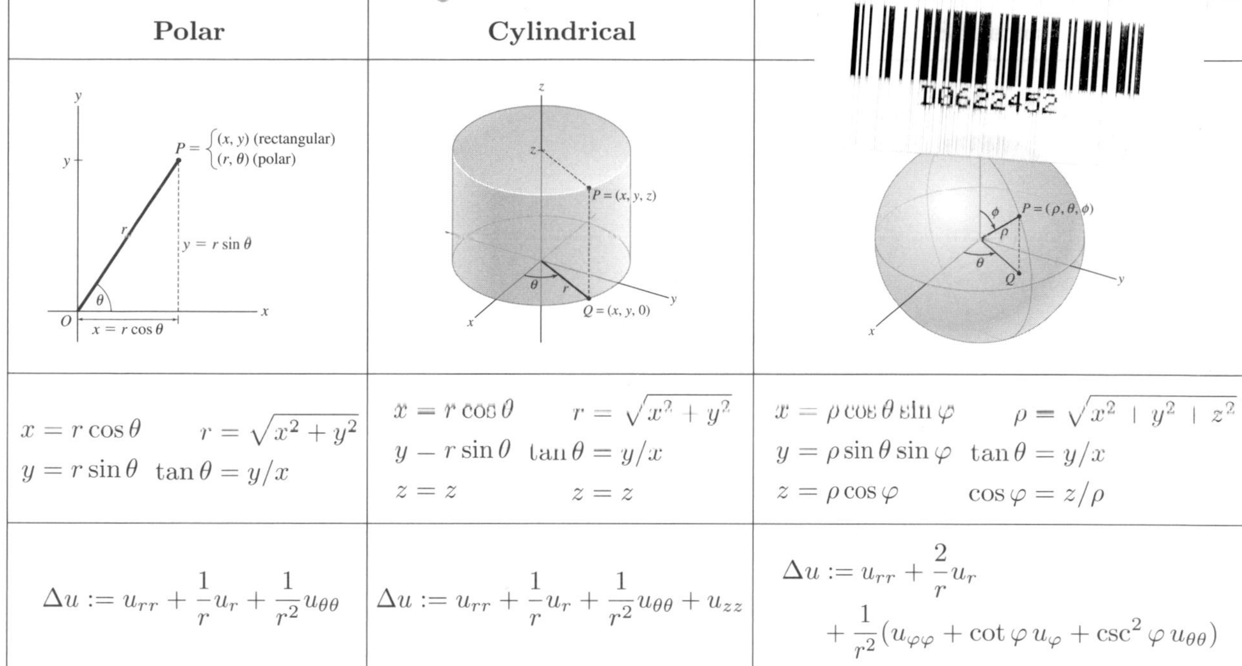

Polar	Cylindrical	
$x = r\cos\theta \quad r = \sqrt{x^2+y^2}$ $y = r\sin\theta \quad \tan\theta = y/x$	$x = r\cos\theta \quad r = \sqrt{x^2+y^2}$ $y = r\sin\theta \quad \tan\theta = y/x$ $z = z \quad z = z$	$x = \rho\cos\theta\sin\varphi \quad \rho = \sqrt{x^2+y^2+z^2}$ $y = \rho\sin\theta\sin\varphi \quad \tan\theta = y/x$ $z = \rho\cos\varphi \quad \cos\varphi = z/\rho$
$\Delta u := u_{rr} + \frac{1}{r}u_r + \frac{1}{r^2}u_{\theta\theta}$	$\Delta u := u_{rr} + \frac{1}{r}u_r + \frac{1}{r^2}u_{\theta\theta} + u_{zz}$	$\Delta u := u_{rr} + \frac{2}{r}u_r + \frac{1}{r^2}(u_{\varphi\varphi} + \cot\varphi\, u_\varphi + \csc^2\varphi\, u_{\theta\theta})$

$\mathbb{R}^2$	$\mathbb{R}^3$
$\nabla := \left\langle \frac{\partial}{\partial x}, \frac{\partial}{\partial y} \right\rangle$ $\operatorname{grad} f := \nabla f := \langle f_x, f_y \rangle$ $\operatorname{div} \mathbf{F} := \nabla \cdot \mathbf{F} := \frac{\partial F_1}{\partial x} + \frac{\partial F_2}{\partial y}$ $\Delta f := \operatorname{div}\operatorname{grad} f := \nabla \cdot \nabla f = f_{xx} + f_{yy}$	$\nabla := \left\langle \frac{\partial}{\partial x}, \frac{\partial}{\partial y}, \frac{\partial}{\partial z} \right\rangle$ $\operatorname{grad} f := \nabla f := \langle f_x, f_y, f_z \rangle$ $\operatorname{div} \mathbf{F} := \nabla \cdot \mathbf{F} := \frac{\partial F_1}{\partial x} + \frac{\partial F_2}{\partial y} + \frac{\partial F_3}{\partial z}$ $\operatorname{curl} \mathbf{F} := \nabla \times \mathbf{F} := \begin{vmatrix} \mathbf{i} & \mathbf{j} & \mathbf{k} \\ \frac{\partial}{\partial x} & \frac{\partial}{\partial y} & \frac{\partial}{\partial z} \\ F_1 & F_2 & F_3 \end{vmatrix}$ $\Delta f := \operatorname{div}\operatorname{grad} f := \nabla \cdot \nabla f = f_{xx} + f_{yy} + f_{zz}$

	Dirichlet	Neumann	Robin
Mathematical Form	$u = T$	$\frac{\partial u}{\partial \mathbf{n}} = R$ $\nabla u \cdot \mathbf{n} = R$	$\frac{\partial u}{\partial \mathbf{n}} + Ku = T$ $\nabla u \cdot \mathbf{n} + Ku = T$
Physical Form	$u = T$	$-\frac{\partial u}{\partial \mathbf{n}} = R$ $-\nabla u \cdot \mathbf{n} = R$	$-\frac{\partial u}{\partial \mathbf{n}} = K(u - T/K),\ K > 0$ $-\nabla u \cdot \mathbf{n} = K(u - T/K),\ K > 0$

Introduction to Applied Partial Differential Equations

Introduction to Applied Partial Differential Equations

John M. Davis

Baylor University

W. H. Freeman and Company
New York

Publisher: Ruth Baruth
Executive Editor: Terri Ward
Executive Marketing Manager: Jennifer Somerville
Associate Editor: Katrina Wilhelm
Editorial Assistant: Tyler Holzer
Marketing Assistant: Joan Rothschild
Photo Editor: Christine Buese
Cover Designer: Diana Blume
Project Editor: Jodi Isman
Illustrations: John M. Davis, Network Graphics
Illustration Coordinator: Janice Donnola
Director of Production: Ellen Cash
Printing and Binding: RR Donnelley

Library of Congress Control Number: 2011944063

ISBN-13: 978-1-4292-7592-7
ISBN-10: 1-4292-7592-8

Printed in the United States of America

First printing

W. H. Freeman and Company
41 Madison Avenue
New York, NY 10010
Houndmills, Basingstoke RG21 6XS, England
www.whfreeman.com

Contents

To Tiffany

Preface

This text is based on my teaching of a one-semester, junior level, partial differential equations course at Baylor University over the last dozen or so years. The typical prerequisites are completion of the calculus sequence, ordinary differential equations, and linear algebra, although the linear algebra concepts here are relatively self-contained. About three-quarters of the students in our course are mechanical or electrical engineering majors with the remainder split between mathematics, physics, and other hard science majors. As such, the vision of the course is equal parts computational proficiency, geometric insight, and physical interpretation of the problems at hand. The goal is to inextricably link all three of those throughout this text.

The method of separation of variables in various coordinate systems naturally receives the most attention, but we also tackle pure initial value problems on unbounded domains using the methods of d'Alembert and Fourier transforms. More theoretical topics such as error analysis, modes of convergence, basic L^2 theory, and Sturm-Liouville theory are dealt with at a level that is hopefully digestable for this audience. This list of topics for a PDEs course is by no means exhaustive, nor is it meant to be. Our mantra is to pick the most salient topics for a reasonably robust set of one-semester courses, then do them consistent with the philosophy summarized below.

The focus is on deeper concepts over mundane calculations. We compute *by hand* selectively and when there is something to be gained from doing so. The appropriate use of technology is encouraged, for example, in computing the integrals arising in Fourier series coefficients or plotting 3D animations of solution surfaces. We use *Mathematica* to accomplish this, but the text is agnostic with respect to the particular software or computer algebra system used. Most importantly, technology only complements the text—it is not the main thrust.

Geometric insight and physical interpretation are emphasized. Obtaining a series or integral representation of a solution is not an end unto itself: we do so in order to gain some understanding of the qualitative properties of the solution and what they tell us about the real world physical model being solved. Computer algebra systems are an excellent tool for visualizing solutions and strengthening these connections. Therefore, we constantly reflect on what the trio of computation, visualization, and physical interpretation reveals to us about the problem.

The language and tools of vector calculus are embraced. The student for whom PDEs is important is precisely the student for whom fluency in vector calculus is necessary. Many concepts from vector calculus need to be solidified in students' minds, and a PDEs course is a natural place for this to happen. We take full advantage of this opportunity, for example, by deriving the multidimensional heat equation and wave equation from a variational viewpoint since these are excellent opportunities to showcase the Divergence Theorem and Stokes' Theorem.

Future analysis topics are foreshadowed. A good PDEs course should serve as a segue to upper-level mathematics courses, insomuch as studying PDEs motivates many questions which are properly answered later in real analysis. This pivotal role of the course should not be underestimated because it provides a frame of reference and a wealth of motivating examples during later studies. Therefore, we purposefully introduce concepts such as vector spaces, inner product spaces, eigenvalue problems, orthogonality, various modes of convergence, and basic L^2 theory at a level appropriate for this audience.

Chapter Overviews

Chapter 1 compares ODEs with PDEs and shows how PDEs and relevant boundary conditions are developed from physical models.

Chapter 2 introduces the method of separation of variables in the context of the heat equation and wave equation in one spatial dimension on bounded domains. Fourier series are viewed as *natural* by-products of this method and the Fourier coefficients are determined by appealing to the orthogonality of the underlying basis functions.

Chapter 3 is a pause from the computational techniques of the previous chapter and aims to put those techniques in a broader, overarching perspective. The convergence of an infinite series of functions—what that even *means* and the role it plays in applications—is often lost on students. We strive to accomplish this by emphasizing the various notions of error in a Fourier series approximation.

Chapter 4 treats the Sturm-Liouville theory succinctly but hopefully puts the completeness of the eigenfunctions used to solve the problems in Chapters 5 and 6 on more solid ground. We accomplish this by arguing for the symmetry of the underlying Sturm-Liouville operator.

Chapter 5 deals with the method of separation of variables in multiple spatial dimensions. This is a natural place to emphasize and reinforce the tools of vector calculus. We emphasize the *big three*—heat, wave, and Laplace's equation—in 2D with various combinations of boundary conditions.

Chapter 6 extends the method of separation of variables to polar, cylindrical, and spherical coordinates. The payoff of the theory from Chapter 4 becomes evident as the underlying eigenvalue problems become more complicated. Moreover, the rich geometry of solutions in these problems makes all that work worthwhile.

Chapter 7 explores problems on unbounded domains, abandoning separation of variables for d'Alembert's solution, characteristics, and transform methods. The geometry here is different than before, but equally rich.

Sample Syllabi and Exercises

Various instructors have taught from this material in a typical, one-semester partial differential equations course using the following sections:

- Techniques Only
  - Sections 1.1–1.4; 2.1–2.5; 5.1–5.6; 6.1, 6.3-6.5; 7.1–7.4
  - 23 sections, approximately 34 hours
- Techniques, Some Theory
  - Sections 1.1–1.4; 2.1–2.5; 3.1, 3.3–3.5; 5.1–5.6; 6.1, 6.3; 7.1–7.4
  - 25 sections, approximately 35 hours
- Techniques, More Theory
  - Sections 1.1–1.4; 2.1–2.5; 3.1, 3.3–3.5; 4.1–4.3; 5.1–5.6; 6.1–6.3; either 6.4, 6.5 or 7.1–7.4
  - 27 sections, approximately 38 hours

The starred sections in the table of contents can be skipped without disrupting the logical flow of the course. I tend to rotate which subset of these enrichment sections I cover depending on the personality of my class that semester. Chapter 7 is independent from Chapters 2–6, so it can be covered early or late depending on the instructor's preference. Finally, at least the high points of Sections 4.1–4.3 are likely needed to do Sections 6.3–6.5 and discuss completeness of the eigenfunctions.

There are more than 340 exercises in the text, ranging from routine calculations to filling in the details of an argument as well as using technology to visualize solutions of PDEs. Answers to most of the odd numbered problems are in the back of the text. Exercises within the same section are referenced simply by their number. A reference to Exercise A.B.C means Chapter A, Section B, Exercise C.

Technology

Modern computer algebra systems such as *Mathematica*, *Maple*, and MATLAB have revolutionized the teaching and understanding of a partial differential equations course. This textbook aims to embrace these tools in a *balanced* way. The goal here is not to avoid computing by hand entirely, but rather to avoid either mundane or intractable computations that significantly compound the complexity of already complicated PDE problems and let a computer algebra system handle those for us so we can focus on more important things. Embracing this technology enables us to visualize solutions to PDEs in much greater depth and discuss geometric interpretations of concepts with more accuracy than ever before. In my own teaching, this has been the most powerful tool in helping my students to understand PDEs. The exercises for which a computer algebra system might be especially beneficial, for example, to compute Fourier series coefficients or to plot a three dimensional animation of a solution, are denoted with this symbol: .

Acknowledgments

I thank Matthew Beauregard, Mariette Maroun, Frank Mathis, and Tim Sheng for teaching from early copies of the manuscript and offering many helpful suggestions.

I also thank the following reviewers whose comments, suggestions, and criticisms significantly improved the final version of the text: John Alford, *Sam Houston State University*; Eduardo Balreira, *Trinity University*; Paul W. Eloe, *University of Dayton*; William D. Emerson, *Metropolitan State College of Denver*; Stephen Goode, *California State University, Fullerton*; Grant Hart, *Brigham Young University*; David Isaacson, *Rensselaer Polytechnic Institute*; Billy Jackson, *University of Northern Colorado*; Jon Jacobsen, *Harvey Mudd College*; Sharon R. Lubkin, *North Carolina State University*; Paul Martin, *Colorado School of Mines*; David Nicholls, *University of Illinois at Chicago*; David Rollins, *University of Central Florida*; Weihua Ruan, *Purdue University-Calumet*; Constance Schober, *University of Central Florida*; Patrick Shipman, *Colorado State University*; Thomas J. Smith, *Manhattan College*; Bertram Zinner, *Auburn University*.

I have learned from and leaned on the accumulated teaching wisdom of so many of my colleagues and former teachers. You all know who you are, but probably don't know how much of an impact you have had on me.

I especially want to thank my friend Matthew Beauregard for cheerfully and carefully reading through and teaching from multiple versions of the manuscript, offering fantastic suggestions and corrections along the way, and accuracy checking the answers to the exercises. The innumerable hours we've spent discussing the pedagogy of PDEs has benefited me greatly. I am indebted to Roger Lipsett for tenaciously working through the text and every single exercise, finding countless errors and polishing many rough edges. Thanks also to Selwyn Hollis for answering my *Mathematica* questions. I am grateful to Johnny Henderson for his guidance and mentorship from the beginning.

Above all, I owe my loving wife and family so much for their sacrifice during this project. I wouldn't have been interested in doing this without their faithful support. I also thank my parents for their unwavering encouragement.

I could not imagine a more supportive editor than Terri Ward, who enthusiastically championed this project from the start and gave me full creative license to pursue my vision. I thoroughly enjoyed working almost daily with my developmental editor, Katrina Wilhelm, who shares my passion for attention to detail and who handled every curve ball I threw at her with the utmost grace and professionalism.

Despite the hard work of so many involved, the errors that remain are solely mine. I plan to keep an updated errata and *Mathematica* notebooks for each section at

`http://homepages.baylor.edu/john_m_davis/PDE/`

Finally, if this book can be used—even in a small way—to grab the mind of one student with the beauty and power of differential equations, I will consider it a success.

John M. Davis
Baylor University
Waco, Texas

Introduction to Applied Partial Differential Equations

Chapter 1

Introduction to PDEs

1.1 ODEs vs. PDEs

An *ordinary differential equation* (ODE) is an equation involving an unknown function (of one independent variable) and at least one of its ordinary derivatives. For example, if we let[1] $y := y(t)$, then

$$y' = -y, \qquad y'' + y = 0, \qquad y'' + 2y' + 3y = \cos t$$

are all familiar ODEs in terms of the unknown function $y(t)$.

All solutions of the first ODE above, $y' = -y$, are of the form $y = Ce^{-t}$, where C is an arbitrary constant. Therefore, $y = Ce^{-t}$ represents an (infinite) family of one-variable functions, each of which is a solution to the given ODE. If, in addition to the ODE, we also require $y(0) = 1$, then this condition forces $C = 1$, thereby selecting precisely one solution from the infinite family of solutions. See Figure 1.1.

On the other hand, a *partial differential equation* (PDE) is an equation involving an unknown function (of more than one independent variable) and at least one of its partial derivatives. For example, if we let $u := u(x,t)$, then

$$\frac{\partial u}{\partial t} + \frac{\partial u}{\partial x} = 0, \qquad \frac{\partial u}{\partial t} = \frac{\partial^2 u}{\partial x^2}, \qquad \frac{\partial^2 u}{\partial t^2} = \frac{\partial^2 u}{\partial x^2} \tag{1.1}$$

are all examples of PDEs in the unknown function $u(x,t)$. However, we may consider as many independent variables for the unknown function as we like. For example,

$$u_{xx} + u_{yy} + u_{zz} = 0 \text{ is a PDE in the unknown } u := u(x,y,z),$$

$$u_{rr} + \frac{1}{r}u_r + \frac{1}{r^2}u_{\theta\theta} = u_t \text{ is a PDE in the unknown } u := u(r,\theta,t),$$

$$\rho\left(\frac{\partial \mathbf{v}}{\partial t} + \mathbf{v}\cdot\nabla\mathbf{v}\right) = -\nabla p + \mu\nabla^2\mathbf{v} + \mathbf{f} \text{ is a PDE in the unknown } \mathbf{v} := \mathbf{v}(x,y,z,t).$$

[1]The notation := means *is defined as* or *is equal by definition.*

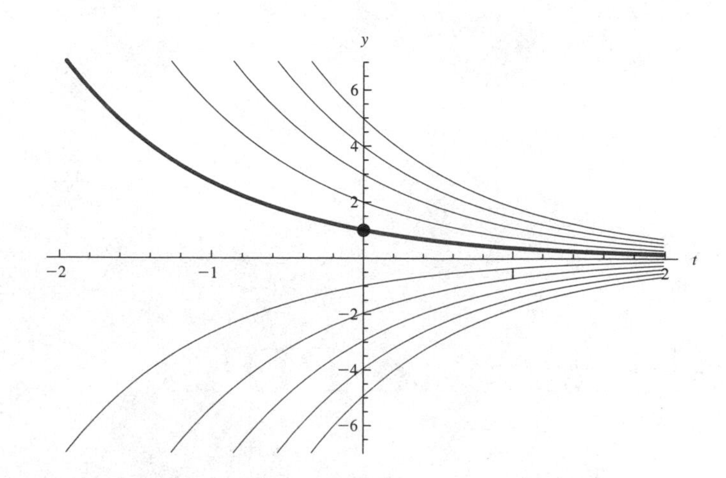

Figure 1.1: Several members of the family $y = Ce^{-t}$ are shown which solve $y' = -y$. Here, the initial condition $y(0) = 1$ specifies the one solution which passes through the point $(0, 1)$.

Throughout this text, we will always assume sufficient smoothness of all functions involved so that mixed partial derivatives are equal, i.e., $u_{xy} = u_{yx}$, etc.

To demonstrate one significant difference between solving ODEs and PDEs, take the first PDE in (1.1),

$$u_t + u_x = 0.$$

We can quickly verify[2] that functions of the form

$$u(x,t) = f(x - t)$$

for an arbitrary continuously differentiable *function* f are indeed solutions of $u_t + u_x = 0$. To see this, use the chain rule:

$$\begin{aligned} u(x,t) = f(x-t) \implies u_t &= f'(x-t) \cdot \frac{\partial}{\partial t}(x-t) = f'(x-t) \cdot (-1) \\ u_x &= f'(x-t) \cdot \frac{\partial}{\partial x}(x-t) = f'(x-t) \cdot 1 \\ u_t + u_x &= -f'(x-t) + f'(x-t) = 0. \end{aligned}$$

For example, if we impose the condition $u(x,0) = \frac{1}{1+x^2}$, then f is completely determined since

$$\begin{aligned} u(x,0) = f(x-0) &= \frac{1}{1+x^2} \\ f(x) &= \frac{1}{1+x^2} \end{aligned}$$

[2]We will learn where this formula came from later.

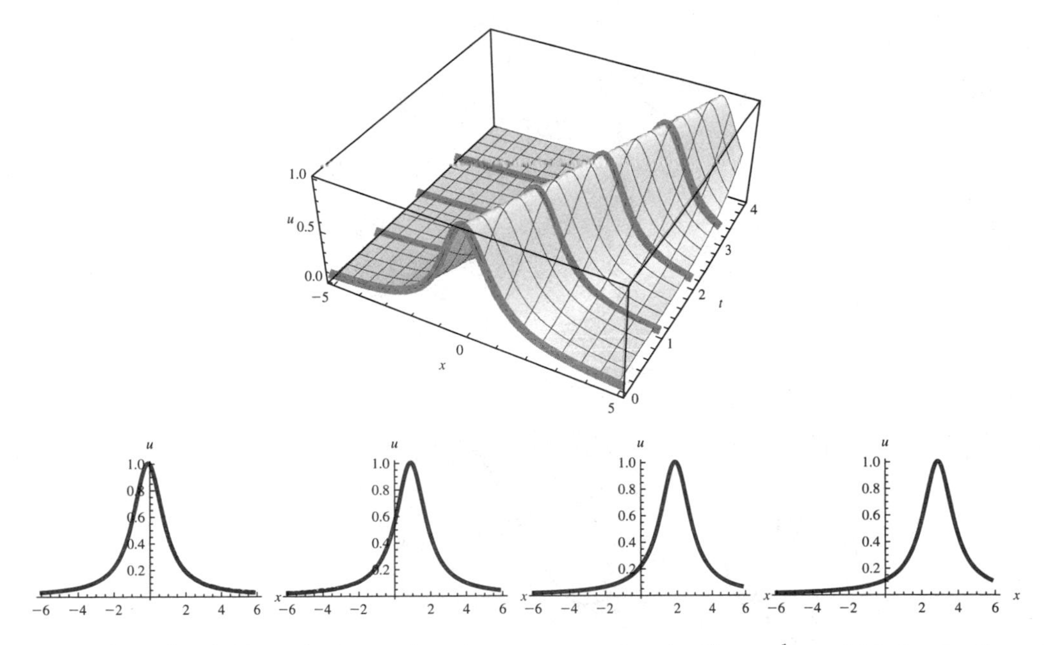

Figure 1.2: (Top) The solution surface for $u_t + u_x = 0,\ u(x,0) = \frac{1}{1+x^2}$, which is $u(x,t) = \frac{1}{1+(x-t)^2}$. The curves traced out at $t = 0, 1, 2, 3$ are highlighted. (Bottom) The "time slices" of the solution surface at $t = 0, 1, 2, 3$ above. In contrast to the ODE problem, solutions to the PDE are surfaces. For an ODE, an initial condition determines the arbitrary *constant(s)* in a solution, but for a PDE, an initial condition determines the arbitrary *function(s)* in a solution.

and therefore the solution to $u_t + u_x = 0,\ u(x,0) = \frac{1}{1+x^2}$ that we seek is

$$u(x,t) = f(x-t) = \frac{1}{1+(x-t)^2}.$$

This type of condition, where we specify the value of the solution at $t = 0$ is called an *initial condition* since we think of it as specifying the value the solution must obtain at the starting time $t = 0$.

On the other hand, if we require $u(0,t) = \sin t$, then

$$\begin{aligned} u(0,t) = f(0-t) &= \sin t \\ f(t) &= -\sin t, \end{aligned}$$

and therefore the solution to $u_t + u_x = 0,\ u(0,t) = \sin t$ that we seek is

$$u(x,t) = f(x-t) = -\sin(x-t).$$

This type of condition, where we specify the value of the solution for a fixed value of a spatial variable, is called a *boundary condition* since we think of this as specifying the

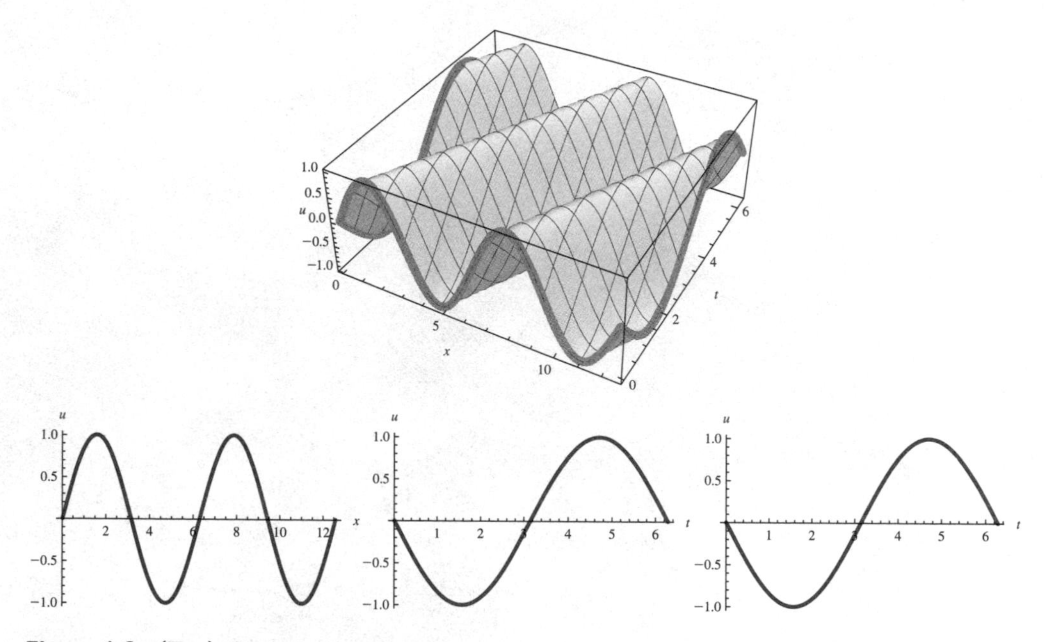

Figure 1.3: (Top) The solution of $u_t + u_x = 0,\ u(0,t) = \sin t$ which is $u(x,t) = -\sin(x-t)$, shown here for $0 < x < 4\pi,\ 0 < t < 2\pi$. The curves highlighted on the left and right correspond to the boundary conditions $u(0,t)$ and $u(4\pi,t)$, respectively. The front curve corresponds to the initial condition $u(x,0)$. (Bottom) The curves prescribed by the initial condition and two boundary conditions.

value the solution must obtain at the physical boundary of the spatial domain of the problem. See Figures 1.2 and 1.3.

Note that in the ODE case, the general solution $y = Ce^{-t}$ was a (one parameter) family of *curves* in the t-y plane and the initial condition selected one particular curve from the family. In the PDE case, the general solution was a (one parameter) family of *surfaces* in x-t-u space and the initial condition selected a particular surface from the family. This indicates how solutions to PDEs are inherently more complicated than solutions to ODEs.

In the next section, we will explore the physical origins of PDEs from mathematical physics and discuss what information is required in conjunction with a PDE to form a well-posed problem. Afterwards, we will turn our attention to techniques for *solving* certain types of PDEs.

Exercises

1. (a) Let $u := u(x,t)$. Solve $u_t + u_x = 0$ subject to the condition $u(x,0) = e^{-|x|}$.

   (b) Plot the solution surface in (a) for $-5 < x < 5$ and $0 < t < 5$.

   (c) Plot curves that $u(x,t)$ traces out when $t = 0, 1, 2, 3, 4$.

2. Discuss whether the highlighted curves on the surfaces below represent initial conditions or boundary conditions.

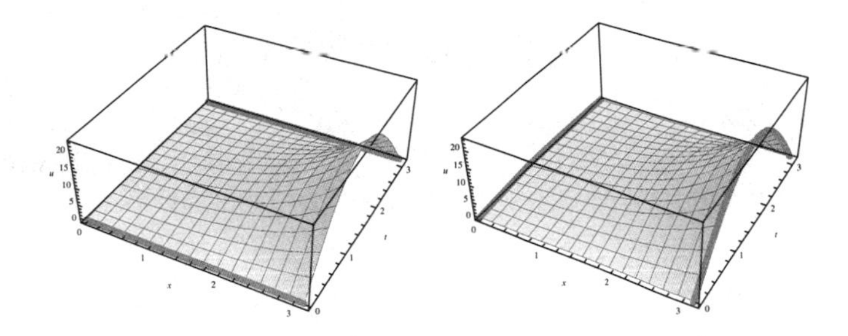

3. Consider the PDE $u_t + cu_x = 0$, where $c > 0$ is a constant.

   (a) Show that $u(x,t) = f(x - ct)$ is a solution of the PDE, where f is any differentiable function of one variable.

   (b) The solutions in (a) are called *right traveling wave solutions.* Explain why.

   (c) Let $f(z) = z^2$ and $c = 1/2$. Plot the surface $u(x,t)$ and the curves in the x-u plane traced out by $u(x,0)$, $u(x,1)$, $u(x,2)$, $u(x,3)$.

   (d) This PDE is called the *transport equation.* Using (c), explain why.

4. (a) The PDE $u_{tt} = u_{xx}$ arises in modeling wave phenomena and is therefore called the *wave equation.* Show that $u(x,t) = (x+t)^4$ is a solution.

   (b) What initial condition does u satisfy?

   (c) Plot the solution surface.

   (d) Using (c), discuss the difference between the conditions $u(x,0)$ and $u(0,t)$.

5. (a) Verify that $u(x,t) = \frac{1}{\sqrt{4\pi kt}} \exp\left(-\frac{x^2}{4kt}\right)$ is a solution of the PDE $u_t = ku_{xx}$, called the *heat equation* because it models heat flow.

   (b) Let $k = 1$. What curve does the surface $u(x,t)$ trace out when $x = 0$ and $0 < t < 1$?

6. Show that $u(x,y) = e^{-y}(\sin x + \cos x)$ satisfies the PDE $u_{xx} + u_{yy} = 0$ as well as the two auxiliary conditions $u(x,0) = \sin x + \cos x$, $u(0,y) = u_x(0,y)$.

7. Consider the PDE in polar coordinates for $u := u(r,\theta)$ given by

$$u_{rr} + \frac{1}{r}u_r + \frac{1}{r^2}u_{\theta\theta} = 0.$$

   (a) Show that $u(r,\theta) = \ln r$ and $u(r,\theta) = r\cos\theta$ are both solutions.

   (b) Plot the solution surfaces in (a) for $0 < r < 1$ and $-\pi < \theta < \pi$.

1.2 How PDEs Are Born: Conservation Laws, Fluids, and Waves

Conservation Laws and Fluids

Consider fluid flow in a one dimensional domain (e.g., heat flow in a very thin rod or the diffusion of dye in a very thin tube of fluid) of length ℓ as shown in Figure 1.4. Let $t > 0$ denote time and $0 < x < \ell$ denote the spatial variable. Define the following quantities:

$u(x,t) :=$ the density of the quantity of interest from fluid dynamics (e.g., heat energy or dye concentration) at the point x and time t,

$\varphi(x,t) :=$ the flux of the quantity at the point x and time t (the amount of the quantity flowing to the right per unit time per unit area through a cross section at x),

$f(x,t) :=$ rate of internal generation of the quantity at the point x and time t (amount per unit volume per unit time),

$A :=$ cross-sectional area of a slice of the domain (assumed to be the same for all x).

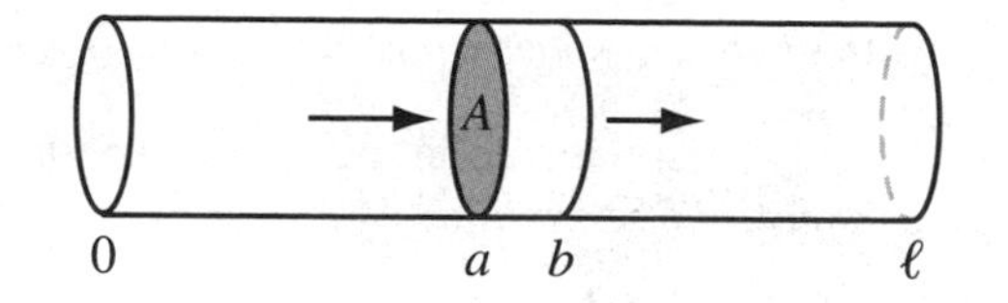

Figure 1.4: Modeling fluid flow in one space dimension.

Consider an arbitrary subinterval $[a, b]$ of the domain $0 < x < \ell$ with (constant) cross-sectional area A. The conservation law here is

$$\begin{array}{c}\text{time rate of change}\\ \text{of total quantity in a slice}\end{array} = \text{rate of inflow} - \text{rate of outflow} + \begin{array}{c}\text{rate of internal}\\ \text{generation.}\end{array}$$

Translating this into symbols,

$$\frac{d}{dt}\int_a^b u(x,t)A\,dx = A\varphi(a,t) - A\varphi(b,t) + \int_a^b f(x,t)A\,dx. \tag{1.2}$$

Applying the Fundamental Theorem of Calculus to the difference on the right-hand side,

$$\int_a^b \frac{\partial u}{\partial t}(x,t)\,dx = -\int_a^b \frac{\partial \varphi}{\partial x}(x,t)\,dx + \int_a^b f(x,t)\,dx.$$

We can rewrite this as the integral equation

$$\int_a^b \left[\frac{\partial u}{\partial t}(x,t) + \frac{\partial \varphi}{\partial x}(x,t)\right] dx = \int_a^b f(x,t)\,dx.$$

Since these integrals are equal for an arbitrary subinterval $[a,b]$ of $0 < x < \ell$, the integrands must be equal, producing the *Fundamental Conservation Law*:

$$\boxed{\frac{\partial u}{\partial t}(x,t) + \frac{\partial \varphi}{\partial x}(x,t) = f(x,t).} \tag{1.3}$$

However, the Fundamental Conservation Law alone is not enough to derive a partial differential equation modeling heat flow or chemical diffusion because there are still *two* unknowns—u and φ (the source term f is typically thought of as given). Thus, we appeal to further physical assumptions on the flux term φ called *constitutive equations.*[3]

The following are some physically realistic constitutive equations that lead to well-studied PDEs.

- $\varphi = cu$, $f \equiv 0$. Then (1.3) becomes

$$\boxed{u_t + cu_x = 0,} \tag{1.4}$$

  called the *transport equation* because it models the transport of a (nondiffusing) substance by a flowing fluid.

- $\varphi = -ku_x$, $f \equiv 0$. Then (1.3) becomes

$$\boxed{u_t - ku_{xx} = 0, \qquad \text{or equivalently} \qquad u_t = ku_{xx},} \tag{1.5}$$

  called the *heat equation* or *diffusion equation* since this constitutive equation is consistent with Fourier's Law of Heat Conduction[4] and Fick's Law of Diffusion.[5]

- $\varphi = cu - ku_x$, $f \equiv 0$. Then (1.3) becomes

$$u_t + cu_x - ku_{xx} = 0,$$

  called the *transport-diffusion equation* because it models the combined dynamics of a diffusing substance in a flowing fluid.

- $\varphi = u^2/2$, $f \equiv 0$. Then (1.3) becomes

$$u_t + uu_x = 0,$$

  called the *inviscid*[6] *Burgers' equation*, which arises in fluid dynamics.

- $\varphi = u^2/2 - ku_x$, $f \equiv 0$. Then (1.3) becomes

$$u_t + uu_x - ku_{xx} = 0,$$

  called the *viscous Burgers' equation* and arises in more advanced fluid dynamics.

[3] Also called *continuity equations* or *state equations.*
[4] Heat flows down the gradient—from areas of high concentration to low.
[5] The time rate of change of a diffusing concentration is proportional to the concentration gradient.
[6] Fluids with negligible viscous (frictional) forces are called *inviscid.*

- $\varphi = -ku_x$, $f(u) = u(1-u)$. Then (1.3) becomes
$$u_t - ku_{xx} = u(1-u), \tag{1.6}$$
called *Fisher's equation.* It arises in mathematical biology because it models logistic-type populations which disperse in space and time.

As these examples demonstrate, the power of the Fundamental Conservation Law lies in how a diverse collection of PDEs follow from it simply by applying various constitutive equations for φ and/or for f.

The Wave Equation in One Space Dimension

Consider a perfectly flexible string of length ℓ whose endpoints have fixed positions. Let $t > 0$ denote time and $0 < x < \ell$ denote the spatial variable. Define the following quantities:

$$\begin{aligned} u(x,t) &:= \text{vertical position of the point on the string } x \text{ units} \\ & \text{ from the left endpoint at time } t, \\ \rho(x) &:= \text{linear density of the string at } x, \\ \mathbf{T}(x,t) &:= \text{tension vector at } x \text{ at time } t. \end{aligned}$$

We will make the following (physically realistic) assumptions in our model:

- The deflections of the string are "small" so that $|u(x,t)|$ and $|u_x(x,t)|$ are "small."
- The only significant motion is vertical; all else is negligible. In particular, the motion is confined to the x-u plane and horizontal motion is disregarded.
- The only force present is due to tension (we ignore friction, external forces, etc.), which is directed tangentially to the string at x.

Consider an arbitrary subinterval $[a,b]$ of the domain $0 < x < \ell$, as shown in Figure 1.5.

First, we resolve the tension $\mathbf{T}(x,t)$ into its horizontal and vertical components[7]:

$$\begin{aligned} \mathbf{T}_{\text{horiz}}(x,t) &= \frac{|\mathbf{T}(x,t)|}{\sqrt{1+u_x^2(x,t)}}, \\ \mathbf{T}_{\text{vert}}(x,t) &= \frac{|\mathbf{T}(x,t)|u_x(x,t)}{\sqrt{1+u_x^2(x,t)}}. \end{aligned}$$

Balancing horizontal forces on the segment $[a,b]$ requires

$$\frac{|\mathbf{T}(b,t)|}{\sqrt{1+u_x^2(b,t)}} - \frac{|\mathbf{T}(a,t)|}{\sqrt{1+u_x^2(a,t)}} = 0, \tag{1.7}$$

[7]Recall that the horizontal component of a vector $\mathbf{v}$ is given by $\mathbf{v}_{\text{horiz}} = |\mathbf{v}|\cos\theta$, while the vertical component is $\mathbf{v}_{\text{vert}} = |\mathbf{v}|\sin\theta$.

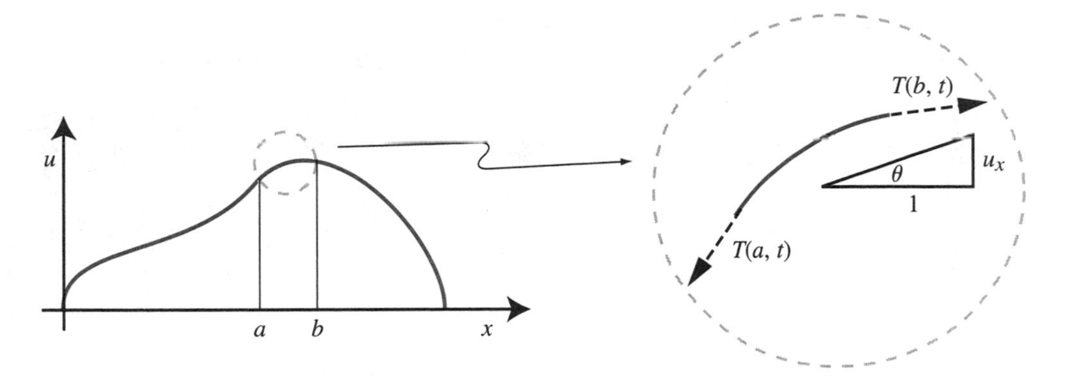

Figure 1.5: Modeling small vertical vibrations in a flexible string. Since the tension is directed tangentially, the slope that the tension vector makes with the horizonal should be u_x; that is, $\tan\theta = \frac{u_x}{1}$. The triangle is then used to compute $\mathbf{T}_{\text{horiz}} = |\mathbf{T}|\cos\theta$ in (1.7).

which says the tension is independent of x (since a and b are arbitrary), so we will now write $|\mathbf{T}(x,t)| = T(t)$.

Balancing the vertical forces on $[a,b]$, Newton's Second Law[8] yields

$$\frac{d}{dt}\int_a^b \rho(x)u_t(x,t)\sqrt{1+u_x^2(x,t)}\,dx = \frac{T(t)u_x(b,t)}{\sqrt{1+u_x^2(b,t)}} - \frac{T(t)u_x(a,t)}{\sqrt{1+u_x^2(a,t)}}. \tag{1.8}$$

Since we assumed the deflections were small, $u_x^2 \approx 0$ and hence $\sqrt{1+u_x^2(x,t)} \approx 1$, so the last equation reduces to

$$\frac{d}{dt}\int_a^b \rho(x)u_t(x,t)\,dx = T(t)[u_x(b,t) - u_x(a,t)].$$

Putting the time derivative inside the left-hand integral and reformulating the difference on the right-hand side as an integral, we have

$$\int_a^b \rho(x)u_{tt}(x,t)\,dx = \int_a^b T(t)u_{xx}(x,t)\,dx.$$

Since these integrals are equal for an arbitrary subinterval $[a,b]$ of $0 < x < \ell$, the integrands must be equal

$$\rho(x)u_{tt}(x,t) = T(t)u_{xx}(x,t). \tag{1.9}$$

If we make the final simplifying assumption that the density of the string doesn't vary with x and the tension doesn't vary with time, i.e., $\rho(x) \equiv \rho$ and $T(t) \equiv T$, then we have the *one dimensional wave equation*

$$\boxed{u_{tt} = c^2 u_{xx}.} \tag{1.10}$$

[8] $F = ma$, but we are using the equivalent statement $F = \frac{d}{dt}(mv)$, i.e., force equals the time rate of change of momentum.

Here, $c = \sqrt{T/\rho}$ is the *wave speed* of the string.

Variations of the wave equation occur by modifying the assumptions involved:

- One way to account for frictional (damping) forces is to assume that the damping force is proportional to the velocity u_t. This yields the *damped wave equation*

$$u_{tt} = c^2 u_{xx} - r u_t, \qquad r > 0.$$

- If there is a restorative, elastic force present which is proportional to the displacement u (e.g., in a spring), we get the equation

$$u_{tt} = c^2 u_{xx} - k u, \qquad k > 0.$$

Exercises

1. (a) In the derivation of the Fundamental Conservation Law, what units do $u(x,t)$, $\varphi(x,t)$, and $f(x,t)$ have?

   (b) Show that each term in (1.2) has units [quantity]/[time].

2. (a) In the derivation of the wave equation, what units do $u(x,t)$, $\rho(x)$, and $\mathbf{T}(x,t)$ have?

   (b) Show that each term in (1.8) has force units.

   (c) Show that c in (1.10) has velocity units.

3. Write each of the following PDEs in the form $u_t + \varphi_x = f$ and identify the flux term in each.

   (a) $u_t + u u_x + u_{xxx} = 0$

   (b) $u_t - u_x e^{-u} = 0$

   (c) $u_t + \frac{u_x}{1+u^2} = |t|$

4. Suppose $u(x,t)$ denotes the density of cars on a certain stretch of highway (in units of cars per mile).

   (a) Explain why $u(x,t)$ obeys the Fundamental Conservation Law.

   (b) If we further assume $f \equiv 0$, what does that mean in this scenario?

   (c) Assume $f \equiv 0$ as in (b). One proposal is to model the flux of cars with the constitutive equation $\varphi = (M - u)u$, where M is the maximum number of cars that can fit on the stretch of road under discussion. Discuss the reasonableness of this flux assumption in terms of a traffic flow model.

   (d) Assume $f \equiv 0$. For the proposed flux in (c), write the PDE modeling the traffic density and simplify.

5. In the heat conduction problem, we assumed the sides (not ends) of the rod were insulated so that no heat flow could occur across this lateral edge. This time, assume the bar does lose heat across this lateral edge at a rate proportional to the temperature at that point $u(x,t)$. Assume no internal generation of heat. Derive the PDE for this situation.

6. When deriving the 1D wave equation, why is $u_x^2 \approx 0$ valid without having $u_x \approx 0$?

7. In this problem, we tweak the vibrating string problem by considering the effect of some extra forces in the model.

   (a) This time account for the force due to gravity weighing on the string. When balancing the vertical forces, show that this results in an extra term of the form $-\int_a^b g\rho(x)\,dx$. Derive the PDE for this situation.

   (b) This time assume the vertical motion is damped by a force proportional to the velocity of the string. When balancing the vertical forces, show that this results in an extra term of the form $-\int_a^b \rho(x)cu_t(x,t)\,dx$. Derive the PDE for this situation.

8. **(Telegraph equations)** Using basic electrical theory, it can be shown that the current $I(x,t)$ and voltage $V(x,t)$ in an electrical transmission line (e.g., a power line or telephone line) at position x in the line and time t obey the equations

$$\begin{aligned} \frac{\partial I}{\partial x} + C\frac{\partial V}{\partial t} + GV = 0, \\ \frac{\partial V}{\partial x} + L\frac{\partial I}{\partial t} + RI = 0, \end{aligned} \qquad (*)$$

   where C is the capacitance per unit length, G is the leakage per unit length, R is the resistance per unit length, and L is the inductance per unit length.

   (a) Assume C, G, R, L are all constants. Eliminate V in the equations above to obtain the PDE

$$\frac{\partial^2 I}{\partial x^2} = CL\frac{\partial^2 I}{\partial t^2} + (CR+GL)\frac{\partial I}{\partial t} + GRI.$$

   (Hint: Differentiate the first equation with respect to x, the second with respect to t, and combine to eliminate the mixed derivative V_{xt} and V_{tx} terms. Use the second equation again to eliminate the remaining V_x term.)

   (b) Similar to (a), determine the PDE for the voltage if I is eliminated.

   (c) If $R = 0$ and $G = 0$, then this idealized case is called a *lossless transmission line.* Show for lossless transmission lines, $(*)$ reduces to the two wave equations

$$\frac{\partial^2 I}{\partial t^2} = c^2\frac{\partial^2 I}{\partial x^2}, \qquad \frac{\partial^2 V}{\partial t^2} = c^2\frac{\partial^2 V}{\partial x^2},$$

   where $c = \sqrt{\frac{1}{CL}}$.

9. **(Classifying Second Order Linear Equations)** The most general form of a second order linear PDE in $u := u(x, y)$ is

$$Au_{xx} + Bu_{xy} + Cu_{yy} + Du_x + Eu_y + Fu = 0. \qquad (**)$$

If A, B, C, D, E, F are constants, we classify $(**)$ according to the sign of the discriminant $B^2 - 4AC$ as follows:

$$B^2 - 4AC \begin{cases} > 0, & \text{then } (**) \text{ is called } \textit{hyperbolic}, \\ = 0, & \text{then } (**) \text{ is called } \textit{parabolic}, \\ < 0, & \text{then } (**) \text{ is called } \textit{elliptic}. \end{cases}$$

Using this scheme, classify these PDEs as hyperbolic, parabolic, or elliptic.

(a) $u_t = ku_{xx}$ (heat equation)

(b) $u_{tt} = c^2 u_{xx}$ (wave equation)

(c) $u_{xx} + u_{yy} = 0$ (Laplace's equation)

(d) $\frac{\partial^2 I}{\partial x^2} = CL\frac{\partial^2 I}{\partial t^2} + (CR+GL)\frac{\partial I}{\partial t} + GRI$ for $C, L, R, G > 0$ (telegraph equation)

1.3 Boundary Conditions in One Space Dimension

Problems in one space dimension have the spatial variable defined on an interval such as $0 < x < \ell$. As such, the boundary of the domain consists of the endpoints of that interval, $x = 0$ and $x = \ell$.

PDE models often require that a solution fulfills further constraints on the boundary of the domain, called *boundary conditions*, which play a crucial role in the overall solution. Let's look at the most common types of boundary conditions that arise in practice.

Dirichlet Boundary Conditions

A *Dirichlet*[9] *boundary condition* (also called a *boundary condition of the first kind*) specifies the value of the unknown function u on the physical boundary of the domain.

For example, a Dirichlet boundary condition at $x = 0$ would have the form $u(0, t) = T$. In the context of the 1D heat equation, the physical interpretation is that the temperature u at the left endpoint of the rod $(x = 0)$ is held fixed at T degrees for all time. For the wave equation, it means the left endpoint of the string is held fixed at a height of T units for all time. This type of boundary condition is named in honor of Gustav Dirichlet. See Figure 1.7.

[9]Pronounced *DEER-ah-shlay* or *DEER-ah-clay*.

Neumann Boundary Conditions

A *Neumann*[10] *boundary condition* (also called a *boundary condition of the second kind*) specifies the value of the spatial derivative $\frac{\partial u}{\partial x}$ on the physical boundary of the domain.

There are two standard ways of writing a Neumann boundary condition—in *mathematical form* or *physical form*. The mathematical form simply specifies $\frac{\partial u}{\partial x}$ on the boundary of the domain, e.g., in the 1D heat equation for a rod of length ℓ, we could write $u_x(0,t) = R$ at the left edge or $u_x(\ell,t) = R$ at the right edge. These are simply mathematical conditions imposed at the boundary of the domain since (as is) they don't claim to represent any physical phenomena at the boundary.

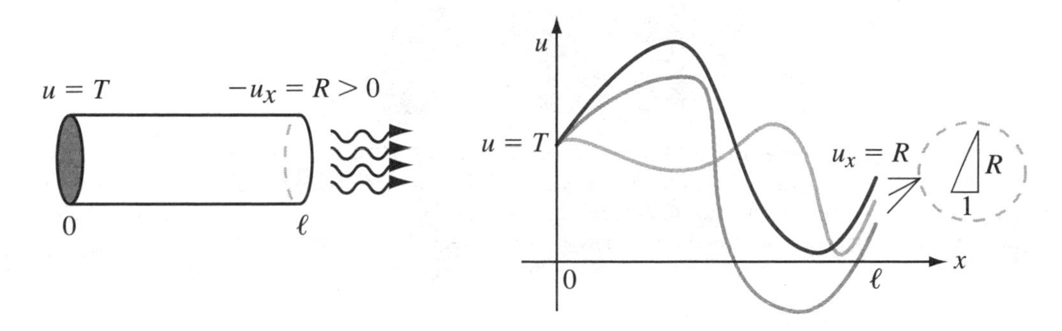

Figure 1.6: (Left) A Dirichlet boundary condition at the left endpoint and a Neumann boundary condition at the right endpoint for the heat conduction problem. (Right) The same boundary conditions for the vibrating string.

On the other hand, the physical form of a Neumann boundary condition in the context of fluid flow (e.g., heat transfer, fluid dynamics, etc.) strives to express the *outward flux* of the quantity normal (perpendicular) to the boundary. For example, in the context of the 1D heat equation on $0 < x < \ell$, specifying an outward flux of R units normal to the boundary at the left edge ($x = 0$), would be written $u_x(0,t) = R > 0$. To see why, recall that the flux term in the constitutive equation involving heat flow is $\varphi = -ku_x$, $k > 0$, and by convention a positive flux means rightward flow. At the left edge ($x = 0$) of the rod, outward flow normal to the boundary would be leftward flow, so we must have $\varphi < 0$ at $x = 0$. But then $-ku_x(0,t) < 0$, so $u_x(0,t) > 0$ represents *outward* flux normal to the boundary at the left edge. Since $u_x(0,t) = R > 0$ has a specific physical interpretation in terms of outward flux, we call this the *physical form* of the Neumann boundary condition and prefer to write our Neumann boundary conditions this way by default.

Note, however, that the physical form of a Neumann boundary condition at the right edge ($x = \ell$) of the rod looks different than at the left edge since the outward normals point in opposite directions. To see why, note that we want rightward flow at the right edge, so $\varphi = -ku_x > 0$ at $x = \ell$. Thus, an outward flux of R units at the right edge would be written $-u_x(\ell,t) = R > 0$.

[10]Pronounced *NOY-mahn* and named in honor of Carl Gottfried Neumann.

Figure 1.7: Johann Peter Gustav Lejeune Dirichlet (1805–1859) of Germany studied under the renowned Fourier, Laplace, Poisson, and Legendre. Dirichlet made significant contributions to analysis, number theory, and mathematical physics. He had a reputation as an excellent teacher and very clear lecturer, but was modest, shy, and reserved.

In the physical form of a Neumann boundary condition, the left-hand side should always represent *outward flux normal to the boundary.*

The mathematical form and physical form can always be translated back and forth easily. For example, a Neumann boundary condition at $x = 0$ could have the form

$$\underbrace{\frac{\partial u}{\partial x}(0,t) = R,}_{\text{mathematical form}} \qquad \text{or equivalently,} \qquad \underbrace{\overbrace{\frac{\partial u}{\partial x}(0,t)}^{\substack{\text{outward}\\ \text{flux}\\ \text{at } x=0}} = R.}_{\text{physical form (preferred)}}$$

On the other hand, a Neumann boundary condition at $x = \ell$ could have the form

$$\underbrace{\frac{\partial u}{\partial x}(\ell,t) = R,}_{\text{mathematical form}} \qquad \text{or equivalently,} \qquad \underbrace{\overbrace{-\frac{\partial u}{\partial x}(\ell,t)}^{\substack{\text{outward}\\ \text{flux}\\ \text{at } x=\ell}} = -R.}_{\text{physical form (preferred)}}$$

Note that any discussion of flux in the context of a wave equation makes no sense. Instead, for the wave equation, a Neumann boundary condition such as $u_x(\ell, t) = R$ means that the slope of the string (rather than its position) is specified. See Figure 1.6.

Robin Boundary Conditions

A *Robin*[11] *boundary condition* (also called a *boundary condition of the third kind*) specifies a linear combination of the spatial derivative $\frac{\partial u}{\partial x}$ and the unknown function u on the physical boundary of the domain.

For example, a Robin boundary condition at $x = \ell$ might take the form

$$\underbrace{\frac{\partial u}{\partial x}(\ell,t) + Ku(\ell,t) = T}_{\text{mathematical form}}, \quad \text{or equivalently,} \quad \underbrace{\overbrace{-\frac{\partial u}{\partial x}}^{\substack{\text{outward}\\ \text{flux}\\ \text{at } x=\ell}}(\ell,t) = K(u(\ell,t) - T/K)}_{\text{physical form (preferred)}},$$

where K is a proportionality constant. For the heat equation, this is a type of Newton's Law of Cooling in action—the rate at which heat energy is exchanged across the right endpoint is proportional to the difference between the temperature at the right endpoint of the rod $u(\ell,t)$ and the temperature T/K of the surrounding medium. For physically realistic heat conduction problems (those that are required to obey Fourier's Law), we must have $K > 0$ in this formulation. See Figure 1.8.

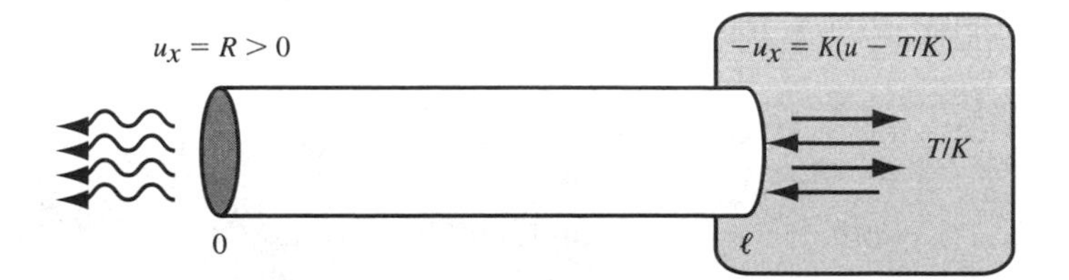

Figure 1.8: In the heat conduction problem, a radiating Neumann boundary condition at the left endpoint signifies a constant flux across the boundary. A Robin boundary condition at the right endpoint indicates convection with an outside medium fixed at T/K degrees.

> In the physical form of a Robin boundary condition, the left-hand side should always represent *outward flux normal to the boundary.* The right-hand side should be a constant times the difference between u and the surrounding medium. The convection is realistic if and only if the proportionality constant K is positive.

For the wave equation, it can be shown that this represents when the right endpoint of the string is attached to a spring-mass system, where the equilibrium position of the mass is a height of T units. Again, $K > 0$ corresponds to a physically realistic spring: one that obeys Hooke's Law.

[11]These boundary conditions were first studied in the context of thermodynamics problems by French applied mathematician Victor Gustave Robin (1855–1897), who was a lecturer in mathematical physics at the Sorbonne in Paris. Unfortunately, no known portraits of Robin exist today.

Although mathematically valid, not all linear combinations of u and u_x at an endpoint are physically realistic. Therefore, we must specify a Robin boundary condition carefully if the focus is on an accurate physical model (as opposed to a purely mathematical exploration).

	Dirichlet	**Neumann**	**Robin**, $K > 0$
mathematical form	$u = T$	$u_x = R$	$u_x + Ku = T$
physical form	$u = T$	left: $u_x = R$ right: $-u_x = R$	left: $u_x = K(u - T/K)$ right: $-u_x = K(u - T/K)$

Table 1.1: Comparison of mathematical versus physical forms for stating boundary conditions. The guiding principle in the physical form of Neumann and Robin boundary conditions is that the left-hand side take the form of outward flux, and for Robin boundary conditions, the right-hand side needs to be stated in a way consistent with Fourier's Law.

Periodic Boundary Conditions

Periodic boundary conditions are a pair of conditions at each endpoint requiring the function u and its spatial derivative $\frac{\partial u}{\partial x}$ to agree at the boundary points; that is,

$$u(0,t) = u(\ell,t) \quad \text{and} \quad u_x(0,t) = u_x(\ell,t).$$

For (1.5), modeling the diffusion of dye in a tube of length ℓ, envision joining the ends of the tube to form a circle. Then the $x = 0$ end physically "matches" the $x = \ell$ end; as such, the value of the solution u must be the same at $x = 0$ as it is at $x = \ell$, and the value of u_x must be the same at $x = 0$ as it is at $x = \ell$. See Figure 1.9.

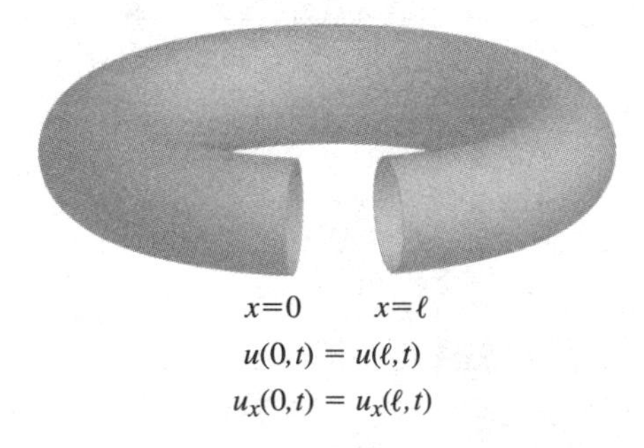

Figure 1.9: Periodic boundary conditions are relevant when modeling the diffusion of dye in a circular tube.

Other Boundary Conditions

Here, we have emphasized the major types of boundary conditions that arise when modeling physically realistic problems with PDEs. Note, however, that there are many other types of boundary conditions possible, both physical and nonphysical ones.

Exercises

1. Consider the 1D heat equation for a rod of length ℓ. Suppose the left endpoint is insulated so that no heat energy flows in or out of that endpoint. Write the boundary condition modeling this scenario.

2. Set up the PDE and boundary conditions that model the temperature $u := u(x,t)$ in a rod of length ℓ, where the left endpoint convects with an outside medium whose temperature is $M(t)$. The right endpoint has a thermostat attached which maintains a temperature $f(t)$.

3. In the context of the 1D heat equation for a rod of length ℓ, match the following boundary conditions with their physical interpretation. Roman numerals can be used more than once if needed.

(a) $u(0,t) = 1$	(I) fixed temperature of 1 unit
(b) $-u(0,t) = 1$	(II) fixed outward flux of 1 unit
(c) $u_x(0,t) = 1$	(III) realistic convection with an outside medium maintained at 1 unit
(d) $-u_x(0,t) = 1$	(IV) unrealistic convection with an outside medium maintained at 1 unit
(e) $u_x(\ell,t) = 1$	(V) fixed temperature of -1 unit
(f) $-u_x(\ell,t) = 1$	(VI) fixed inward flux of 1 unit
(g) $u_x(0,t) = u(0,t) - 1$	(VII) none of the above
(h) $u_x(\ell,t) = u(\ell,t) - 1$	

4. Consider the 1D heat equation for a rod of length ℓ.

   (a) Explain why the Neumann boundary condition $u_x(0,t) = R > 0$ is called a *radiating* condition, while $u_x(\ell,t) = R > 0$ is called an *absorbing* condition.

   (b) Is $u_x(0,t) = R < 0$ radiating or absorbing? Explain.

   (c) Is $u_x(\ell,t) = R < 0$ radiating or absorbing? Explain.

5. Without solving a PDE—using just physical intuition—answer the following:

   (a) Find $\lim_{t\to\infty} u(x,t)$ for $u_t = u_{xx}$, $u(0,t) = T_1$, $u_x(\ell,t) = T_2$.

   (b) Find $\lim_{t\to\infty} u(x,t)$ for $u_t = u_{xx}$, $u_x(0,t) = 0$, $-u_x(\ell,t) = u(\ell,t) - 100$.

6. Interpret the meaning of periodic boundary conditions for the 1D wave equation.

7. Consider the Robin boundary condition $\frac{\partial u}{\partial x}(0,t) + u(0,t) = -100$. Is this a physically realistic Robin boundary condition? Explain.

1.4 ODE Solution Methods

In this section, we give a brief review of some concepts and techniques from calculus and/or a first course in ODEs needed to study PDEs.

Linear Operators and Linear Equations

Linearity is a central concept in mathematics, and plays an important role throughout the calculus. We begin our study of linearity with a seemingly abstract definition; however, we will soon recognize it as a familiar idea.

An *operator* is a mapping between function spaces; the input is a function and the output is another function. For example, let $\mathscr{L} := \frac{d}{dx}$ so that $\mathscr{L}$ is the operator whose action is differentiation with respect to x. The input to $\mathscr{L}$ must be a differentiable function (else $\mathscr{L}$ cannot act on it) and the output of $\mathscr{L}$ is another function (possibly differentiable, but maybe not). $\mathscr{L}$ is an operator because it is a mapping between these spaces of functions. We could describe this operator in two ways:

$$\mathscr{L} := \frac{d}{dx} \qquad \text{or} \qquad \mathscr{L}u := \frac{du}{dx}.$$

Both say the same thing: the first tells us the action that is to be performed; the second demonstrates the action on a particular input function u. For the latter, we may write $\mathscr{L}u$ or $\mathscr{L}(u)$ or even $\mathscr{L}[u]$; these all denote the same thing.

Example 1.4.1. Let's look at some examples of operators. Here, $u := u(x)$.

1. Let $\mathscr{L}u := \frac{du}{dx}$. Then the action of $\mathscr{L}$ on an input u is differentiation with respect to x. For example, $\mathscr{L}(\sin x) = \cos x$ and $\mathscr{L}(\ln x) = \frac{1}{x}$.

2. Let $\mathscr{L}u := \int u(x)\,dx$. Then the action of $\mathscr{L}$ on an input u is antidifferentiation. For example, $\mathscr{L}(x^2) = \frac{x^3}{3} + C$ and $\mathscr{L}(\pi) = \pi x + C$.

3. Let $\mathscr{L}\mathbf{x} := A\mathbf{x}$, where $\mathbf{x}$ is an $n \times 1$ vector and A an $n \times n$ matrix. The action of $\mathscr{L}$ on an input vector $\mathbf{x}$ is matrix multiplication by A.

4. Let $\mathscr{L}u := uu'$. The action of $\mathscr{L}$ is multiplication of the input by its derivative.

Several other examples are explored in the exercises. ◊

Definition 1.1. We say that $\mathscr{L}$ is a *linear operator* if

$$\mathscr{L}(f+g) = \mathscr{L}f + \mathscr{L}g \quad \text{and} \quad \mathscr{L}(cf) = c\mathscr{L}f \tag{1.11}$$

hold for all functions f, g in the domain of $\mathscr{L}$ and all constants c. An operator that is not linear is called *nonlinear* .

Example 1.4.2. Let's revisit the previous example to see which operators are in fact linear.

1. $\mathscr{L}u := \frac{du}{dx}$ is a linear operator because

$$\begin{aligned}\mathscr{L}(f+g) &= \frac{d}{dx}(f+g) = \frac{df}{dx} + \frac{dg}{dx} = \mathscr{L}f + \mathscr{L}g,\\ \mathscr{L}(cf) &= \frac{d}{dx}(cf) = c\frac{df}{dx} = c\mathscr{L}f\end{aligned}$$

both hold for all valid inputs f, g and constants c. This restates what we learned in calculus: differentiation is a *linear* operation.

2. $\mathscr{L}u := \int u(x)\,dx$ is a linear operator because

$$\begin{aligned}\mathscr{L}(f+g) &= \int (f+g)\,dx = \int f(x)\,dx + \int g(x)\,dx = \mathscr{L}f + \mathscr{L}g,\\ \mathscr{L}(cf) &= \int cf(x)\,dx = c\int f(x)\,dx = c\mathscr{L}f\end{aligned}$$

both hold for all valid inputs f, g and constants c. This restates what we learned in calculus: antidifferentiation is a *linear* operation.

3. $\mathscr{L}\mathbf{x} := A\mathbf{x}$ is a linear operator because

$$\begin{aligned}\mathscr{L}(\mathbf{u}+\mathbf{v}) &= A(\mathbf{u}+\mathbf{v}) = A\mathbf{u} + A\mathbf{v} = \mathscr{L}\mathbf{u} + \mathscr{L}\mathbf{v},\\ \mathscr{L}(c\mathbf{u}) &= A(c\mathbf{u}) = cA\mathbf{u} = c\mathscr{L}\mathbf{u}\end{aligned}$$

both hold for all valid inputs $\mathbf{u}$, $\mathbf{v}$ and constants c. This restates the linear algebra fact that matrix multiplication is a *linear* operation.

4. $\mathscr{L}u := uu'$ is *not* a linear operator because

$$\mathscr{L}(f+g) = (f+g)(f+g)' = ff' + fg' + gf' + gg' \quad \text{while} \quad \mathscr{L}f + \mathscr{L}g = ff' + gg',$$

so the first part of (1.11) fails. ◇

We can form equations with operators. For example, the differential equation $y'' + 2y' + 3y = 0$ can be expressed as the operator equation $\mathscr{L}u = 0$ by defining $\mathscr{L}u := u'' + 2u' + 3u$. This abstraction leads us to an important classification system for differential equations.

Definition 1.2. A differential equation is *linear* if it can be written in operator form $\mathscr{L}u = f$, where $\mathscr{L}$ is a linear differential operator and f does not depend on u but only on the independent variable(s). A differential equation which is not linear is called *nonlinear*. For linear equations, if $f \equiv 0$, we say the equation is *homogeneous*; otherwise, the equation is *nonhomogeneous*.

A powerful consequence for linear differential equations is the following.

Theorem 1.1 (Superposition Principle).
If $\{u_1, u_2, \dots, u_n\}$ all satisfy the linear equation $\mathscr{L}u = 0$, then any linear combination

$$c_1 u_1 + c_2 u_2 + \cdots + c_n u_n$$

also satisfies $\mathscr{L}u = 0$.

Proof. Since $\mathscr{L}$ is a linear operator,

$$\begin{aligned}
\mathscr{L}(c_1 u_1 + c_2 u_2 + \cdots + c_n u_n) &= \mathscr{L}(c_1 u_1) + \mathscr{L}(c_2 u_2) + \cdots + \mathscr{L}(c_n u_n) \\
&= c_1 \mathscr{L} u_1 + c_2 \mathscr{L} u_2 + \cdots + c_n \mathscr{L} u_n \\
&= c_1 \cdot 0 + c_2 \cdot 0 + \cdots + c_n \cdot 0 \\
&= 0.
\end{aligned}$$

□

Next, we review some basic ODE solution techniques that will be used throughout the text.

First Order Linear Equations

Consider the homogeneous first order linear ODE

$$y'(t) + p(t)y(t) = 0. \tag{1.12}$$

This equation is separable and can be solved by direct integration:

$$\begin{aligned}
\frac{dy}{dt} &= -p(t)y \\
\frac{dy}{y} &= -p(t)\,dt \\
\int \frac{dy}{y} &= -\int p(t)\,dt \\
\ln|y| + C &= -\int p(t)\,dt \\
y(t) &= De^{-\int p(t)\,dt}.
\end{aligned}$$

If $p(t) \equiv p$ (p is a constant instead of a function of t), then (1.12) takes on a particularly familiar form:

$$y' + py = 0 \qquad \text{or} \qquad \frac{dy}{dt} = -py.$$

This is the most common ODE from calculus since it is the exponential growth/decay model (the time rate of change of y is proportional to y). The general solution is $y(t) = Ce^{-pt}$, where C is an arbitrary constant.

On the other hand, consider the nonhomogeneous case,

$$y'(t) + p(t)y(t) = f(t). \tag{1.13}$$

Here, the *integrating factor* $m(t) = e^{\int p(t)\,dt}$ must be employed: multiply both sides of (1.13) by $m(t)$. A little calculus will show (1.13) reduces to

$$\begin{aligned} [m(t)y(t)]' &= m(t)f(t) \\ m(t)y(t) &= \int m(t)f(t)\,dt + C \\ y(t) &= \frac{\int m(t)f(t)\,dt + C}{m(t)}. \end{aligned}$$

Homogeneous Second Order Linear Equations with Constant Coefficients

Consider the prototype equation

$$ay''(t) + by'(t) + cy(t) = 0, \qquad a, b, c \in \mathbb{R}. \tag{1.14}$$

We look for solutions of the form $y(t) = e^{rt}$, where r is a parameter since a linear combination of an exponential function together with its first and second derivatives has a reasonable chance of combining to give 0. If $y(t) = e^{rt}$, then $y'(t) = re^{rt}$ and $y''(t) = r^2e^{rt}$. Substituting these into (1.14), we obtain

$$ar^2e^{rt} + bre^{rt} + ce^{rt} = 0.$$

Since e^{rt} is never zero (r is a real number here), we can divide by it to obtain the *characteristic equation* for (1.14):

$$ar^2 + br + c = 0. \tag{1.15}$$

Stop for a moment to compare (1.14) with (1.15). The simplicity (and beauty!) of this method is that we have now reduced solving the apparently complicated *differential* equation (1.14) to solving the very elementary *algebraic* equation (1.15).

The nature of the solutions to (1.14) is dictated by the nature of the roots of the quadratic equation (1.15), which has solutions

$$r_{1,2} = \frac{-b \pm \sqrt{b^2 - 4ac}}{2a}.$$

There are three cases based on the sign of the discriminant $b^2 - 4ac$.

- Case 1: $b^2 - 4ac > 0$, a pair of distinct real roots. Then—as expected—the second order equation (1.14) has two linearly independent solutions given by e^{r_1t} and e^{r_2t}. The general solution to (1.14) in this case is

$$y(t) = c_1e^{r_1t} + c_2e^{r_2t}.$$

- CASE 2: $b^2 - 4ac = 0$, A REPEATED REAL ROOT. Here, $r_1 = r_2 = -b/2a$, so one solution to (1.14) is $e^{r_1 t}$, but we need a second linearly independent solution in order to write down the general solution. Using methods from your ODE course, the second linearly independent solution we seek is given by $te^{r_1 t}$. Therefore, the general solution to (1.14) in this case is

$$y(t) = c_1 e^{r_1 t} + c_2 t e^{r_1 t}.$$

- CASE 3: $b^2 - 4ac < 0$, A PAIR OF COMPLEX CONJUGATE ROOTS. Here, $r_{1,2} = \alpha \pm \beta i$, where α and β are determined by the values of a, b, c, so the two solutions are $e^{(\alpha+\beta i)t}$ and $e^{(\alpha-\beta i)t}$. Using Euler's formula $e^{i\theta} = \cos\theta + i\sin\theta$, and methods from your ODE course, these produce the linearly independent solutions $e^{\alpha t}\cos\beta t$ and $e^{\alpha t}\sin\beta t$. The general solution to (1.14) in this case is

$$\begin{aligned} y(t) &= c_1 e^{\alpha t}\cos\beta t + c_2 e^{\alpha t}\sin\beta t \\ &= e^{\alpha t}(c_1\cos\beta t + c_2\sin\beta t). \end{aligned}$$

Initial Value Problems: General and Particular Solutions

In many ODE courses, only *initial value problems* (IVPs) are studied. For example,

$$y''(t) + 2y'(t) + 5y(t) = 0, \qquad y(0) = 1,\ y'(0) = 0, \tag{1.16}$$

specifies two conditions on the solution $y(t)$ of the given ODE. These are *initial conditions* since both are specified at the *same* point ($t = 0$ here). The word choice comes from the fact that the variable is often thought of as time and the point where the conditions are specified is thought of as the starting or initial time.

Solving the IVP (1.16) consists of first finding the *general solution* using the methods of the previous section:

$$\begin{aligned} y'' + 2y' + 5y &= 0 \\ r^2 + 2r + 5 &= 0 && \leftarrow \text{characteristic equation} \\ r &= -1 \pm 2i \\ y(t) &= e^{-t}(c_1\cos 2t + c_2\sin 2t). && \leftarrow \text{general solution} \end{aligned}$$

To find the *particular solution* that satisfies $y(0) = 1$, $y'(0) = 0$, we use the general solution:

$$\begin{aligned} y(0) &= e^0(c_1\cos 0 + c_2\sin 0) = 1 \implies c_1 = 1 \\ y'(t) &= -e^{-t}(c_1\cos 2t + c_2\sin 2t) + e^{-t}(-2c_1\sin 2t + 2c_2\cos 2t) \\ y'(0) &= -e^0(c_1\cos 0 + c_2\sin 0) + e^0(-2c_1\sin 0 + 2c_2\cos 0) = 0 \implies c_2 = 1/2 \\ y(t) &= e^{-t}(\cos 2t + \frac{1}{2}\sin 2t). \qquad \leftarrow \text{particular solution} \end{aligned}$$

Different sets of initial conditions yield different values of c_1, c_2 and, hence, different particular solutions. The key here is they are all fruit from the same tree; that tree being the general solution of the differential equation at hand.

Boundary Value Problems

In contrast to (1.16), consider the *boundary value problem* (BVP)

$$y''(x) + 2y'(x) + 5y(x) = 0, \qquad y(0) = 1,\ y(4) = 0.$$

Here, the two conditions the solution $y(x)$ must satisfy are specified at $x = 0$ and $x = 4$. These are called *boundary conditions* since we often think of only looking for the solution at values of x between these two extremes, i.e., solving for $0 < x < 4$.

Finding the general solution has nothing to do with initial or boundary conditions, only the differential equation itself, so in this case there is nothing new to do:

$$y(x) = e^{-x}(c_1 \cos 2x + c_2 \sin 2x).$$

To find the particular solution satisfying $y(0) = 1$, $y(4) = 0$, apply those conditions and solve for c_1, c_2:

$$\begin{aligned}
y(0) &= e^0(c_1 \cos 0 + c_2 \sin 0) = 1 && \Longrightarrow && c_1 = 1\\
y(4) &= e^{-4}(\cos 8 + c_2 \sin 8) = 0 && \Longrightarrow && c_2 = -\cot 8\\
y(x) &= e^{-x}(\cos 2x - \cot 8 \sin 2x) && && \leftarrow \text{particular solution}
\end{aligned}$$

Obtaining a unique solution to a boundary value problem is a much more delicate matter than for an initial value problem. The structure of solutions to an ODE can vary dramatically as different boundary conditions are imposed. For example, the relatively simple equation $y''(x) + y(x) = 0$:

- subject to the boundary conditions $y(0) = 2$, $y(\pi) = 0$ has no solution.
- subject to the boundary conditions $y(0) = 2$, $y(\pi) = -2$ has infinitely many solutions.
- subject to the boundary conditions $y(0) = 2$, $y(\pi/2) = 0$ has exactly one solution.

See Exercise 12.

Cauchy-Euler Equations

The natural next step from (1.14) is to consider nonconstant coefficient equations of this type, i.e., equations of the form $a(x)y''(x) + b(x)y'(x) + c(x)y(x) = 0$. However, this is a tall task in general, even for "tame" linear equations. Instead, we focus our energies on a particular class of nonconstant coefficient equations called *Cauchy-Euler equations* (named in honor of Augustin-Louis Cauchy and Leonhard Euler; see Figure 1.10) which have the standard form[12]

$$ax^2y''(x) + bxy'(x) + cy(x) = 0, \qquad a, b, c \in \mathbb{R},\ a \neq 0. \tag{1.17}$$

[12]For convenience, we will assume $x > 0$.

Figure 1.10: Augustin-Louis Cauchy (1789–1857) of France was among the most impactful mathematicians of all time, having published 789 papers. However, he was known for a coarse personality. In a letter, Abel remarked that Cauchy is mad and that nothing that could be done about him, although, right now, he is the only one who knows how mathematics is supposed to be done.

Two avenues are reasonable to pursue when deriving the general solution to (1.17). One is to look for solutions of a particular form, like we did to solve (1.14). This time, the form we choose is $y(x) = x^p$, so that $y'(x) = px^{p-1}$ and $y''(x) = p(p-1)x^{p-2}$. Substituting these into (1.17),

$$\begin{aligned} ax^2p(p-1)x^{p-2} + bxpx^{p-1} + cx^p &= 0 \\ ap(p-1)x^p + bpx^p + cx^p &= 0 \\ x^p(ap^2 + (b-1)p + c) &= 0. \end{aligned}$$

Since $x^p \neq 0$, we can divide both sides by it to obtain the *Cauchy-Euler characteristic equation*:

$$ap^2 + (b-1)p + c = 0. \tag{1.18}$$

Similar to our analysis of (1.16), the nature of the solutions to (1.17) is determined by the roots of (1.18).

- Case 1: a pair of distinct real roots. If p_1, p_2 are distinct real roots, the general solution to (1.17) is $y(x) = c_1x^{p_1} + c_2x^{p_2}$.
- Case 2: a repeated real root. If p_1 is a repeated real root, the general solution to (1.17) is $y(x) = c_1x^{p_1} + c_2x^{p_1}\ln x$.
- Case 3: a pair of complex conjugate roots. If $\alpha \pm \beta i$ are roots, the general solution to (1.17) is $y(x) = c_1x^\alpha \cos(\beta \ln x) + c_2x^\alpha \sin(\beta \ln x)$.

As mentioned, a second method for solving (1.17) is possible. If we make the change of variables $x = e^t$, (1.17) becomes an equation of the form

$$ay''(t) + (b-1)y'(t) + cy(t) = 0.$$

The three cases of the form of the general solution for this equation can be translated to the solutions in the three cases above via $x = e^t$. See Exercise 14.

The Hyperbolic Trigonometric Functions

The hyperbolic trigonometric functions will be useful to use when solving boundary value problems. We begin with their definitions:

$$\cosh x := \frac{e^x + e^{-x}}{2}, \qquad \sinh x := \frac{e^x - e^{-x}}{2}. \tag{1.19}$$

The calculus of these functions is strikingly similar to that of the usual trig functions:

$$\tanh x = \frac{\sinh x}{\cosh x}$$
$$\cosh(-x) = \cosh(x), \quad \sinh(-x) = -\sinh(x), \quad \tanh(-x) = -\tanh(x)$$
$$\cosh(0) = 1, \quad \sinh(0) = 0$$
$$\frac{d}{dx}\cosh x = \sinh x, \quad \frac{d}{dx}\sinh x = \cosh x.$$

Note that $\cosh x \neq 0$ always, while $\sinh x = 0$ and $\tanh x = 0$ only when $x = 0$. See Figure 1.11.

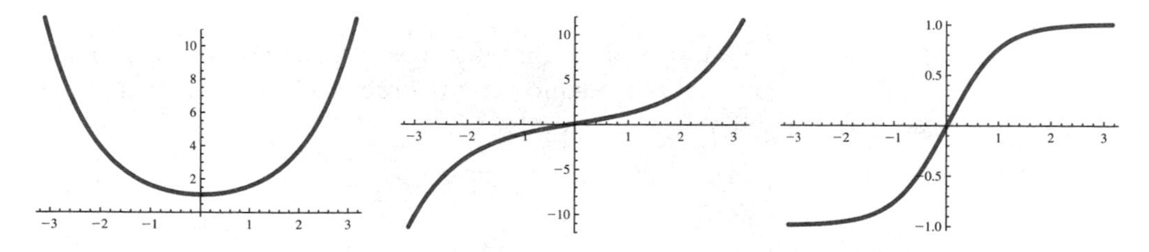

Figure 1.11: The graphs of $\cosh x$, $\sinh x$, and $\tanh x$.

The main utility of the hyperbolic functions will be in solving certain boundary value problems. Consider, for example, $y''(x) - 4y(x) = 0$. From before, the characteristic equation is $r^2 - 4 = 0$, so $r = \pm 2$, and hence the general solution is

$$y(x) = c_1 e^{2x} + c_2 e^{-2x}. \tag{1.20}$$

By using (1.19), we could also write the general solution as

$$y(x) = c_3 \cosh(2x) + c_4 \sinh(2x). \tag{1.21}$$

When solving boundary value problems, it will often be computationally easier to work with (1.21) rather than (1.20); see Exercise 15. It behooves us to become comfortable with the hyperbolic trig functions.

Exercises

1. Determine whether the following operators are linear or nonlinear. Prove your assertion.

   (a) $\mathscr{L}u := au'' + bu' + cu$, where $a, b, c \in \mathbb{R}$

   (b) $\mathscr{L}u := |u|$

   (c) $\mathscr{L}u := \int_0^\infty e^{-st}u(t)\,dt$ (the Laplace transform operator)

   (d) $\mathscr{L}\mathbf{x} := \mathbf{x}^T\mathbf{x}$ where $\mathbf{x}^T$ denotes the transpose of the $n \times 1$ vector $\mathbf{x}$

2. Write each of the PDEs (1.4)–(1.6) in operator form. Determine which ones are linear/nonlinear.

3. Write each of these differential equations in operator form. Determine whether it is a linear or nonlinear equation. If the equation is linear, determine whether it is homogeneous or nonhomogeneous. Explain your answer.

   (a) $y'' + y = y^2$

   (b) $\frac{dy}{dt} = ky$, k constant

   (c) $a(x)y''(x) + b(x)y'(x) + c(x)y(x) = \sin(x)$

   (d) $\frac{dy}{dx} = \frac{x}{y}$

4. (a) Let $\mathscr{L}$ be a linear operator. Show that if φ solves the homogeneous problem $\mathscr{L}u = 0$ and ψ solves the nonhomogeneous problem $\mathscr{L}u = f$, then $\varphi + \psi$ also solves the nonhomogeneous problem $\mathscr{L}u = f$.

   (b) Is the same result true if $\mathscr{L}$ is not a linear operator? Explain why or why not.

5. Consider $\frac{dy}{dt} = y$.

   (a) Find the general solution.

   (b) Find the particular solution satisfying the initial condition $y(1) = 2$.

6. Consider $y' + 2ty = t$.

   (a) Find the general solution.

   (b) Find the particular solution satisfying the initial condition $y(0) = \sqrt{2}$.

7. Consider $y' - \frac{2}{x}y = x^2 \cos x$, $x > 0$.

   (a) Find the general solution.

   (b) Find the particular solution satisfying the initial condition $y(\pi) = 1$.

8. Consider $y'' + 4y' - 5y = 0$.

   (a) Find the general solution.

   (b) Find the particular solution satisfying the initial conditions $y(0) = 1$, $y'(0) = 0$.

   (c) Find the particular solution satisfying the boundary conditions $y(0) = 0$, $y'(1) = 1$.

9. Consider $y'' + 6y' + 9y = 0$.

   (a) Find the general solution.

   (b) Find the particular solution satisfying the initial conditions $y(0) = 1$, $y'(0) = -1$.

   (c) Find the particular solution satisfying the boundary conditions $y(0) = 1$, $y(1) = 0$.

10. Consider $y'' - 4y' + 5y = 0$.

    (a) Find the general solution.

    (b) Find the particular solution satisfying the initial conditions $y(0) = 1$, $y'(0) = 1$.

    (c) Find the particular solution satisfying the boundary conditions $y'(0) = 1$, $y(\pi/2) = 0$.

11. Consider $y'' + 16y = 0$.

    (a) Find the general solution.

    (b) Find the particular solution satisfying the initial conditions $y(0) = 1$, $y'(0) = 1$.

    (c) Find the particular solution satisfying the boundary conditions $y'(0) = -4$, $y'(1) = 0$.

12. Consider $y'' + y = 0$.

    (a) Show that this equation together with the boundary conditions $y(0) = 2$, $y(\pi) = 0$ has no solution.

    (b) Show that this equation together with the boundary conditions $y(0) = 2$, $y(\pi) = -2$ has infinitely many solutions.

    (c) Show that this equation together with the boundary conditions $y(0) = 2$, $y(\pi/2) = 0$ has exactly one solution.

13. Find the general solution for $x > 0$:

    (a) $x^2y''(x) + 2xy'(x) - 6y(x) = 0$

    (b) $x^2y''(x) + 9xy'(x) + 17y(x) = 0$

14. This exercise will show the connection between the three forms of general solution of (1.14) and the three forms of the general solution of (1.17).

    (a) Let $y(x)$ be a solution of (1.17), and consider the substitution $x = e^t$. Show $\frac{dy}{dx} = e^{-t}\frac{dy}{dt}$ and $\frac{d^2y}{dx^2} = \left(\frac{d^2y}{dt^2} - \frac{dy}{dt}\right)e^{-2t}$. (Hint: By the chain rule, $\frac{dy}{dx} = \frac{dy}{dt}\frac{dt}{dx}$.)

    (b) Substitute these into (1.17) to obtain

    $$y''(t) + (b-1)y'(t) + cy(t) = 0. \qquad (*)$$

    That is, the Cauchy-Euler equation (1.17) has been transformed into a standard constant coefficient equation of the form (1.14).

    (c) Suppose the characteristic equation for $(*)$ has distinct real roots r_1, r_2. Then the general solution is $y(t) = c_1e^{r_1t} + c_2e^{r_2t}$. Use $x = e^t$ to transform this general solution to one of the form $y(x) = c_1x^{r_1} + c_2x^{r_2}$.

    (d) Suppose the characteristic equation for $(*)$ has a repeated real root r_1. Then the general solution is $y(t) = c_1e^{r_1t} + c_2te^{r_1t}$. Use $x = e^t$ to transform this general solution to one of the form $y(x) = c_1x^{r_1} + c_2x^{r_1}\ln x$.

    (e) Suppose the characteristic equation for $(*)$ has a complex conjugate pair of roots $\alpha \pm \beta i$. Then the general solution is $y(t) = e^{\alpha t}(c_1 \cos \beta t + c_2 \sin \beta t)$. Use $x = e^t$ to transform this general solution to one of the form $y(x) = c_1x^{\alpha}\cos(\beta \ln x) + c_2x^{\alpha}\sin(\beta \ln x)$.

15. Consider $y'' - 25y = 0$, $y(0) = 0$, $y'(1) = 1$.

    (a) Write the general solution as a linear combination of exponential functions.

    (b) Solve for the constants in the general solution in (a) by applying the boundary conditions.

    (c) Write the general solution as a linear combination of hyperbolic cosine and hyperbolic sine.

    (d) Solve for the constants in the general solution in (c) by applying the boundary conditions.

    (e) Which was easier: (b) or (d)?

Chapter 2

Fourier's Method: Separation of Variables

2.1 Linear Algebra Concepts

The Geometry of Vector Spaces

There are many parallels between linear algebra and differential equations because they are based upon similar theoretical foundations. The setting for studying linear algebra is the *vector space.* When the vector space has an associated *inner product* or *dot product*, there is a concept of "size" or "length" of a vector arrived at by computing its *norm* defined in terms of the inner product:

$$\|v\| := \sqrt{\langle \mathbf{v}, \mathbf{v} \rangle} = \sqrt{\mathbf{v} \cdot \mathbf{v}}.$$

The inner product can also be used to determine when two vectors are *orthogonal* (perpendicular):

$$\mathbf{u} \perp \mathbf{v} \Longleftrightarrow \langle \mathbf{u}, \mathbf{v} \rangle = 0,$$

that is, two vectors are orthogonal when their inner product is zero.

Let $S := \{\mathbf{v}_1, \ldots, \mathbf{v}_n\}$ be a set of vectors in the vector space V. S is *linearly independent* if

$$c_1\mathbf{v}_1 + \cdots + c_n\mathbf{v}_n = 0 \Longleftrightarrow c_1 = \cdots = c_n = 0.$$

S is *linearly dependent* if it is not linearly independent. S *spans* V if every vector in V can be written as a linear combination of elements of S.

B is a *basis* for the vector space V if B is a linearly independent set that spans V. The *dimension* of V is the number of elements in B. Although there may be many bases for V, they must all have the same dimension.

Example 2.1.1. We demonstrate these concepts with the prototypical vector space from linear algebra: $V = \mathbb{R}^n$. Consider the vectors $\mathbf{u} = (u_1, u_2, \ldots, u_n)$ and $\mathbf{v} = (v_1, v_2, \ldots, v_n)$ from $\mathbb{R}^n$.

- The inner product is the usual dot product: $\langle \mathbf{u}, \mathbf{v} \rangle = \mathbf{u} \cdot \mathbf{v} = u_1 v_1 + \cdots + u_n v_n$.
- The norm of $\mathbf{u}$ is given by $\|\mathbf{u}\| = \sqrt{\langle \mathbf{u}, \mathbf{u} \rangle} = \sqrt{u_1^2 + \cdots + u_n^2}$.
- $\mathbf{u}$ and $\mathbf{v}$ are orthogonal (in symbols, $\mathbf{u} \perp \mathbf{v}$) when $\langle \mathbf{u}, \mathbf{v} \rangle = u_1 v_1 + \cdots + u_n v_n = 0$.
- Let $\mathbf{e}_i$ denote the vector with 1 in the ith entry and 0 everywhere else. Then $B := \{\mathbf{e}_1, \mathbf{e}_2, \ldots, \mathbf{e}_n\}$ is a basis for $\mathbb{R}^n$ because (i) the set is linearly independent, and (ii) the set spans $\mathbb{R}^n$. Since B has n elements, $\mathbb{R}^n$ has dimension n.
- Let $S := \{\mathbf{e}_1, \mathbf{e}_2, \ldots, \mathbf{e}_{n-1}\}$. S is a linearly independent set, but does not span $\mathbb{R}^n$ so S is *not* a basis for $\mathbb{R}^n$. (It is, however, a basis for $\mathbb{R}^{n-1}$). ◊

Example 2.1.2. Consider functions $y(x)$ defined on the interval $[\alpha, \beta]$, and let

$$V := \{y : ay'' + by' + cy = 0, \text{ where } a, b, c \in \mathbb{R}, \ a \neq 0\}.$$

- By the Superposition Principle, V is a vector space under the usual operations of addition and scalar multiplication of functions.
- If we define the inner product

$$\langle u, v \rangle := \int_\alpha^\beta u(x) v(x)\, dx, \tag{2.1}$$

  then we have a method for determining the "size" of elements in V by their norm:

$$\|u\| = \sqrt{\langle u, u \rangle} = \left(\int_\alpha^\beta |u(x)|^2\, dx \right)^{1/2}.$$

  We denote the set of all continuous functions on $[\alpha, \beta]$ by $C[\alpha, \beta]$; that is,

$$C[\alpha, \beta] := \{f : [\alpha, \beta] \to \mathbb{R} \;\big|\; f \text{ is continuous}\}.$$

  We will often (but not always) use the inner product (2.1) for $C[\alpha, \beta]$.
- u and v are orthogonal (in symbols, $u \perp v$) when $\langle u, v \rangle = \int_\alpha^\beta u(x) v(x)\, dx = 0$.
- Let $B := \{y_1(x), y_2(x)\}$ be a fundamental set of solutions[1] for $ay'' + by' + cy = 0$; that is, the general solution to $ay'' + by' + cy = 0$ can be written in the form $y(x) = c_1 y_1(x) + c_2 y_2(x)$. Then B is a basis for V since (i) it is a linearly independent set (the Wronskian[2] is nonzero), and (ii) any element of V (any solution) can be written as a linear combination of members of B. Since B has 2 elements, the vector space V has dimension 2. ◊

[1] Also called a *fundamental system.*

[2] The *Wronskian* of $\{y_1(x), y_2(x)\}$ is defined as $W(y_1, y_2) := \begin{vmatrix} y_1(x) & y_2(x) \\ y_1'(x) & y_2'(x) \end{vmatrix}$.

Eigenvalue Problems

Recall the eigenvalue problem from linear algebra: find values of the scalar λ such that

$$A\mathbf{x} = \lambda\mathbf{x}, \tag{2.2}$$

where A is an $n \times n$ matrix, and $\mathbf{x} \neq 0$ is an $n \times 1$ vector. Any λ satisfying (2.2) is called an *eigenvalue* and the corresponding vector $\mathbf{x}$ is called the associated *eigenvector*.

On the other hand, consider the boundary value problem $y'' + \lambda y = 0$, $y'(0) = 0$, $y(1) = 0$. If we set $\mathscr{L}u := -u''$, the problem can be rewritten in the form

$$\mathscr{L}u = \lambda u, \quad u'(0) = 0, \; u(1) = 0. \tag{2.3}$$

Compare (2.3) and (2.2): they are very similar! Both problems seek to find values of the parameter λ (eigenvalues) such that the action of the linear operator on a nontrivial vector/function (eigenvector/eigenfunction) is equal to scalar multiplication of that vector/function by λ.

Example 2.1.3. Consider the eigenvalue problem on the interval $0 < x < 1$:

$$y''(x) + \lambda y(x) = 0, \quad y'(0) = 0, \; y(1) = 0. \tag{2.4}$$

First, we can rewrite the differential equation in operator form by defining $\mathscr{L}u := -u''$ so that $y'' + \lambda y = 0$ becomes $\mathscr{L}u = \lambda u$, while the boundary conditions are unchanged:

$$\mathscr{L}u = \lambda u, \quad u'(0) = 0, \; u(1) = 0. \tag{2.5}$$

Note that (2.4) and (2.5) are simply two ways to state the same problem. The upshot to (2.5) is twofold: it is reminiscent of eigenvalue problems in linear algebra[3] and we can immediately identify the underlying operator. You should be equally comfortable with both forms.

Next, we examine three cases for the parameter λ:

- Case 1: $\lambda = 0$. The eigenvalue problem (2.4) reduces to $y''(x) = 0$, $y'(0) = y(1) = 0$. The general solution is $y(x) = Ax + B$ so $y'(x) = A$. Applying the boundary conditions, $y'(0) = A = 0$, so $A = 0$ and $y(1) = B = 0$ so $B = 0$. This says $y(x) \equiv 0$, i.e., y is the trivial solution. Therefore, $\lambda = 0$ is not an eigenvalue (since eigenvalues by definition must yield *nontrivial* eigenvectors).

- Case 2: $\lambda < 0$. Let's say $\lambda = -p^2 < 0$. The eigenvalue problem (2.4) becomes $y''(x) - p^2 y(x) = 0$, $y'(0) = y(1) = 0$. The general solution is $y(x) = A\cosh(px) + B\sinh(px)$ so $y'(x) = Ap\sinh(px) + Bp\cosh(px)$. Applying the first boundary condition, $y'(0) = Ap\sinh(0) + Bp\cosh(0) = 0$ so either $p = 0$ or $B = 0$. But $p \neq 0$ since $-p^2 < 0$ by assumption, so it must be that $B = 0$. Applying the second boundary condition, $y(1) = A\cosh p = 0$ so $A = 0$ or $\cosh p = 0$. However, $\cosh p \neq 0$, so it must be that $A = 0$. Since $A = B = 0$, the solution is $y(x) \equiv 0$, the trivial solution. Since this case produces only the trivial solution, there are no negative eigenvalues.

[3] In the sense that both have the form "operator applied to input equals scalar times input."

- CASE 3: $\lambda > 0$. Let's say $\lambda = p^2 > 0$. The eigenvalue problem (2.4) becomes $y''(x) + p^2 y(x) = 0$, $y'(0) = y(1) = 0$. The general solution is

$$y(x) = A\cos(px) + B\sin(px), \tag{2.6}$$

so that $y'(x) = -Ap\sin(px) + Bp\cos(px)$. The first boundary condition implies $y'(0) = Bp = 0$ so either $B = 0$ or $p = 0$. But $p \neq 0$ since $p^2 > 0$ by assumption, so it must be that $B = 0$. On the other hand, the second boundary condition implies $y(1) = A\cos p = 0$ so either $A = 0$ or $\cos p = 0$. Since $B = 0$ already, then $A = 0$ would yield the trivial solution $y(x) \equiv 0$, so instead we consider $\cos p = 0$, which has solutions $p = (2n-1)\frac{\pi}{2}$, $n = 0, \pm 1, \pm 2, \ldots$. Thus, the eigenvalues are

$$\lambda_n = p_n^2 = \left[(2n-1)\frac{\pi}{2}\right]^2, \quad n = 1, 2, \ldots.$$

To determine the eigenfunctions, we look back at the general solution (2.6) and see which terms were not forced to vanish in the above analysis. Since the first boundary condition forced $B = 0$, the eigenfunctions are

$$y_n(x) = A\cos(p_n x), \quad \text{or, equivalently,} \quad y_n(x) = A\cos(\sqrt{\lambda_n}\, x), \quad n = 1, 2, \ldots.$$

In this particular example, all eigenvalues were of one sign. Although this does not always happen, we will explore this issue in more detail later. ◊

Exercises

1. Define the vectors $\mathbf{u} = (1, 5)$ and $\mathbf{v} = (-\sqrt{2}, 2)$ in $\mathbb{R}^2$.

   (a) Find $\|\mathbf{u}\|$ and $\|\mathbf{v}\|$.

   (b) Are $\mathbf{u}$ and $\mathbf{v}$ orthogonal? Why or why not?

   (c) Are $\mathbf{u}$ and $\mathbf{v}$ linearly dependent or linearly independent? Explain.

   (d) Does $\{\mathbf{u}, \mathbf{v}\}$ form a basis for $\mathbb{R}^2$? Explain.

2. Let $C[0,\pi] := \{f : [0,\pi] \to \mathbb{R} \mid f \text{ is continuous}\}$. With addition and scalar multiplication defined in the usual way, this is a vector space. Let the inner product on $C[0,\pi]$ be defined analogous to (2.1), that is, $\langle u, v\rangle := \int_0^\pi u(x)v(x)\,dx$.

   (a) Let $f(x) = \sin x$ and $g(x) = x^2$. Which is "bigger": f or g?

   (b) Is $f \perp g$? Explain.

   (c) Find a nontrivial function in $C[0,\pi]$, which is orthogonal to f.

   (d) Find a nontrivial function in $C[0,\pi]$, which is orthogonal to g.

   (e) Make a conjecture on the dimension of $C[0,\pi]$.

3. Consider the vector space $C[0,1]$, with the standard norm arising from the inner product $\langle u, v\rangle := \int_0^1 u(x)v(x)\,dx$. Let $f(x) = \sqrt{x}$ and $g(x) = x$.
   (a) Compute $\|f\|$.
   (b) Is $f \perp g$ in this space? Justify your answer.

4. Consider the vector space $C[0,\ell]$ for $\ell > 0$, with the standard norm arising from the inner product $\langle u, v\rangle := \int_0^\ell u(x)v(x)\,dx$. Let $f(x) = x$ and $g(x) = 1 - x^2$.
   (a) Find ℓ such that f and g are orthogonal on the interval $[0,\ell]$.
   (b) Using the value of ℓ you found above, compute $\|f\|$ on the interval $[0,\ell]$.

5. Consider $C[-1,1] := \{f : [-1,1] \to \mathbb{R} \mid f \text{ is continuous}\}$ with the inner product given by $\langle u, v\rangle := \int_{-1}^1 u(x)v(x)\,dx$.
   (a) Find (by trial and error) a nontrivial function in $C[-1,1]$ which is orthogonal to $f(t) = t$.
   (b) Find a second degree polynomial, $p(t) = at^2 + bt + c$, that is orthogonal to $f(t) = t$ in this space. How many such $p(t)$ are there?
   (c) Find some $p(t)$ from (b) with $\|p\| = 1$.

6. Consider the boundary value problem $y''(x) + \lambda y(x) = 0$, $y(0) = 0$, $y(1) = 0$.
   (a) Set this up as an eigenvalue problem of the form $\mathscr{L}u = \lambda u$ with the required boundary conditions.
   (b) Is $\lambda = 0$ an eigenvalue? If so, find all eigenfunctions in this case. If not, explain why not.
   (c) Suppose $\lambda < 0$. Find all eigenvalues and eigenfunctions in this case.
   (d) Suppose $\lambda > 0$. Find all eigenvalues and eigenfunctions in this case.

7. Consider the boundary value problem $y''(x) + \lambda y(x) = 0$, $y'(0) = 0$, $y'(1) = 0$.
   (a) Set this up as an eigenvalue problem of the form $\mathscr{L}u = \lambda u$ with the required boundary conditions.
   (b) Is $\lambda = 0$ an eigenvalue? If so, find all eigenfunctions in this case. If not, explain why not.
   (c) Suppose $\lambda < 0$. Find all eigenvalues and eigenfunctions in this case.
   (d) Suppose $\lambda > 0$. Find all eigenvalues and eigenfunctions in this case.

8. Consider the boundary value problem $x^2y''(x) + xy'(x) + \lambda y(x) = 0$, $y(1) = 0$, $y(5) = 0$.
   (a) Set this up as an eigenvalue problem of the form $\mathscr{L}u = \lambda u$ with the required boundary conditions.

(b) Is $\lambda = 0$ an eigenvalue? If so, find all eigenfunctions in this case. If not, explain why not.

(c) Suppose $\lambda < 0$. Find all eigenvalues and eigenfunctions in this case.

(d) Suppose $\lambda > 0$. Find all eigenvalues and eigenfunctions in this case.

9. Consider the boundary value problem $y''(x) + y'(x) - \lambda y(x) = 0$, $y(0) = 0$, $y(1) = 0$.

   (a) Set this up as an eigenvalue problem of the form $\mathscr{L}u = \lambda u$ with the required boundary conditions.

   (b) The motivation for considering $\lambda = 0$, $\lambda > 0$, and $\lambda < 0$ in the previous problems was because the discriminant of the characteristic equation was simply λ. Show that the discriminant of this problem is $1 + 4\lambda$.

   (c) Consider the case when the discriminant $1 + 4\lambda = 0$; that is, $\lambda = -1/4$. Is $\lambda = -1/4$ an eigenvalue? If so, find all eigenfunctions in this case. If not, explain why not.

   (d) Consider the case when the discriminant $1 + 4\lambda > 0$; that is, $\lambda > -1/4$. Find all eigenvalues and eigenfunctions in this case.

   (e) Consider the case when the discriminant $1 + 4\lambda < 0$; that is, $\lambda < -1/4$. Find all eigenvalues and eigenfunctions in this case.

10. Show that the eigenfunctions in Example 2.1.3 are orthogonal on the interval $0 < x < 1$ with respect to the standard inner product.

11. A *norm* does not have to arise from an inner product. In fact, a norm $\|\cdot\|$ is defined as a function from a vector space V to the nonnegative real numbers which satisfies

   (N1) $\|x\| \geq 0$ for all $x \in V$

   (N2) $\|x\| = 0$ if and only if $x = 0$

   (N3) $\|cx\| = |c|\,\|x\|$ for all $x \in V$ and $c \in \mathbb{R}$

   (N4) $\|x + y\| \leq \|x\| + \|y\|$ for all $x, y \in V$

   (a) Show that the norm arising from the inner product $\langle u, v \rangle := \int_a^b u(x)v(x)\,dx$ on the vector space $C[a, b]$ satisfies (N1)–(N4).

   (b) Show that the norm $\|u\| := \max_{a \leq x \leq b} |u(x)|$ on the vector space $C[a, b]$ satisfies (N1)–(N4).

2.2 The General Solution via Eigenfunctions

The goal of this section is to find the general solution of the one dimensional heat and wave equations subject to a variety of physically relevant boundary conditions. Given the complicated nature of PDEs studied in the last chapter, it should not be surprising that it will take some work to accomplish this. However, history will be our guide: we will use the method of separation of variables due to Joseph Fourier.

Example 2.2.1. Consider the initial-boundary value problem for the heat equation,

$$u_t = ku_{xx}, \qquad 0 < x < \ell,\ t > 0, \tag{2.7a}$$

$$u(0,t) = u(\ell,t) = 0, \quad t > 0, \tag{2.7b}$$

$$u(x,0) = f(x), \qquad 0 < x < \ell. \tag{2.7c}$$

Fourier looked for solutions which are separated in space and time, i.e., solutions of the form $u(x,t) = X(x)T(t)$, called *product solutions* or *separated solutions.* Since $u(x,t)$ is assumed to be a solution, the PDE implies $X(x)T'(t) = kX''(x)T(t)$. After dividing by $kX(x)T(t)$, we can rewrite this as

$$\frac{1}{k}\frac{T'(t)}{T(t)} = \frac{X''(x)}{X(x)} = -\lambda.$$

Here, λ is a constant (to be determined) and the minus sign is simply for convenience. These are two ODEs—one in the time variable t and one in the space variable x:

$$T'(t) + \lambda k T(t) = 0 \qquad \text{and} \qquad X''(x) + \lambda X(x) = 0. \tag{2.8}$$

The time equation is first order in t and can be solved using the methods of Section 1.4. The general solution is $T(t) = Ce^{-\lambda kt}$, where C is an arbitrary constant.

However, the space equation is a second order ODE, so we expect a two parameter family of solutions. The first boundary condition in (2.7b) yields

$$u(0,t) = 0 \implies X(0)T(t) = 0 \implies X(0) = 0 \ \text{ or } \ T(t) = 0.$$

But the previous work reveals $T(t)$ is never zero unless $C = 0$, but $C = 0$ yields a *trivial solution* (i.e., $T(t) \equiv 0$), and we seek nontrivial[4] solutions. It must be that $X(0) = 0$. A similar argument shows $X(\ell) = 0$. Therefore, the X problem in (2.8) becomes an ODE boundary value problem—an eigenvalue problem[5] actually—in the x variable:

$$X''(x) + \lambda X(x) = 0, \qquad X(0) = X(\ell) = 0. \tag{2.9}$$

To solve for $X(x)$, we must consider three cases based on the possible sign of λ.

[4]Trivial solutions (ones that are identically zero) of course satisfy the PDE, but won't satisfy any interesting initial conditions.

[5]To see this, set $\mathscr{L}X := -X''$. Then (2.9) can be rewritten $\mathscr{L}X = \lambda X$, $X(0) = 0$, $X(\ell) = 0$, which is consistent with (2.3).

- CASE 1: $\lambda = 0$. The eigenvalue problem (2.9) reduces to $X''(x) = 0$, $X(0) = X(\ell) = 0$. The general solution is $X(x) = Ax + B$. Applying the boundary conditions, $X(0) = A \cdot 0 + B = 0$ so $B = 0$ and $X(\ell) = A\ell + B = 0$ so $A = 0$. This says $X(x) \equiv 0$, i.e., X is the trivial solution. Therefore, $\lambda = 0$ is not an eigenvalue.

- CASE 2: $\lambda < 0$. Let's say $\lambda = -p^2$. The eigenvalue problem (2.9) becomes $X''(x) - p^2 X(x) = 0$, $X(0) = X(\ell) = 0$. The general solution[6] is $X(x) = A\cosh(px) + B\sinh(px)$. Applying the boundary conditions, $X(0) = A\cosh(0) + B\sinh(0) = 0$, so $A = 0$ and $X(\ell) = A\cosh(p\ell) + B\sinh(p\ell) = 0$ implies $B\sinh(p\ell) = 0$. $B = 0$ results in the trivial solution and $\sinh(p\ell) = 0$ only has solution $p = 0$, but we assumed $\lambda = -p^2 < 0$. Since this case produces only the trivial solution, there are no negative eigenvalues.

- CASE 3: $\lambda > 0$. Let's say $\lambda = p^2$. The eigenvalue problem (2.9) becomes $X''(x) + p^2 X(x) = 0$, $X(0) = X(\ell) = 0$. The general solution is $X(x) = A\cos(px) + B\sin(px)$. Applying the boundary conditions,

$$\begin{aligned} X(0) = A\cos(0) + B\sin(0) = 0 \quad &\Longrightarrow \quad A = 0, \\ X(\ell) = A\cos(p\ell) + B\sin(p\ell) = 0 \quad &\Longrightarrow \quad B\sin(p\ell) = 0. \end{aligned}$$

  Either $B = 0$ or $\sin(p\ell) = 0$. Since $A = 0$, allowing $B = 0$ would make X the trivial solution. So it must be that $\sin(p\ell) = 0$; that is, $p\ell = n\pi$, $n = 1, 2, 3, \ldots$. Hence, the eigenvalues for (2.9) are $\lambda_n = p_n^2 = (n\pi/\ell)^2$, $n = 1, 2, 3, \ldots$, with corresponding eigenfunctions $X_n(x) = \sin(n\pi x/\ell)$, $n = 1, 2, 3, \ldots$.

Summarizing,

$$\lambda_n = (n\pi/\ell)^2, \quad X_n(x) = \sin(n\pi x/\ell), \quad T_n(t) = \exp(-(n\pi/\ell)^2 kt), \quad n = 1, 2, 3, \ldots,$$

where $X_n(x)$ and $T_n(t)$ are only specified up to an arbitrary multiplicative constant. By the Superposition Principle (Theorem 1.1), any linear combination of solutions is again a solution, and therefore the solution to the original initial-boundary value problem (2.7) is

$$\boxed{u(x,t) = \sum_{n=1}^{\infty} c_n X_n(x) T_n(t) = \sum_{n=1}^{\infty} c_n \sin(n\pi x/\ell)\exp(-(n\pi/\ell)^2 kt).}$$

A more succinct way to write this is

$$\boxed{u(x,t) = \sum_{n=1}^{\infty} c_n \sin(\sqrt{\lambda_n}\, x)\exp\left(-\lambda_n kt\right),}$$

which is fine as long as we clearly state what the eigenvalues λ_n are. The only thing left to do is determine the coefficients c_n. We will tackle this a little later using some other methods due to Fourier. See Figure 2.1. ◊

Figure 2.1: Jean Baptiste Joseph Fourier (1768–1830) of France was greatly influenced by the teachings of Lagrange (his dissertation advisor) and Laplace. They served as referees for his 1807 memoir *On the Propagation of Heat in Solid Bodies,* in which Fourier introduced the notion of expanding a function as an infinite series of trigonometric functions. Later, Fourier codirected Dirichlet's dissertation and Dirichlet carried on the work.

Example 2.2.2. Consider the initial-boundary value problem for the wave equation,

$$\begin{aligned}
u_{tt} &= c^2 u_{xx}, && 0 < x < \ell,\ t > 0, && (2.10a)\\
u(0,t) &= u(\ell,t) = 0, && t > 0, && (2.10b)\\
u(x,0) &= f(x), && 0 < x < \ell, && (2.10c)\\
u_t(x,0) &= g(x), && 0 < x < \ell. && (2.10d)
\end{aligned}$$

Let $u(x,t) = X(x)T(t)$. The PDE implies $X(x)T''(t) = c^2X''(x)T(t)$. Dividing by $c^2X(x)T(t)$, we can rewrite this as

$$\frac{1}{c^2}\frac{T''(t)}{T(t)} = \frac{X''(x)}{X(x)} = -\lambda.$$

This yields two ODEs both of which are second order:

$$T''(t) + \lambda c^2 T(t) = 0 \qquad \text{and} \qquad X''(x) + \lambda X(x) = 0.$$

Since the spatial problem dictates the eigenvalues/eigenfunctions, we begin there. Translating (2.10b),

$$X''(x) + \lambda X(x) = 0, \qquad X(0) = X(\ell) = 0.$$

[6]It can also be written as $X(x) = Ce^{-px} + De^{px}$, but the hyperbolic trig function form is easier to use in boundary value problems, as discussed in Section 1.4, Exercise 15.

We already solved this eigenvalue problem in Example 2.2.1. The eigenvalues and eigenfunctions are

$$\lambda_n = \left(\frac{n\pi}{\ell}\right)^2, \qquad X_n(x) = \sin(\sqrt{\lambda_n}\,x), \qquad n = 1, 2, 3, \ldots.$$

We now return to the time problem, $T''(t) + \lambda_n c^2 T(t) = 0$, which has general solution

$$T_n(t) = A\cos(\sqrt{\lambda_n}\,ct) + B\sin(\sqrt{\lambda_n}\,ct), \quad n = 1, 2, 3, \ldots.$$

Since the coefficients could vary with n, it is appropriate to write this as

$$T_n(t) = a_n\cos(\sqrt{\lambda_n}\,ct) + b_n\sin(\sqrt{\lambda_n}\,ct), \quad n = 1, 2, 3, \ldots.$$

By the Superposition Principle, any linear combination of solutions is again a solution; therefore,

$$\boxed{u(x,t) = \sum_{n=1}^{\infty} c_n X_n(x) T_n(t) = \sum_{n=1}^{\infty} \Big(a_n\cos(\sqrt{\lambda_n}\,ct) + b_n\sin(\sqrt{\lambda_n}\,ct)\Big)\sin(\sqrt{\lambda_n}\,x)}$$

is the general solution to the original problem (2.10). Note that there are two families of coefficients (a_n and b_n) in this general solution. ◇

Example 2.2.3. Consider the heat equation with homogeneous Neumann-Neumann boundary conditions,

$$\begin{aligned} u_t &= ku_{xx}, && 0 < x < \ell,\ t > 0, && (2.11a)\\ u_x(0,t) &= u_x(\ell,t) = 0, && t > 0, && (2.11b)\\ u(x,0) &= f(x), && 0 < x < \ell. && (2.11c) \end{aligned}$$

Let $u(x,t) = X(x)T(t)$. The PDE implies $X(x)T'(t) = kX''(x)T(t)$. Dividing by $kX(x)T(t)$,

$$\frac{1}{k}\frac{T'(t)}{T(t)} = \frac{X''(x)}{X(x)} = -\lambda,$$

which results in two ODEs:

$$X''(x) + \lambda X(x) = 0 \qquad \text{and} \qquad T'(t) + \lambda k T(t) = 0. \tag{2.12}$$

The X problem together with the boundary conditions (2.11b) forms an eigenvalue problem in the x variable:

$$X''(x) + \lambda X(x) = 0, \qquad X'(0) = X'(\ell) = 0. \tag{2.13}$$

Note carefully how the boundary conditions here are different from (2.9). To solve for $X(x)$, we must again consider three cases based on the possible sign of λ.

- CASE 1: $\lambda = 0$. Then (2.13) reduces to $X''(x) = 0$, $X'(0) = X'(\ell) = 0$. The general solution is $X(x) = Ax + B$. The boundary conditions force $A = 0$, but leave B arbitrary. This says $\lambda_0 = 0$ is an eigenvalue with eigenfunction $X_0(x) = B$, where B is an arbitrary constant.

- CASE 2: $\lambda < 0$. Let's say $\lambda = -p^2$. Then (2.13) becomes $X''(x) - p^2 X(x) = 0$, $X'(0) = X'(\ell) = 0$. The general solution is $X(x) = A\cosh(px) + B\sinh(px)$. The boundary conditions force $A = B = 0$, i.e., X is the trivial solution. (Be sure you understand why.)

- CASE 3: $\lambda > 0$. Let's say $\lambda = p^2$. Then (2.13) becomes $X''(x) + p^2 X(x) = 0$, $X'(0) = X'(\ell) = 0$. The general solution is $X(x) = A\cos(px) + B\sin(px)$. Then $X'(x) = -Ap\sin(px) + Bp\cos(px)$, and applying the boundary conditions,

$$\begin{aligned} X'(0) &= -Ap\sin(0) + Bp\cos(0) = 0 &\implies& \quad B = 0, \\ X'(\ell) &= -Ap\sin(p\ell) + Bp\cos(p\ell) = 0 &\implies& \quad -Ap\sin(p\ell) = 0. \end{aligned}$$

  Either $A = 0$, $p = 0$, or $\sin(p\ell) = 0$. If $A = 0$, we already had $B = 0$, so $X(x) \equiv 0$. On the other hand, $p \neq 0$ since we assumed $\lambda = p^2 > 0$ for this case. It must be that $\sin(p\ell) = 0$; that is, $p\ell = n\pi$, $n = 1, 2, 3, \ldots$. Hence, the eigenvalues in this case are $\lambda = \lambda_n = p^2 = (n\pi/\ell)^2$, $n = 1, 2, 3, \ldots$, with corresponding eigenfunctions $X_n(x) = \cos(\sqrt{\lambda_n}\, x)$, $n = 1, 2, 3, \ldots$.

We now return to the time problem in (2.12). Based on the work above, there are two cases:

$$\begin{aligned} \lambda = 0: \qquad & T'(t) = 0 &\implies& \quad T(t) = C, \\ \lambda > 0: \qquad & T'(t) + \lambda_n k T(t) = 0 &\implies& \quad T_n(t) = C\exp(-\lambda_n k t), \end{aligned}$$

where C is an arbitrary constant.

We summarize the eigenvalues, eigenfunctions, and time functions below.

$\lambda = 0$ case	$\lambda > 0$ case $(n = 1, 2, \ldots)$
$\lambda_0 = 0$	$\lambda_n = (n\pi/\ell)^2$
$X_0(x) = B$	$X_n(x) = \cos(\sqrt{\lambda_n}\, x)$
$T_0(t) = C$	$T_n(t) = \exp(-\lambda_n\, kt)$

Finally, we apply the Superposition Principle for all the cases in the table above to obtain

$$\begin{aligned} u(x,t) &= \underbrace{c_0 X_0(x) T_0(t)}_{\lambda=0 \text{ case}} + \underbrace{\sum_{n=1}^{\infty} c_n X_n(x) T_n(t)}_{\lambda>0 \text{ case}} \\ &= c_0 \cdot B \cdot C + \sum_{n=1}^{\infty} c_n \cos(\sqrt{\lambda_n}\, x) \exp(-\lambda_n\, kt). \end{aligned}$$

Since c_0, B, and C are all arbitrary constants, we condense them into one so that the general solution to (2.11) takes the form

$$\boxed{u(x,t) = c_0 + \sum_{n=1}^{\infty} c_n \cos(\sqrt{\lambda_n}\, x) \exp(-\lambda_n\, kt).}$$

Note that there is one more coefficient, c_0, to find in this case. We will determine the coefficients c_n, $n = 0, 1, 2, \ldots$, in the next section. ♢

Exercises

1. What role did the initial condition(s) play in any of the examples in this section?

2. (a) Explain why the final solution of Examples 2.2.1 and 2.2.3 involved only one family of constants, whereas in Example 2.2.2, there were two families of constants.

   (b) Explain why it is important for these constants to depend on n.

3. Consider the heat equation with homogeneous Dirichlet-Neumann boundary conditions:

$$\begin{aligned} u_t &= k u_{xx}, && 0 < x < \ell,\ t > 0, \\ u(0,t) &= u_x(\ell,t) = 0, && t > 0, \\ u(x,0) &= f(x), && 0 < x < \ell. \end{aligned}$$

   (a) Give a physical interpretation for each line in the problem above.

   (b) State the eigenvalue problem for X (eigenvalue problems require an ODE plus boundary conditions) and the ODE for T.

   (c) Analyzing the three cases for the sign of λ, determine the eigenvalues and eigenfunctions for the X problem.

   (d) For the λ in (c), solve the T problem.

   (e) Use the Superposition Principle to obtain the general solution of the given initial-boundary value problem as an infinite series.

4. Consider the wave equation with homogeneous Neumann-Dirichlet boundary conditions:

$$\begin{aligned} u_{tt} &= c^2 u_{xx}, && 0 < x < \ell,\ t > 0, \\ u_x(0,t) &= u(\ell,t) = 0, && t > 0, \\ u(x,0) &= f(x), && 0 < x < \ell, \\ u_t(x,0) &= g(x), && 0 < x < \ell. \end{aligned}$$

   (a) Give a physical interpretation for each line in the problem above.

(b) State the eigenvalue problem for X (eigenvalue problems require an ODE plus boundary conditions) and the ODE for T.

(c) Analyzing the three cases for the sign of λ, determine the eigenvalues and eigenfunctions for the X problem.

(d) For the λ in (c), solve the T problem.

(e) Use the Superposition Principle to obtain the general solution of the given initial-boundary value problem as an infinite series.

5. Consider the heat equation with periodic boundary conditions:

$$\begin{aligned} u_t &= ku_{xx}, & -\ell < x < \ell,\ t > 0, \\ u(-\ell, t) &= u(\ell, t), & t > 0, \\ u_x(-\ell, t) &= u_x(\ell, t), & t > 0, \\ u(x, 0) &= f(x), & -\ell < x < \ell. \end{aligned}$$

(a) Give a physical interpretation for each line in the problem above.

(b) State the eigenvalue problem for X (eigenvalue problems require an ODE plus boundary conditions) and the ODE for T.

(c) Analyzing the three cases for the sign of λ, determine the eigenvalues and eigenfunctions for the X problem.

(d) For the λ in (c), solve the T problem.

(e) Use the Superposition Principle to obtain the general solution of the given initial-boundary value problem as an infinite series.

6. Consider the *damped* wave equation with homogeneous Dirichlet-Dirichlet boundary conditions:

$$\begin{aligned} u_{tt} &= u_{xx} - u_t, & 0 < x < 1,\ t > 0, \\ u(0, t) &= u(1, t) = 0, & t > 0, \\ u(x, 0) &= f(x), & 0 < x < 1, \\ u_t(x, 0) &= g(x), & 0 < x < 1. \end{aligned}$$

(a) Give a physical interpretation for each line in the problem above.

(b) State the eigenvalue problem for X (eigenvalue problems require an ODE plus boundary conditions) and the ODE for T.

(c) Analyzing the three cases for the sign of λ, determine the eigenvalues and eigenfunctions for the X problem.

(d) For the λ in (c), solve the T problem.

(e) Use the Superposition Principle to obtain the general solution of the given initial-boundary value problem as an infinite series.

2.3 The Coefficients via Orthogonality

In this section, we revisit each of the examples from Section 2.2. The goal is to determine the coefficients in the general solution from the given initial condition(s).

Fourier Sine Series

Example 2.3.1. In Section 2.2, we solved the initial-boundary value problem

$$u_t = ku_{xx}, \qquad 0 < x < \ell,\ t > 0, \tag{2.14a}$$

$$u(0,t) = u(\ell,t) = 0, \qquad t > 0, \tag{2.14b}$$

$$u(x,0) = f(x), \qquad 0 < x < \ell, \tag{2.14c}$$

to obtain the general solution

$$u(x,t) = \sum_{n=1}^{\infty} b_n \sin(\sqrt{\lambda_n}\, x) \exp\left(-\lambda_n kt\right),$$

where $\lambda_n = (n\pi/\ell)^2$. To find the coefficients b_n, we apply the initial condition to obtain

$$u(x,0) = \sum_{n=1}^{\infty} b_n \sin(\sqrt{\lambda_n}\, x) = f(x), \quad 0 < x < \ell. \tag{2.15}$$

This infinite series expansion of $f(x)$ on the interval $0 < x < \ell$ in terms of sines of various frequencies is called a *Fourier sine series* in honor of Joseph Fourier.

The key fact in discovering the formula for the coefficients lies in the *orthogonality* of the underlying family of eigenfunctions $\{\sin(n\pi x/\ell)\}_{n=1}^{\infty}$ on the interval $0 < x < \ell$:

$$\int_0^{\ell} \sin(n\pi x/\ell)\sin(m\pi x/\ell)\, dx = \begin{cases} 0, & n \neq m, \\ \ell/2, & n = m. \end{cases} \tag{2.16}$$

We use the term orthogonality because (2.16) can be expressed in terms of the inner product (2.1) as

$$\langle \sin(n\pi x/\ell), \sin(m\pi x/\ell)\rangle = 0, \quad n \neq m.$$

To find the formula for the coefficients b_n above, we choose an arbitrary (but fixed) positive integer m, select the mth member from the orthogonal family, $\sin(m\pi x/\ell)$, multiply both sides of (2.15) by $\sin(m\pi x/\ell)$, and integrate over $0 < x < \ell$:

$$\sum_{n=1}^{\infty} b_n \sin(n\pi x/\ell) = f(x)$$

$$\sum_{n=1}^{\infty} b_n \sin(n\pi x/\ell)\sin(m\pi x/\ell) = f(x)\sin(m\pi x/\ell)$$

$$\int_0^{\ell} \sum_{n=1}^{\infty} b_n \sin(n\pi x/\ell)\sin(m\pi x/\ell)\, dx = \int_0^{\ell} f(x)\sin(m\pi x/\ell)\, dx.$$

Interchanging the summation and integration (there is some mathematical delicacy here, but it is justified), the last line becomes

$$\sum_{n=1}^{\infty} b_n \int_0^{\ell} \sin(n\pi x/\ell)\sin(m\pi x/\ell)\,dx = \int_0^{\ell} f(x)\sin(m\pi x/\ell)\,dx.$$

By (2.16), only the $n = m$ term survives out of the infinite series on the left-hand side, enabling us to solve for the coefficients:

$$\begin{aligned}
\sum_{n=1}^{\infty} b_n \underbrace{\int_0^{\ell} \sin(n\pi x/\ell)\sin(m\pi x/\ell)\,dx}_{=0 \text{ except when } n=m} &= \int_0^{\ell} f(x)\sin(m\pi x/\ell)\,dx \\
b_m \int_0^{\ell} \sin(m\pi x/\ell)\sin(m\pi x/\ell)\,dx &= \int_0^{\ell} f(x)\sin(m\pi x/\ell)\,dx \\
b_m \cdot \frac{\ell}{2} &= \int_0^{\ell} f(x)\sin(m\pi x/\ell)\,dx \\
b_m &= \frac{2}{\ell}\int_0^{\ell} f(x)\sin(m\pi x/\ell)\,dx.
\end{aligned}$$

Since m was an arbitrary integer (just a dummy variable), we can replace m with n and view the result as a formula for all the coefficients in the Fourier sine series of $f(x)$ on $0 < x < \ell$. Therefore, the solution of (2.14) is

$$\boxed{u(x,t) = \sum_{n=1}^{\infty} b_n \sin(n\pi x/\ell)\exp(-(n\pi/\ell)^2 kt), \quad b_n = \frac{2}{\ell}\int_0^{\ell} f(x)\sin(n\pi x/\ell)\,dx.}$$

See Figure 2.2. ◊

Fourier Cosine Series

In Section 2.2, we saw examples where the eigenfunctions involved cosines rather than sines. It is natural to ask whether eigenfunction families of the form $\{\cos(n\pi x/\ell)\}_{n=0}^{\infty}$ also share this very useful orthogonality property on $0 < x < \ell$. Fortunately, the answer is yes (see Exercise 1):

$$\int_0^{\ell} \cos(n\pi x/\ell)\cos(m\pi x/\ell)\,dx = \begin{cases} 0, & n \neq m, \\ \ell, & n = m = 0, \\ \ell/2, & n = m \neq 0. \end{cases} \tag{2.17}$$

This allows us to repeat the orthogonality argument from earlier to derive formulas for the coefficients in a *Fourier cosine series* expansion.

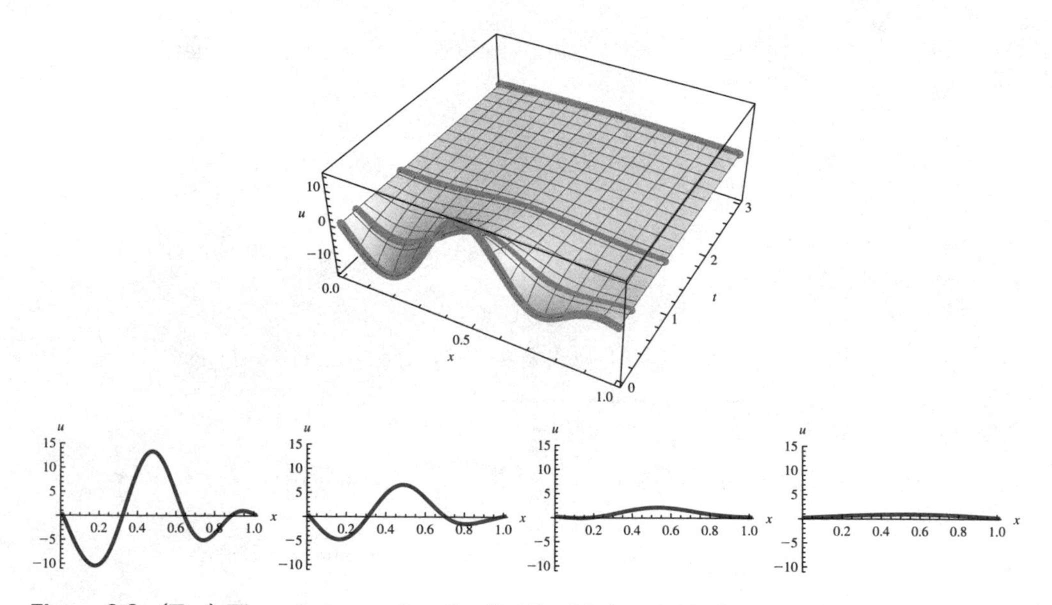

Figure 2.2: (Top) The solution surface for (2.14) with $k = 0.03$, $\ell = 1$, and initial temperature distribution $f(x) = -100x(1-x)^2 \sin(10x)$ using the first 5 terms of the Fourier sine series solution. (Bottom) Time snapshots (slices of the surface) at $t = 0$, $t = 0.25$, $t = 1$, $t = 3$.

Example 2.3.2. In Section 2.2, we considered the 1D heat equation with homogeneous Neumann-Neumann boundary conditions,

$$
\begin{aligned}
u_t &= k u_{xx}, && 0 < x < \ell,\ t > 0, && (2.18a)\\
u_x(0,t) &= u_x(\ell,t) = 0, && t > 0, && (2.18b)\\
u(x,0) &= f(x), && 0 < x < \ell, && (2.18c)
\end{aligned}
$$

and found the general solution to be[7]

$$u(x,t) = \frac{1}{2}a_0 + \sum_{n=1}^{\infty} a_n \cos(\sqrt{\lambda_n}\, x) \exp(-\lambda_n k t),$$

where $\lambda_n = \left(\frac{n\pi}{\ell}\right)^2$, $n = 1, 2, \ldots$. Applying the initial condition,

$$u(x,0) = \frac{1}{2}a_0 + \sum_{n=1}^{\infty} a_n \cos(n\pi x/\ell) = f(x), \quad 0 < x < \ell.$$

We call this a *Fourier cosine series* for $f(x)$ on $0 < x < \ell$ since it expresses $f(x)$ as an infinite sum of cosines of varying frequencies on the interval $0 < x < \ell$.

[7] The factor of $1/2$ on the a_0 term is to account for the factor of $1/2$ missing in the $n = m = 0$ case of (2.17).

To find the coefficients a_n, $n = 0, 1, 2, \ldots$, we repeat the "multiply-and-integrate" orthogonality argument from the last example in conjunction with (2.17) (see Exercise 2), to find

$$a_n = \frac{2}{\ell}\int_0^\ell f(x)\cos(n\pi x/\ell)\,dx, \quad n = 0, 1, 2, \ldots. \tag{2.19}$$

This one compact formula (which includes the $n = 0$ case in it) is very similar to the formula for the Fourier sine coefficients. Therefore, the solution of (2.18) is

$$u(x,t) = \frac{1}{2}a_0 + \sum_{n=1}^{\infty} a_n \cos(\sqrt{\lambda_n}\,x)\exp(-\lambda_n kt), \qquad \begin{aligned} \lambda_n &= (n\pi/\ell)^2, \\ a_n &= \frac{2}{\ell}\int_0^\ell f(x)\cos(n\pi x/\ell)\,dx. \end{aligned}$$

See Figure 2.3. ◊

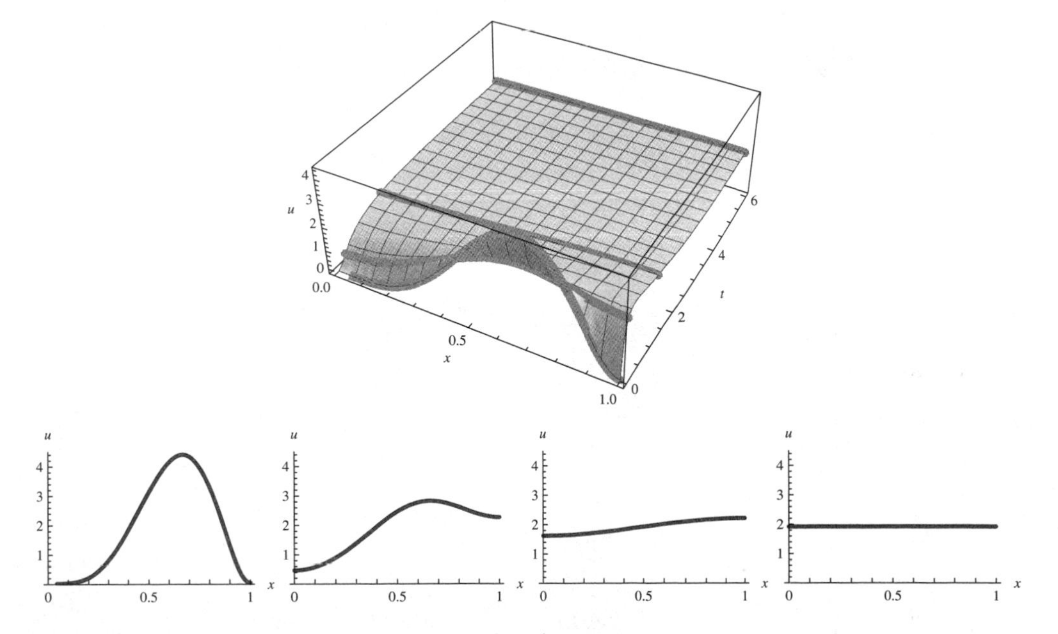

Figure 2.3: (Top) The solution surface for (2.18) with $k = 0.1$, $\ell = 1$, and initial temperature distribution $f(x) = 200x^4(1-x)^2$ using the first 10 terms of the Fourier cosine series solution. (Bottom) Time snapshots at $t = 0$, $t = 0.25$, $t = 1.5$, $t = 6$.

Full Fourier Series

On the other hand, we know from Section 2.2 it is possible that *both* sines and cosines appear as eigenfunctions. Following the theme of this section, the hope is that when this does happen that the underlying family of eigenfunctions,

$$\{1, \cos(n\pi x/\ell), \sin(n\pi x/\ell)\}_{n=1}^{\infty},$$

is orthogonal. Fortunately, this is true on the interval $-\ell < x < \ell$ since

$$\left.\begin{aligned} \int_{-\ell}^{\ell} 1 \cdot \cos(n\pi x/\ell)\,dx &= 0, \\ \int_{-\ell}^{\ell} 1 \cdot \sin(n\pi x/\ell)\,dx &= 0, \\ \int_{-\ell}^{\ell} \cos(n\pi x/\ell)\cos(m\pi x/\ell)\,dx &= 0, \quad n \neq m, \\ \int_{-\ell}^{\ell} \sin(n\pi x/\ell)\sin(m\pi x/\ell)\,dx &= 0, \quad n \neq m, \\ \int_{-\ell}^{\ell} \cos(n\pi x/\ell)\sin(m\pi x/\ell)\,dx &= 0. \end{aligned}\right\} \tag{2.20}$$

See Exercise 3. In the next example, we will see how the orthogonality relations in (2.20) can be used to find the two families of coefficients in a *full Fourier series.*

Example 2.3.3. The 1D heat equation with periodic boundary conditions,

$$u_t = ku_{xx}, \qquad -\ell < x < \ell,\ t > 0, \tag{2.21a}$$

$$u(-\ell, t) = u(\ell, t), \qquad t > 0, \tag{2.21b}$$

$$u_x(-\ell, t) = u_x(\ell, t), \qquad t > 0, \tag{2.21c}$$

$$u(x, 0) = f(x), \qquad -\ell < x < \ell. \tag{2.21d}$$

has general solution

$$u(x,t) = \frac{1}{2}a_0 + \sum_{n=1}^{\infty} \left[a_n \cos(\sqrt{\lambda_n}\, x) + b_n \sin(\sqrt{\lambda_n}\, x)\right] \exp(-\lambda_n kt),$$

where $\lambda_n = \left(\frac{n\pi}{\ell}\right)^2$, $n = 1, 2, \ldots$. Applying the initial condition,

$$u(x,0) = \frac{1}{2}a_0 + \sum_{n=1}^{\infty} [a_n \cos(n\pi x/\ell) + b_n \sin(n\pi x/\ell)] = f(x), \quad -\ell < x < \ell.$$

This is called a *full Fourier series* of $f(x)$ on the interval $-\ell < x < \ell$ since it involves both sine and cosine terms in the infinite series. We have to find formulas for two sets of coefficients, but this follows directly from a (now familiar) orthogonality argument (see Exercise 4):

$$\begin{aligned} a_n &= \frac{1}{\ell}\int_{-\ell}^{\ell} f(x)\cos(n\pi x/\ell)\,dx, \quad n = 0, 1, 2, \ldots, \\ b_n &= \frac{1}{\ell}\int_{-\ell}^{\ell} f(x)\sin(n\pi x/\ell)\,dx, \quad n = 1, 2, 3, \ldots. \end{aligned} \tag{2.22}$$

Therefore, the solution of (2.21) is

$$u(x,t) = \frac{1}{2}a_0 + \sum_{n=1}^{\infty}\Big[a_n\cos(\sqrt{\lambda_n}\,x) + b_n\sin(\sqrt{\lambda_n}\,x)\Big]\exp(-\lambda_n kt),$$
$$\lambda_n = (n\pi/\ell)^2, \quad a_n, b_n \text{ are given by (2.22).}$$

See Figure 2.4. ◊

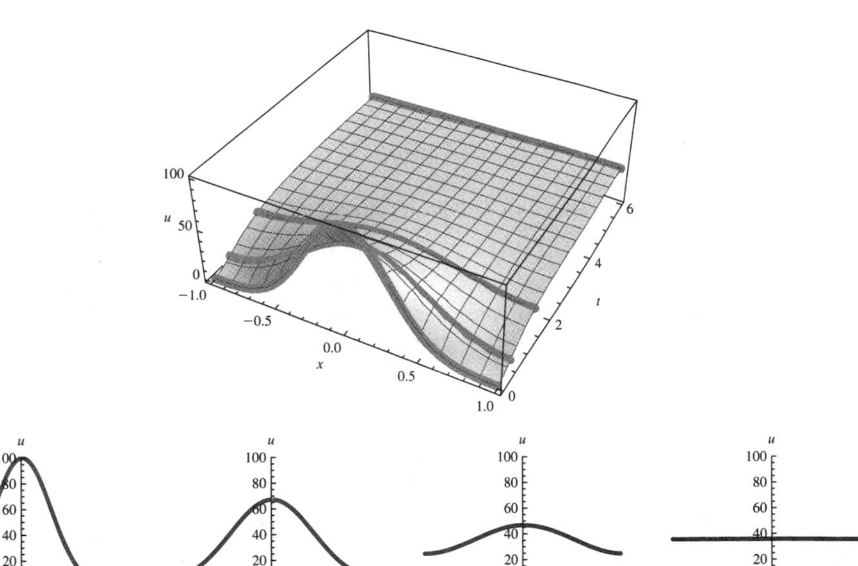

Figure 2.4: (Top) The solution surface for (2.21) with $k = 0.1$, $\ell = 1$, and $f(x) = 100(1 - x^2)\exp(-5x^2)$, using the first 5 terms of the full Fourier series solution. (Bottom) Time snapshots at $t = 0$, $t = 0.25$, $t = 1.5$, $t = 6$.

Exercises

1. (a) Use the trig identity $\sin a \sin b = \frac{1}{2}\cos(a-b) - \frac{1}{2}\cos(a+b)$ and then integrate to verify the orthogonality relation (2.16) for the family $\{\sin(n\pi x/\ell)\}_{n=1}^{\infty}$ on the interval $0 < x < \ell$.

   (b) Use the trig identity $\cos a \cos b = \frac{1}{2}\cos(a+b) + \frac{1}{2}\cos(a-b)$ and then integrate to verify the orthogonality relation (2.17) for the family $\{\cos(n\pi x/\ell)\}_{n=0}^{\infty}$ on the interval $0 < x < \ell$.

2. Use an orthogonality argument to derive the formula (2.19) for the coefficients in a Fourier cosine series.

3. Use your work from Exercise 1 and the trig identity $\sin a \cos b = \frac{1}{2}\sin(a+b) + \frac{1}{2}\sin(a-b)$ to verify each of the orthogonality relations in (2.20).

4. Use an orthogonality argument to derive the formula (2.22) for the coefficients in a full Fourier series.

5. (a) Find the full Fourier series for $f(x) = \begin{cases} 0, & -1 < x < 0, \\ x, & 0 < x < 1. \end{cases}$

   Compute the coefficients first by hand and then in *Mathematica* to check your answer.

   (b) Plot $f(x)$ and the first 5, 10, 30 terms of the full Fourier series on the same coordinate plane. (Do each comparison in a separate plot for clarity.)

6. (a) Find the full Fourier series for $f(x) = \begin{cases} -1, & -1 < x < 0, \\ 1, & 0 < x < 1. \end{cases}$

   Compute the coefficients first by hand and then in *Mathematica* to check your answer.

   (b) Plot $f(x)$ and the first 5, 10, 30 terms of the full Fourier series on the same coordinate plane. (Do each comparison in a separate plot for clarity.)

7. (a) In Section 2.2, we found the general solution of the 1D wave equation with homogeneous Dirichlet-Dirichlet boundary conditions:

$$\begin{aligned} u_{tt} &= c^2 u_{xx}, & 0 < x < \ell,\ t > 0, \\ u(0,t) = u(\ell,t) &= 0, & t > 0, \\ u(x,0) &= f(x), & 0 < x < \ell, \\ u_t(x,0) &= g(x), & 0 < x < \ell. \end{aligned}$$

   Use the initial conditions to find the coefficients a_n, b_n, $n = 1, 2, \ldots$.

   (b) Suppose $c = 1$, $\ell = 1$, $f(x) = 180x^4(1-x)$, and $g(x) = 1$. Using the first 10 terms in the series, plot the solution surface and enough time snapshots to display the dynamics of the solution.

   (c) What happens to the solution as $t \to \infty$? Explain your answer in light of (a) and the physical interpretation of the problem. Does (b) reflect this?

8. (a) In Exercise 2.2.3, we found the general solution of

$$\begin{aligned} u_t &= k u_{xx}, & 0 < x < \ell,\ t > 0, \\ u(0,t) = u_x(\ell,t) &= 0, & t > 0, \\ u(x,0) &= f(x), & 0 < x < \ell. \end{aligned}$$

   Find the coefficients in the general solution.

   (b) Suppose $k = 0.2$, $\ell = 1$, and $f(x) = 180x(1-x)^4$. Using the first 10 terms in the series, plot the solution surface and enough time snapshots to display the dynamics of the solution.

(c) What happens to the solution as $t \to \infty$? Explain your answer in light of (a) and the physical interpretation of the problem. Does (b) reflect this?

9. (a) In Exercise 2.2.4, we found the general solution of

$$\begin{aligned} u_{tt} &= c^2 u_{xx}, && 0 < x < \ell,\ t > 0, \\ u_x(0,t) &= u(\ell,t) = 0, && t > 0, \\ u(x,0) &= f(x), && 0 < x < \ell, \\ u_t(x,0) &= g(x), && 0 < x < \ell. \end{aligned}$$

Find the coefficients in the general solution.

❄ (b) Suppose $c = 1$, $\ell = 1$, $f(x) = -100x^3(1-x)(x-0.75)$, and $g(x) = 1 - x$. Using the first 10 terms in the series, plot the solution surface and enough time snapshots to display the dynamics of the solution.

(c) What happens to the solution as $t \to \infty$? Explain your answer in light of (a) and the physical interpretation of the problem. Does (b) reflect this?

10. (a) In Exercise 2.2.6, we found the general solution of

$$\begin{aligned} u_{tt} &= u_{xx} - u_t, && 0 < x < 1,\ t > 0, \\ u(0,t) &= u(1,t) = 0, && t > 0, \\ u(x,0) &= f(x), && 0 < x < 1, \\ u_t(x,0) &= g(x), && 0 < x < 1. \end{aligned}$$

Find the coefficients in the general solution.

❄ (b) Suppose $f(x) = -100x^3(1-x)(x-0.75)$, and $g(x) = 1 - x$. Using the first 10 terms in the series, plot the solution surface and enough time snapshots to display the dynamics of the solution.

(c) What happens to the solution as $t \to \infty$? Explain your answer in light of (a) and the physical interpretation of the problem. Does (b) reflect this?

11. What is the physical interpretation of the $\frac{1}{2}a_0$ term in the Fourier cosine series?

12. Look back at the work from various problems in this section and the last section and compile a table outlining all eigenvalues and eigenfunctions for $X''(x) + \lambda X(x) = 0$ subject to the following boundary conditions:

(a) Dirichlet-Dirichlet: $X(0) = 0$, $X(\ell) = 0$

(b) Dirichlet-Neumann: $X(0) = 0$, $X'(\ell) = 0$

(c) Neumann-Dirichlet: $X'(0) = 0$, $X(\ell) = 0$

(d) Neumann-Neumann: $X'(0) = 0$, $X'(\ell) = 0$

(e) Periodic: $X(-\ell) = X(\ell)$, $X'(-\ell) = X'(\ell)$

2.4 Consequences of Orthogonality

In Section 2.1, we discussed how a vector space of functions on $a < x < b$ can be equipped with the inner product

$$\langle u, v \rangle := \int_a^b u(x)v(x)\,dx. \tag{2.23}$$

Analogous to linear algebra on $\mathbb{R}^n$, the functions u and v are orthogonal on $a < x < b$, provided their inner product is zero. This concept can be extended to a (possibly infinite) family of functions as follows.

Definition 2.1. A family of functions $\{f_n(x)\}_{n=1}^{\infty}$, $a < x < b$, none of which is identically zero, is an *orthogonal family on* $a < x < b$ if

$$\langle f_i, f_j \rangle = 0, \quad i \neq j,$$

where the inner product is $\langle u, v \rangle := \int_a^b u(x)v(x)\,dx$.

Example 2.4.1. $\{\sin(n\pi x/\ell)\}_{n=1}^{\infty}$ is an orthogonal family on $0 < x < \ell$ since

$$\langle \sin(n\pi x/\ell), \sin(m\pi x/\ell) \rangle = \int_0^\ell \sin(n\pi x/\ell)\sin(m\pi x/\ell)\,dx = 0,\ n \neq m.$$

This orthogonal family forms the basis of Fourier sine series on $0 < x < \ell$. ◊

Example 2.4.2. $\{\cos(n\pi x/\ell)\}_{n=0}^{\infty}$, or equivalently, $\{1, \cos(n\pi x/\ell)\}_{n=1}^{\infty}$ is an orthogonal family on $0 < x < \ell$ since

$$\langle \cos(n\pi x/\ell), \cos(m\pi x/\ell) \rangle = \int_0^\ell \cos(n\pi x/\ell)\cos(m\pi x/\ell)\,dx = 0,\ n \neq m.$$

This orthogonal family forms the basis of Fourier cosine series on $0 < x < \ell$. ◊

Example 2.4.3. $\{1, \cos(n\pi x/\ell), \sin(n\pi x/\ell)\}_{n=1}^{\infty}$ is an orthogonal family on $-\ell < x < \ell$ due to the calculations in (2.20). This orthogonal family forms the basis of full Fourier series on $-\ell < x < \ell$. ◊

Orthogonality was the key to determining the coefficients in the Fourier series expansions in Section 2.3—and this was no accident, as we now show.

Consider $X'' + \lambda X = 0$ with Dirichlet, Neumann, Robin, or periodic boundary conditions. Let λ_1 be an eigenvalue for this problem with eigenfunction X_1, and λ_2 be a different eigenvalue with corresponding eigenfunction X_2. Then

$$-X_1'' = \lambda_1 X_1, \quad -X_2'' = \lambda_2 X_2, \quad \lambda_1 \neq \lambda_2, \tag{2.24}$$

plus the boundary conditions. Next, consider the identity

$$-X_1''X_2 + X_1X_2'' = [-X_1'X_2 + X_1X_2']'$$

(this is just the product rule) and then integrate both sides over $a < x < b$ to obtain

$$\begin{aligned}\int_a^b (-X_1''X_2 + X_1X_2'')\,dx &= \int_a^b [-X_1'X_2 + X_1X_2']'\,dx \\ &= [-X_1'X_2 + X_1X_2']_a^b . \end{aligned} \tag{2.25}$$

Applying (2.24) to the left-hand side of the equation above and expanding the right-hand side, we get

$$\int_a^b (\lambda_1 X_1 X_2 - X_1 \lambda_2 X_2)\,dx = -X_1'(b)X_2(b) + X_1(b)X_2'(b) + X_1'(a)X_2(a) - X_1(a)X_2'(a).$$

Rearranging the left-hand side,

$$\begin{aligned}(\lambda_1 - \lambda_2)\int_a^b X_1X_2\,dx = &-X_1'(b)X_2(b) + X_1(b)X_2'(b) \\ &+ X_1'(a)X_2(a) - X_1(a)X_2'(a).\end{aligned} \tag{2.26}$$

Let's investigate (2.26) with the four main types of boundary conditions we discussed in Section 1.3.

- **Dirichlet:** Suppose X_1 and X_2 both satisfy the Dirichlet boundary conditions $X(a) = X(b) = 0$. Then the right-hand side of (2.26) collapses to 0.
- **Neumann:** Suppose X_1 and X_2 both satisfy the Neumann boundary conditions $X'(a) = X'(b) = 0$. Then the right-hand side of (2.26) collapses to 0.
- **Robin:** Suppose X_1 and X_2 both satisfy the Robin boundary conditions $X'(a) + c_1X(a) = 0$, $X'(b) + c_2X(b) = 0$. Then the right-hand side of (2.26) collapses to 0. (See Exercise 1.)
- **Periodic:** Suppose X_1 and X_2 both satisfy the periodic boundary conditions $X(a) = X(b)$, $X'(a) = X'(b)$. Then the right-hand side of (2.26) collapses to 0. (See Exercise 2.)

The conclusion is that in any of the four cases above (but not just any set of random boundary conditions—see Exercise 3), the right-hand side of (2.26) vanishes:

$$(\lambda_1 - \lambda_2)\int_a^b X_1X_2\,dx = 0. \tag{2.27}$$

Thus, either $\lambda_1 = \lambda_2$—which contradicts (2.24)—or $\int_a^b X_1X_2\,dx = 0$. Therefore, it must be that X_1 and X_2 are orthogonal on $a < x < b$.

Definition 2.2. Boundary conditions of the form

$$\begin{aligned} \alpha_1 X(a) + \beta_1 X(b) + \gamma_1 X'(a) + \delta_1 X'(b) = 0, \\ \alpha_2 X(a) + \beta_2 X(b) + \gamma_2 X'(a) + \delta_2 X'(b) = 0, \end{aligned} \tag{2.28}$$

are called *symmetric boundary conditions* if

$$[f'(x)g(x) - f(x)g'(x)]_a^b = 0, \tag{2.29}$$

for any pair of functions f and g which satisfies (2.28).

The argument above showed that the four main types of boundary conditions are in fact symmetric boundary conditions for pairs of eigenfunctions. This leads us to the following important theorem.

Theorem 2.1 (Orthogonality of Eigenfunctions for Symmetric BVPs). *Consider $X'' + \lambda X = 0$ with any type of symmetric boundary conditions.*

(a) *The sequence of eigenfunctions $\{X_n(x)\}_{n=1}^{\infty}$ forms an orthogonal family.*

(b) *Suppose $\int_a^b f^2(x)\,dx < \infty$. If f is expanded in an infinite series of these eigenfunctions,*

$$f(x) = \sum_{n=1}^{\infty} c_n X_n(x), \quad a < x < b, \tag{2.30}$$

then the coefficients are given by

$$c_n = \frac{\langle f, X_n \rangle}{\langle X_n, X_n \rangle} = \frac{\int_a^b f(x) X_n(x)\,dx}{\int_a^b X_n^2(x)\,dx}, \quad n = 1, 2, \ldots. \tag{2.31}$$

Proof. (a) Suppose X_1 and X_2 are distinct eigenfunctions of $X'' + \lambda X = 0$ that both satisfy the same set of symmetric boundary conditions. From (2.29), we see (2.25) is zero and hence (2.27) holds. Since λ_1 and λ_2 are distinct, it must be that X_1 and X_2 are orthogonal on $a < x < b$.

(b) To obtain the coefficients, we use the orthogonality of the eigenfunctions:

$$\begin{aligned} f(x) &= \sum_{n=1}^{\infty} c_n X_n(x), \quad a < x < b, \\ \langle f, X_m \rangle &= \sum_{n=1}^{\infty} c_n \langle X_n, X_m \rangle \\ \langle f, X_m \rangle &= c_m \langle X_m, X_m \rangle \\ c_m &= \frac{\langle f, X_m \rangle}{\langle X_m, X_m \rangle} \quad \text{or} \quad c_m = \frac{\langle f, X_m \rangle}{\|X_m\|^2}, \end{aligned}$$

since $\|u\| = \langle u, u \rangle^{1/2}$. The inner products above are from (2.23). □

The first part of this very powerful theorem reveals that it wasn't just a calculus identity for sines and cosines which allowed us to solve for the coefficients via the orthogonality of the eigenfunctions—this is an inherent characteristic of eigenvalue problems which arises in conjunction with symmetric boundary conditions. This is useful because now we don't have to verify orthogonality relations for the eigenfunctions arising in every single initial-boundary value problem we want to solve. Instead, Theorem 2.1 guarantees that they are orthogonal.

The second part of Theorem 2.1 shows that the "multiply and integrate" orthogonality arguments from Section 2.3 carry over to general *orthogonal expansions* or *eigenfunction expansions*, as (4.9) are called. Compare the succinct coefficient formula (4.10) to the formulas for the coefficients in Section 2.3: they are equivalent.

We have not, however, discussed any type of convergence issues, i.e., in what sense the equality between $f(x)$ and $\sum_{n=1}^{\infty} c_n X_n(x)$ is true on $a < x < b$. We will address this in Chapter 3.

Also, we have dodged the issue of complex eigenvalues/eigenfunctions—which we know are possible in linear algebra on $\mathbb{R}^n$. However, the next theorem assures us we did not miss any relevant eigenvalues in our analysis.

Theorem 2.2 (Eigenvalues for Symmetric BVPs are Real). *Consider $X'' + \lambda X = 0$ with any type of symmetric boundary conditions. All of the eigenvalues are real and the corresponding eigenfunctions can be chosen to be real-valued.*

Proof. Consider $X'' + \lambda X = 0$ with symmetric boundary conditions of the form (2.28). Taking the complex conjugate[8] of both sides of the ODE, we see $\overline{X}'' + \overline{\lambda}\,\overline{X} = 0$. Thus, λ is an eigenvalue of the original problem with corresponding eigenfunction X, while $\overline{\lambda}$ is an eigenvalue of the conjugated problem with corresponding eigenfunction $\overline{X}$.

Applying (2.25) for X, $\overline{X}$ and using the symmetry of the boundary conditions, we get an analogue of (2.27):

$$(\lambda - \overline{\lambda}) \int_a^b X\,\overline{X}\,dx = 0.$$

However, $X\,\overline{X} = |X|^2 \geq 0$. X cannot be identically zero since it is an eigenfunction, so the integral cannot vanish. It must be that $\lambda - \overline{\lambda} = 0$, i.e., $\lambda = \overline{\lambda}$ and therefore is real.

To show that the complex eigenfunction X corresponding to the real eigenvalue λ can be chosen to be real-valued, write an eigenfunction in the form $X(x) = A(x) + iB(x)$, where $A(x)$, $B(x)$ are real-valued functions. Then

$$X'' + \lambda X = (A'' + iB'') + \lambda(A + iB) = A'' + \lambda A + i(B'' + \lambda B).$$

[8]If $z = a + bi$ is complex, then the *complex conjugate* of z is $\overline{z} = a - bi$.

Since $X'' + \lambda X = 0$, the real and imaginary parts must be zero: $A'' + \lambda A = 0$ and $B'' + \lambda B = 0$. Therefore, the real eigenvalue λ has real eigenfunctions A and B. □

The last consequence of orthogonality that we highlight in this section is a result that allows us to exclude the possibility of negative eigenvalues in certain situations.

Theorem 2.3 (Ruling Out Negative Eigenvalues).
Consider $X'' + \lambda X = 0$ with any type of symmetric boundary conditions. If the eigenfunctions satisfy

$$X'(x)X(x)\Big|_a^b \leq 0,$$

then there are no negative eigenvalues.

Proof. Integration by parts yields the identity

$$\int_a^b f''(x)g(x)\,dx = f'(x)g(x)\Big|_a^b - \int_a^b f'(x)g'(x)\,dx. \tag{2.32}$$

Take $f(x) = g(x) = X(x)$, where $X(x)$ is an eigenfunction of $X'' + \lambda X = 0$ with symmetric boundary conditions. Then (2.32) becomes

$$\begin{aligned}\int_a^b X''X\,dx &= X'X\Big|_a^b - \int_a^b X'X'\,dx \\ &= X'X\Big|_a^b - \int_a^b (X')^2\,dx \\ &\leq 0.\end{aligned} \tag{2.33}$$

Since $X'' + \lambda X = 0$, the left-hand side above can also be written

$$\int_a^b X''X\,dx = \int_a^b -\lambda XX\,dx = -\lambda \int_a^b X^2\,dx. \tag{2.34}$$

Combining (2.33) and (2.34), and solving for $-\lambda$, we conclude

$$-\lambda = \frac{-\int_a^b (X')^2\,dx + X'X\Big|_a^b}{\int_a^b X^2\,dx} \leq 0.$$

Therefore, $\lambda \geq 0$. □

Notice that each of the boundary conditions in Section 2.2 met the conditions of Theorem 2.3. In the future, this theorem can be used to rule out negative eigenvalues very quickly.

Exercises

1. Verify that the right-hand side of (2.26) is zero for Robin boundary conditions.

2. Verify that the right-hand side of (2.26) is zero for periodic boundary conditions.

3. Show that for boundary conditions of the form $X(a) = X(b)$, $X'(a) = -X'(b)$, the right-hand side of (2.26) is *not* zero. This shows that the right-hand side of (2.26) isn't necessarily zero for every set of boundary conditions.

4. (a) Show that $f_1(x) = x$ and $f_2(x) = x^2$ are orthogonal on $-2 < x < 2$.

   (b) Find values of c_1 and c_2 such that $f_3(x) = x + c_1x^2 + c_2x^3$ is orthogonal to both $f_1(x)$ and $f_2(x)$ on $-2 < x < 2$.

5. (a) Does $\{\sin x, \sin 3x, \sin 5x, \dots\}$ form an orthogonal family on $0 < x < \pi/2$? Explain why or why not.

   (b) Does $\{\cos x, \cos 3x, \cos 5x, \dots\}$ form an orthogonal family on $0 < x < \pi/2$? Explain why or why not.

6. Let $P_0(x) = 1$, $P_1(x) = x$, and $P_2(x) = \frac{1}{2}(3x^2 - 1)$. In applied mathematics, these are referred to as the first few *Legendre polynomials*, a special family of orthogonal polynomials.

   (a) Show by direct calculation that $\{P_0(x), P_1(x), P_2(x)\}$ forms an orthogonal family on the interval $-1 < x < 1$.

   (b) Plot these three Legendre polynomials all together on $-1 < x < 1$.

   (c) Use Theorem 2.1 to compute the coefficients in the orthogonal expansion

   $$x \sin x \approx c_0P_0(x) + c_1P_1(x) + c_2P_2(x), \quad -1 < x < 1.$$

   Plot $x \sin x$ and its three term Legendre polynomial expansion on the same coordinate plane.

   (d) Repeat (c) using the function $x\sin(2x)$ on $-1 < x < 1$.

   (e) Repeat (c) using the function $x\sin(3x)$ on $-1 < x < 1$.

   (f) Explain the difference in the plots of the Legendre polynomial approximations in (c)–(e).

7. Let $x_n = \frac{n}{10}$ for $n = 0, 1, 2, \dots, 10$, and consider the family of "square waves" on the interval $0 \le x \le 1$:

   $$\varphi_n(x) = \begin{cases} 1, & x_{n-1} < x < x_n, \\ 0, & \text{otherwise.} \end{cases}$$

   (a) Show by direct calculation that $\{\varphi_n(x)\}_{n=1}^{10}$ forms an orthogonal family on $0 \le x \le 1$.

(b) Compute the coefficients in the orthogonal expansion

$$xe^{-5x} \approx \sum_{n=1}^{10} c_n \varphi_n(x), \qquad 0 \leq x \leq 1.$$

(c) Plot xe^{-5x} and $\sum_{n=1}^{10} c_n \varphi_n(x)$ on the same coordinate plane.

8. Use Theorem 2.3 to answer the following:

(a) Show that there are no negative eigenvalues for $X'' + \lambda X = 0$, $0 < x < \ell$, when the boundary conditions are homogeneous Dirichlet-Dirichlet or Neumann-Neumann type.

(b) Show that there are no negative eigenvalues for $X'' + \lambda X = 0$, $-\ell < x < \ell$, with the periodic boundary conditions $X(-\ell) = X(\ell)$, $X'(-\ell) = X'(\ell)$.

9. Let $\{f_k(x)\}$, $k = 1, 2, \ldots$ be an orthogonal family on $a < x < b$. Show that $\|f_n + f_m\|^2 = \|f_n\|^2 + \|f_m\|^2$, for all $n \neq m$.

10. We can modify the concept of orthogonality slightly as follows. We say $u(x)$ and $v(x)$ are *orthogonal with respect to the weight function* $w(x)$ *on* $a < x < b$ if

$$\int_a^b u(x)v(x)w(x)\,dx = 0.$$

We can formulate this in terms of inner products by defining the *inner product with weight* w by

$$\langle u, v \rangle_w := \int_a^b u(x)v(x)w(x)\,dx.$$

Then $u \perp_w v \iff \langle u, v \rangle_w = 0$. Our encounters with orthogonality and inner products up until now have always had weight function $w(x) \equiv 1$, but there are times when a different weight function is needed.

(a) Let $H_0(x) = 1$, $H_1(x) = 2x$, and $H_2(x) = 4x^2 - 2$. In applied mathematics, these are referred to as the first few *Hermite polynomials*, a special family of orthogonal polynomials. Show by direct calculation that $\{H_0(x), H_1(x), H_2(x)\}$ forms an orthogonal family with respect to the weight function $w(x) = e^{-x^2}$ on the interval $(-\infty, \infty)$.

(b) Plot these three Hermite polynomials on a single coordinate plane.

(c) Let $L_0(x) = 1$, $L_1(x) = 1 - x$, and $L_2(x) = \frac{1}{2}(x^2 - 4x + 2)$. In applied mathematics, these are referred to as the first few *Laguerre polynomials*, a special family of orthogonal polynomials. Show by direct calculation that $\{L_0(x), L_1(x), L_2(x)\}$ forms an orthogonal family with respect to the weight function $w(x) = e^{-x}$ on the interval $(0, \infty)$.

(d) Plot these three Laguerre polynomials on a single coordinate plane.

11. **(Complex Form of Fourier Series)** If the underlying family of functions is complex-valued rather than real-valued, another modification of the standard inner product is

$$\langle u, v \rangle := \int_a^b u(x)\overline{v(x)}\, dx, \qquad (*)$$

where the bar denotes the complex conjugate, i.e., if $z = a + bi$, then $\overline{z} = a - bi$, where $i := \sqrt{-1}$. This notion leads to the alternate (but equivalent) complex form for a full Fourier series.

(a) Show that $\{e^{in\pi x/\ell}\}$, $n = 0, \pm 1, \pm 2, \ldots$ is an orthogonal family on $-\ell < x < \ell$ with respect to the complex inner product $(*)$. (Hint: Use Euler's formula $e^{i\theta} = \cos\theta + i\sin\theta$.)

(b) Use an orthogonality argument to show that the coefficients c_n, $n = 0, \pm 1, \pm 2, \ldots$ in the complex Fourier series

$$f(x) = \sum_{n=-\infty}^{\infty} c_n e^{in\pi x/\ell}, \qquad -\ell < x < \ell,$$

are given by

$$c_n = \frac{1}{2\ell}\int_{-\ell}^{\ell} f(x) e^{-in\pi x/\ell}\, dx, \qquad n = 0, \pm 1, \pm 2, \ldots.$$

(c) Show that the series in (b) is equivalent to the full Fourier series found in Section 2.3 and their coefficients are related via

$$c_n = \begin{cases} \frac{1}{2}a_0, & n = 0, \\ \frac{1}{2}(a_n - ib_n), & n > 0, \\ \frac{1}{2}(a_{-n} + ib_{-n}), & n < 0, \end{cases}$$

or, equivalently,

$$a_n = 2\mathrm{Re}\,(c_n),$$

$$b_n = \begin{cases} 0, & n = 0, \\ -2\mathrm{Im}\,(c_n) = 2\mathrm{Im}\,(c_{-n}), & n \neq 0. \end{cases}$$

(Hint: Euler's formula yields $\cos\theta = \frac{1}{2}(e^{i\theta} + e^{-i\theta})$ and $\sin\theta = \frac{1}{2i}(e^{i\theta} - e^{-i\theta})$.)

✿ 12. (a) Use the results of Exercise 11 to compute the complex Fourier series for $f(x) = x$, $-1 < x < 1$. Do this first by hand and then check your calculations in *Mathematica*.

(b) Compute the real Fourier series using the methods of Section 2.3. Compare your answer with (a).

2.5 Robin Boundary Conditions

In Section 2.2, the Dirichlet, Neumann, and periodic boundary conditions all led to problems where the infinite sequence of eigenvalues could be expressed with a simple explicit formula such as

$$\lambda_n = \left(\frac{n\pi}{\ell}\right)^2 \quad \text{or} \quad \lambda_n = \left(\frac{(2n-1)\pi}{2\ell}\right)^2.$$

This is not always the case, as we see in the next example.

Example 2.5.1. Consider the problem for the 1D heat equation:

$$u_t = u_{xx}, \qquad 0 < x < 1,\ t > 0, \tag{2.35a}$$

$$u_x(0,t) - u(0,t) = 0, \qquad t > 0, \tag{2.35b}$$

$$u(1,t) = 0, \qquad t > 0, \tag{2.35c}$$

$$u(x,0) = f(x), \qquad 0 < x < 1. \tag{2.35d}$$

The Robin boundary condition (2.35b) at the left endpoint has the equivalent form

$$u_x(0,t) = u(0,t) - 0.$$

Since $K > 0$, the convection obeys Newton's Law of Cooling, and hence is physically realistic. The Dirichlet boundary condition at the right endpoint, $u(1,t) = 0$, indicates a fixed temperature of zero there.

To solve this problem, let $u(x,t) = X(x)T(t)$. Separation of variables leads to

$$X'' + \lambda X = 0, \quad X'(0) - X(0) = 0,\ X(1) = 0, \tag{2.36}$$

$$T' + \lambda T = 0. \tag{2.37}$$

We must again consider three cases based on the possible sign of λ.

- Case 1: $\lambda = 0$. Then (2.36) reduces to $X''(x) = 0$, $X'(0) - X(0) = 0$, $X(1) = 0$. The general solution is $X(x) = Ax + B$. The boundary conditions force $A = B = 0$, i.e., X is the trivial solution.

- Case 2: $\lambda < 0$. Let's say $\lambda = -p^2$. Then (2.36) becomes $X''(x) - p^2 X(x) = 0$, $X'(0) - X(0) = 0$, $X(1) = 0$. The general solution is $X(x) = A\cosh(px) + B\sinh(px)$. The boundary conditions force $A = B = 0$, i.e., X is the trivial solution. (Be sure you understand why.)

- Case 3: $\lambda > 0$. Let's say $\lambda = p^2$. Then (2.36) becomes $X''(x) + p^2 X(x) = 0$, $X'(0) - X(0) = 0$, $X(1) = 0$. The general solution is $X(x) = A\cos(px) + B\sin(px)$. Then $X'(x) = -Ap\sin(px) + Bp\cos(px)$, and applying the boundary conditions,

$$X'(0) - X(0) = Bp - A = 0 \quad \Longrightarrow \quad A = Bp,$$

$$X(1) = A\cos p + B\sin p = 0 \quad \Longrightarrow \quad Bp\cos p + B\sin p = 0.$$

Since $B \neq 0$ (otherwise, $A = 0$ too), dividing by B, the last equation becomes $p \cos p = -\sin p$ or equivalently, $\tan p = -p$. Although this equation has no closed-form expression for its solutions, we can still find the nth solution numerically for any n that we please; see Figure 2.5. Therefore, we simply record the eigenvalues as $\lambda_n = p_n^2$, $n = 1, 2, \ldots$, where p_n is the nth positive root of the equation $\tan p = -p$. The corresponding eigenfunctions are

$$\begin{aligned} X_n(x) &= a_n \cos(\sqrt{\lambda_n}\, x) + b_n \sin(\sqrt{\lambda_n}\, x) \\ &= a_n \cos(\sqrt{\lambda_n}\, x) + \frac{a_n}{\sqrt{\lambda_n}} \sin(\sqrt{\lambda_n}\, x) \\ &= a_n \left[\cos(\sqrt{\lambda_n}\, x) + \frac{1}{\sqrt{\lambda_n}} \sin(\sqrt{\lambda_n}\, x)\right], \quad n = 1, 2, \ldots. \end{aligned}$$

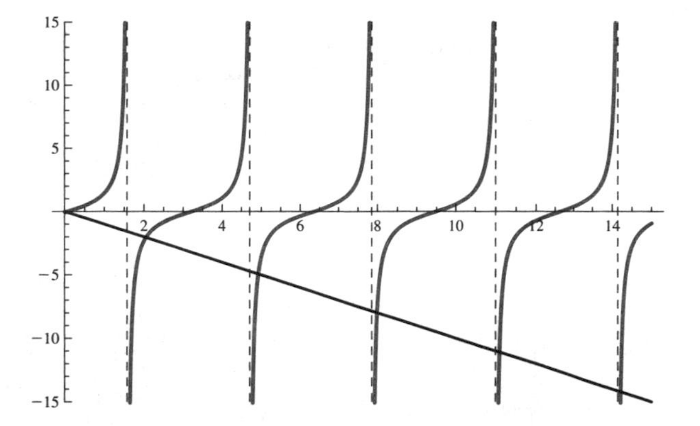

Figure 2.5: Numerically solving for the intersection points of $y = \tan p$ and $y = -p$.

On the other hand, the time problem, $T' + \lambda_n T = 0$, has solution $T_n(t) = C \exp(-\lambda_n t)$, $n = 1, 2, \ldots$. By the Superposition Principle, the general solution is

$$\begin{aligned} u(x,t) &= \sum_{n=1}^{\infty} c_n X_n(x) T_n(t) \\ &= \sum_{n=1}^{\infty} a_n \left[\cos(\sqrt{\lambda_n}\, x) + \frac{1}{\sqrt{\lambda_n}} \sin(\sqrt{\lambda_n}\, x)\right] \exp(-\lambda_n t). \end{aligned}$$

By Theorem 2.1, the eigenfunctions are orthogonal, and the coefficients are given by

$$a_n = \frac{\langle f, X_n \rangle}{\langle X_n, X_n \rangle} = \frac{\int_0^1 f(x) \left[\cos(\sqrt{\lambda_n}\, x) + \frac{1}{\sqrt{\lambda_n}} \sin(\sqrt{\lambda_n}\, x)\right] dx}{\int_0^1 \left[\cos(\sqrt{\lambda_n}\, x) + \frac{1}{\sqrt{\lambda_n}} \sin(\sqrt{\lambda_n}\, x)\right]^2 dx}, \quad n = 1, 2, \ldots.$$

See Figure 2.6. ◊

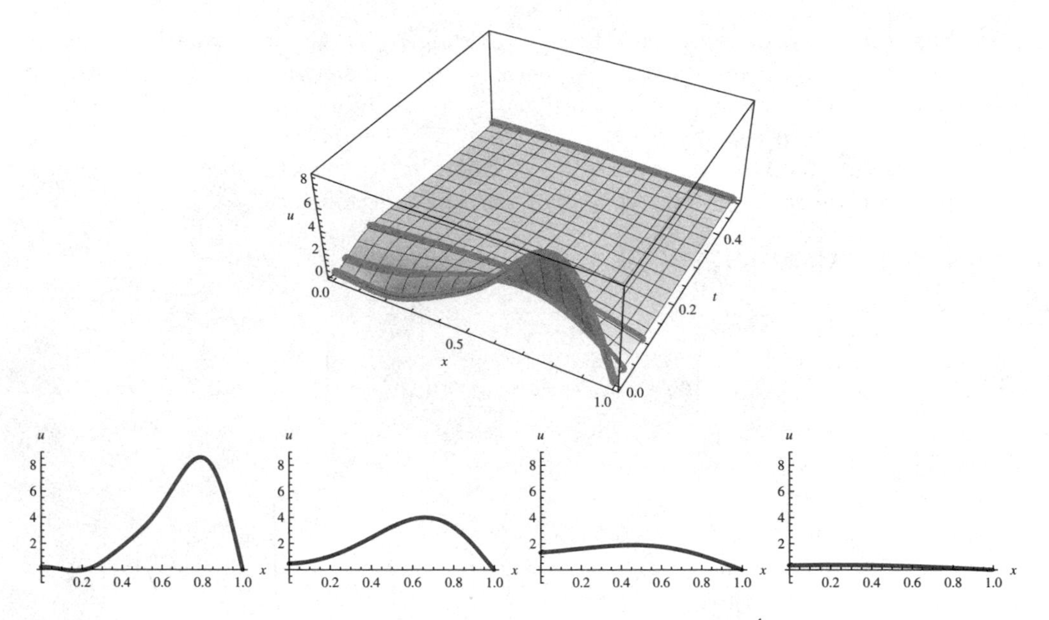

Figure 2.6: (Top) The solution surface for (2.35) with $f(x) = 100x^4(1-x)$ using the first 5 terms of the generalized Fourier series solution. (Bottom) Time snapshots at $t = 0, 0.03, 0.1, 0.5$.

Exercises

1. Consider the Robin-Dirichlet boundary value problem on $0 < x < 1$:

$$X'' + \lambda X = 0, \quad X'(0) + X(0) = 0, \quad X(1) = 0.$$

   (a) What is the physical interpretation of these boundary conditions?

   (b) Show that $\lambda = 0$ is an eigenvalue. Find the corresponding eigenfunction $X_0(x)$.

   (c) Show that there are no negative eigenvalues.

   (d) Find an equation for all the positive eigenvalues.

   (e) Show graphically that there are infinitely many positive solutions to the eigenvalue equation.

   (f) Compute (to 3 decimal places), the numerical values of the first 5 positive eigenvalues. Find the corresponding eigenfunctions.

2. Consider the initial-boundary value problem

$$\begin{aligned} u_t &= u_{xx}, && 0 < x < 1,\ t > 0, \\ u_x(0,t) - u(0,t) &= 0, && t > 0, \\ u_x(1,t) + u(1,t) &= 0, && t > 0, \\ u(x,0) &= f(x), && 0 < x < 1. \end{aligned}$$

(a) Give the physical interpretation of each line in this problem.

(b) Using the method of separation of variables, set up the eigenvalue problem in X and the temporal problem in T.

(c) Explain why all of the eigenvalues for this problem must be positive.

(d) Find an equation for these positive eigenvalues. Find the corresponding eigenfunctions.

(e) Solve the T problem.

(f) Use the Superposition Principle to write the general solution.

(g) Use Theorem 2.1 to set up integral formulas for the coefficients in (f).

❄ (h) Plot the solution surface using the $N = 3$ partial sum of the Fourier series solution when $f(x) = -200x + 100 + 30\cos(x)$. Animate the dynamics of the diffusion using an appropriate set of time slices.

3. Consider the initial-boundary value problem

$$\begin{aligned} u_{tt} &= u_{xx}, && 0 < x < 1,\ t > 0, \\ u_x(0,t) &= u(0,t), && t > 0, \\ u(1,t) &= 0, && t > 0, \\ u(x,0) &= f(x), && 0 < x < 1, \\ u_t(x,0) &= g(x), && 0 < x < 1. \end{aligned}$$

(a) Give the physical interpretation of each line in this problem.

(b) Using the method of separation of variables, set up the eigenvalue problem in X and the temporal problem in T.

(c) Find all the eigenvalues and corresponding eigenfunctions.

(d) Solve the T problem.

(e) Use the Superposition Principle to write the general solution.

(f) Use Theorem 2.1 to set up integral formulas for the coefficients in (e).

❄ (g) Plot the solution surface using the $N = 5$ partial sum of the Fourier series solution when $f(x) = \sin(10x)$ and $g(x) = 0$. Animate the dynamics using an appropriate set of time slices.

4. Consider the initial-boundary value problem

$$\begin{aligned} \frac{\partial u}{\partial t} &= \frac{\partial^2 u}{\partial x^2}, && 0 < x < 1,\ t > 0, \\ u_x(0,t) &= 0, && t > 0, \\ u_x(1,t) + u(1,t) &= 0, && t > 0, \\ u(x,0) &= f(x), && 0 < x < 1. \end{aligned}$$

(a) Give the physical interpretation of each line in this problem.

(b) Using the method of separation of variables, set up the eigenvalue problem in X and the temporal problem in T.

(c) Find all the eigenvalues and corresponding eigenfunctions.

(d) Solve the T problem.

(e) Use the Superposition Principle to write the general solution.

(f) Use Theorem 2.1 to set up integral formulas for the coefficients in the solution from (e).

(g) Plot the solution surface using the $N = 3$ partial sum of the Fourier series solution when $f(x) = \sin(10x)$. Animate the dynamics of the diffusion using an appropriate set of time slices.

5. Consider the initial-boundary value problem on the interval $0 < x < 1$:

$$\begin{aligned} u_t &= u_{xx}, && 0 < x < 1,\ t > 0, \\ u_x(0,t) &= -u(0,t), && t > 0, \\ u_x(1,t) &= -u(1,t), && t > 0, \\ u(x,0) &= f(x), && 0 < x < 1. \end{aligned}$$

(a) Give the physical interpretation of each line in this problem.

(b) Using the method of separation of variables, set up the eigenvalue problem in X and the temporal problem in T.

(c) Show that $\lambda = 0$ is *not* an eigenvalue.

(d) Show that there is exactly one negative eigenvalue. Find it and the corresponding eigenfunction.

(e) Find the infinite sequence of positive eigenvalues and corresponding eigenfunctions.

(f) Solve the T problem.

(g) Use the Superposition Principle to write the general solution.

(h) Use Theorem 2.1 to set up integral formulas for the coefficients in (g).

(i) Plot the solution surface using the $N = 3$ partial sum of the Fourier series solution (i.e., sum up to and including $N = 3$) when $f(x) = \sin(10x)$. Animate the dynamics of the diffusion on the time interval using an appropriate set of time slices.

6. Consider the Robin-Robin boundary value problem on the interval $0 < x < \ell$:

$$X'' + \lambda X = 0, \quad X'(0) - a_0 X(0) = 0, \quad X'(\ell) + a_\ell X(\ell) = 0.$$

(a) Show that $\lambda = 0$ is an eigenvalue if and only if $a_0 + a_\ell = -a_0 a_\ell \ell$. Interpret this condition physically.

(b) Find the eigenfunction(s) corresponding to the $\lambda = 0$ case. (Hint: They are *not* sines or cosines.)

2.6 Nonzero Boundary Conditions: Steady-States and Transients

Intuitively, solutions of the heat equation behave very differently at the beginning of the diffusion process (over a short initial period of time) than they do in the long run (as $t \to \infty$). This was demonstrated in the animations of diffusion dynamics in the last few sections. This physical phenomenon translates into an effective mathematical tool for solving initial-boundary value problems where the boundary conditions are nonhomogeneous (not zero).

The idea is to write the solution $u(x,t)$ of a given problem for the heat equation as the sum of the solution $v(x)$ of an associated *steady-state problem* plus the solution $w(x,t)$ of an associated *transient problem.*

Consider the initial-boundary value problem for the heat equation with nonhomogeneous boundary conditions:

$$u_t = ku_{xx}, \qquad 0 < x < \ell, \ t > 0, \tag{2.38a}$$

$$u(0,t) = T_0, \qquad t > 0, \tag{2.38b}$$

$$u(\ell,t) = T_1, \qquad t > 0, \tag{2.38c}$$

$$u(x,0) = f(x), \qquad 0 < x < \ell. \tag{2.38d}$$

The goal is to decompose this u problem into two subproblems, each of which sheds light on the long-term and short-term behavior of the diffusion process, but together comprise the overall behavior.

The Steady-State Problem

As $t \to \infty$, the temperature in the rod will achieve a *steady-state* or *equilibrium state* which, by definition, is independent of time t. That is,

$$\lim_{t\to\infty} u(x,t) = v(x),$$

for some (yet to be determined) function $v(x)$. Since $v(x)$ is a bona fide solution to (2.38), we can translate this u problem into a simpler *steady-state problem* in v:

$$v''(x) = 0, \qquad 0 < x < \ell,$$

$$v(0) = T_0,$$

$$v(\ell) = T_1.$$

The solution to this ODE boundary value problem is

$$v(x) = \frac{T_1 - T_0}{\ell}x + T_0.$$

Convince yourself that this makes sense in the context of steady-state heat distribution in a bar with the given boundary conditions: it is just a linear function from one endpoint to the other which connects the fixed temperatures at each endpoint.

The Transient Problem

We still have not accounted for the short-time or *transient* behavior of the solution. The key lies in writing

$$\boxed{u(x,t) = \underbrace{v(x)}_{\text{steady-state}} + \underbrace{w(x,t).}_{\text{transient}}}$$

Translating the pieces of the u problem, we obtain the *transient problem* in w:

$$\begin{aligned} w_t &= kw_{xx}, && 0 < x < \ell,\ t > 0, \\ w(0,t) &= 0, && t > 0, \\ w(\ell,t) &= 0, && t > 0, \\ w(x,0) &= f(x) - v(x), && 0 < x < \ell. \end{aligned}$$

Unlike the v problem, this problem still involves a PDE rather than an ODE, but at least the boundary conditions are homogeneous. Notice the variation in the initial condition: since it involves $v(x)$, we must have the solution $v(x)$ before we can solve the transient problem.

However, this initial-boundary value problem in w has homogeneous boundary conditions so it can be solved using the methods of the previous sections. Once that is completed, the solution to (2.38a)–(2.38d) is given by $u(x,t) = v(x) + w(x,t)$. See Figure 2.7.

This process of converting an initial-boundary value problem with nonhomogeneous boundary conditions into two problems—an ODE boundary value problem with nonhomogeneous boundary conditions and a PDE initial-boundary value problem with homogeneous boundary conditions—is referred to as *homogenizing the boundary conditions.* Even when the physical interpretation of a steady-state and transient part of a solution isn't applicable—for example in a wave equation with no damping—this technique can still be used to solve problems with nonhomogeneous boundary conditions.

Consider the initial-boundary value problem

$$\begin{aligned} u_{tt} &= c^2 u_{xx}, && 0 < x < \ell,\ t > 0, \\ u(0,t) &= T_0, && t > 0, \\ u(\ell,t) &= T_1, && t > 0, \\ u(x,0) &= f(x), && 0 < x < \ell, \\ u_t(x,0) &= g(x), && 0 < x < \ell. \end{aligned}$$

There is no *steady-state* solution to an undamped wave equation since the solution varies with time even as $t \to \infty$. However, the ever vibrating string does have a time-independent *equilibrium state* $v(x)$ that can be found just as before:

$$\begin{aligned} v''(x) &= 0, && 0 < x < \ell, \\ v(0) &= T_0, \\ v(\ell) &= T_1. \end{aligned}$$

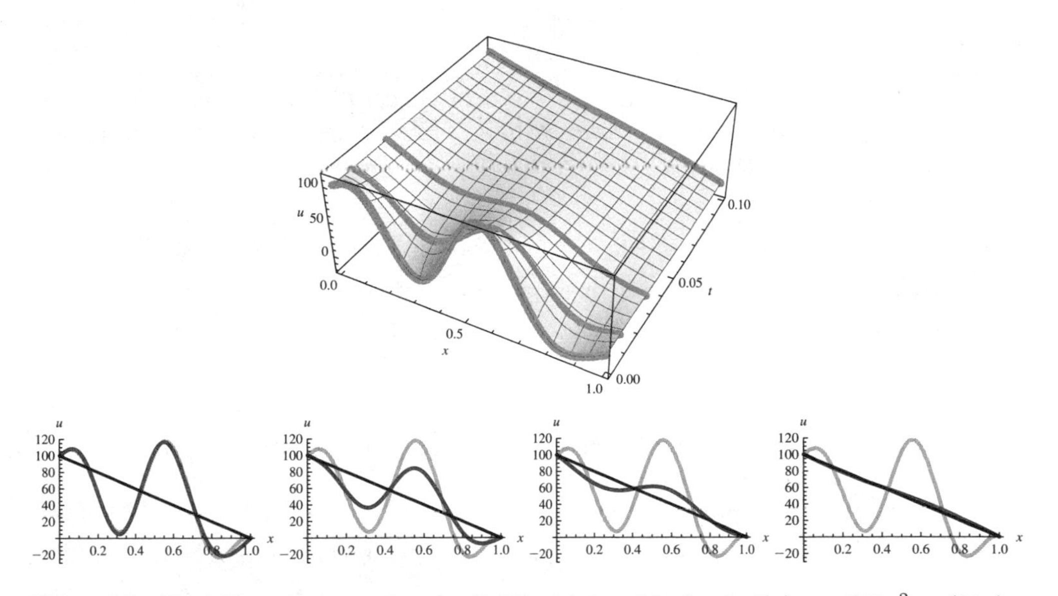

Figure 2.7: (Top) The solution surface for (2.38) with $k = 0.5$, $\ell = 1$, $f(x) = -300x^2\cos(11x) - 100x + 300x\cos(11x) + 100$, $T_0 = 100$, and $T_1 = 0$, using the first 5 terms of the Fourier series solution. (Bottom) Time snapshots at $t = 0, 0.01, 0.03, 0.1$ show the evolution from the initial state to the steady-state. The initial temperature distribution $f(x)$ is light blue, the steady-state $v(x)$ is black, and the solution is dark blue.

The solution to this ODE boundary value problem is

$$v(x) = \frac{T_1 - T_0}{\ell}x + T_0.$$

Indeed, if the left endpoint of the string is fixed at height T_0 while the right endpoint is fixed at height T_1, then the physical equilibrium state of the string is a line connecting those heights.

As before, if we set $u(x,t) := v(x) + w(x,t)$ so that $w(x,t) = u(x,t) - v(x)$, then we get a corresponding initial-boundary value problem in w, but with homogeneous boundary conditions:

$$\begin{aligned}
w_{tt} &= c^2 w_{xx}, && 0 < x < \ell,\ t > 0,\\
w(0,t) &= 0, && t > 0,\\
w(\ell,t) &= 0, && t > 0,\\
w(x,0) &= f(x) - v(x), && 0 < x < \ell,\\
w_t(x,0) &= g(x), && 0 < x < \ell.
\end{aligned}$$

This problem is solved using the standard methods of the previous sections. Finally, the solution $u(x,t)$ to (2.38) is assembled by $u(x,t) = v(x) + w(x,t)$.

Exercises

1. Consider the 1D heat equation in a rod of length ℓ with diffusion constant k. Suppose the left endpoint is fixed at $100°$, while the right endpoint is insulated, and the initial temperature distribution in the rod is $f(x) = 1000x^3(1-x)+100$, $0 < x < \ell$.

   (a) Set up the initial-boundary value problem modeling this scenario.

   (b) Set up and solve the steady-state problem for (a).

   (c) Set up and solve the transient problem for (a).

   (d) Solve the problem in (a).

   ❄ (e) Take $k = \ell = 1$. Plot the solution surface from (d) using the first 5 terms of the Fourier series. On a single coordinate plane, plot the initial temperature distribution and the steady-state solution and animate the dynamics of the solution with an appropriately chosen set of time slices.

   (f) Is the "steady-state" solution a physical steady-state here? Explain.

2. Consider the 1D heat equation in a rod of length ℓ with diffusion constant k. Suppose the left endpoint is convecting (in obedience to Newton's Law of Cooling with proportionality constant $K = 1$) with an outside medium which is $500°$, while the right endpoint is insulated. The initial temperature distribution in the rod is given by $f(x) = 2000|x - 0.65| - 300$, $0 < x < \ell$.

   (a) Set up the initial-boundary value problem modeling this scenario.

   (b) Set up and solve the steady-state problem for (a).

   (c) Set up and solve the transient problem for (a).

   (d) Solve the problem in (a).

   ❄ (e) Take $k = \ell = 1$. Plot the solution surface from (d) using the first 5 terms of the Fourier series. On a single coordinate plane, plot the initial temperature distribution and the steady-state solution and animate the dynamics of the solution with an appropriately chosen set of time slices.

   (f) Is the "steady-state" solution a physical steady-state here? Explain.

3. Consider the 1D heat equation in a rod of length ℓ with diffusion constant k. Suppose the left endpoint is fixed at $100°$, while the right endpoint is convecting (in obedience to Newton's Law of Cooling with proportionality constant $K = 1$) with an outside medium which is $500°$. The initial temperature distribution in the rod is given by $f(x) = 1000x^3(1 - x) + 100$, $0 < x < \ell$.

   (a) Set up the initial-boundary value problem modeling this scenario.

   (b) Set up and solve the steady-state problem for (a).

   (c) Set up and solve the transient problem for (a).

   (d) Solve the problem in (a).

(e) Take $k = \ell = 1$. Plot the solution surface from (d) using the first 5 terms of the Fourier series. On a single coordinate plane, plot the initial temperature distribution and the steady-state solution and animate the dynamics of the solution with an appropriately chosen set of time slices..

(f) Is the "steady-state" solution a physical steady-state here? Explain.

4. Consider the 1D wave equation for a string of length ℓ and with wave speed c. Suppose the left endpoint of the string is fixed at a height of 1, while the right endpoint is attached to a mechanical device in such a way that a slope of zero is maintained at all times. The string has initial shape given by $f(x) = \cos(2\pi x)$, $0 < x < \ell$, and is released with initial velocity $g(x) \equiv 1$, $0 < x < \ell$.

   (a) Set up the initial-boundary value problem modeling this scenario.

   (b) Set up and solve the equilibrium state problem for (a).

   (c) Set up and solve the transient problem for (a).

   (d) Solve the problem in (a).

   (e) Take $c = \ell = 1$. Plot the solution surface from (d) using the first 5 terms of the Fourier series. On a single coordinate plane, plot the initial displacement and the equilibrium solution and animate the dynamics of the solution with an appropriately chosen set of time slices..

   (f) What is the physical interpretation of the solution in (b)?

5. **(Time-Dependent Boundary Conditions)** The methods of this section can be extended to transform nonhomogeneous, time-dependent boundary conditions to homogeneous boundary conditions. The trade-off is that the nonhomogeneity is transferred from the boundary conditions to the PDE.

   Consider the problem

$$\begin{aligned} u_t &= k u_{xx}, && 0 < x < \ell,\ t > 0, \\ u(0,t) &= f_1(t), && t > 0, \\ u_x(\ell,t) + u(\ell,t) &= f_2(t), && t > 0, \\ u(x,0) &= g(x), && 0 < x < \ell. \end{aligned}$$

   (a) Give the physical interpretation of each line in this problem.

   (b) Since the boundary conditions (BCs) are time dependent, the terms "steady-state" and "transient" are no longer appropriate. Instead, we will look to write the solution in the form

$$u(x,t) := \underbrace{v(x,t)}_{\text{nonhomogeneous BCs}} + \underbrace{w(x,t)}_{\text{homogeneous BCs}}.$$

   Let $v(x,t) := v_1(t)(1 - \frac{x}{\ell}) + v_2(t)(x/\ell)$. Find $v_1(t)$, $v_2(t)$ such that $v(x,t)$ satisfies the given nonhomogeneous BCs:

$$v(0,t) = f_1(t), \quad v_x(\ell,t) + v(\ell,t) = f_2(t).$$

(c) What initial condition does $v(x,t)$ satisfy?

(d) With the choice of $v(x,t)$ in (b), show that $w(x,t)$ solves the problem

$$\begin{aligned} w_t &= kw_{xx} - v_t, && 0 < x < \ell,\ t > 0, \\ w(0,t) &= 0, && t > 0, \\ w_x(\ell,t) + w(\ell,t) &= 0, && t > 0, \\ w(x,0) &= g(x) - v(x,0), && 0 < x < \ell. \end{aligned}$$

Since $v(x,t)$ can be determined from (b) and (c), this is a problem in the unknown $w(x,t)$. Although the boundary conditions are now homogeneous, there is now a (known) source term in the PDE.

Chapter 3

Fourier Series Theory

3.1 Fourier Series: Sine, Cosine, and Full

In Chapter 2, we saw how Fourier series—sine, cosine, and full—all arise naturally from the method of separation of variables with each type of series corresponding to different orthogonal families of eigenfunctions. Motivated by this, we define the following.

Definition 3.1. The *Fourier sine series* for $f(x)$ on $0 < x < \ell$ is given by

$$f(x) = \sum_{n=1}^{\infty} b_n \sin(n\pi x/\ell), \qquad b_n = \frac{2}{\ell}\int_0^{\ell} f(x)\sin(n\pi x/\ell)\,dx, \tag{3.1}$$

and the *Fourier cosine series* for $f(x)$ on $0 < x < \ell$ is given by

$$f(x) = \frac{1}{2}a_0 + \sum_{n=1}^{\infty} a_n \cos(n\pi x/\ell), \qquad a_n = \frac{2}{\ell}\int_0^{\ell} f(x)\cos(n\pi x/\ell)\,dx. \tag{3.2}$$

The *full Fourier series* for $f(x)$ on $-\ell < x < \ell$ is given by

$$\begin{aligned} f(x) &= \frac{1}{2}a_0 + \sum_{n=1}^{\infty}[a_n \cos(n\pi x/\ell) + b_n \sin(n\pi x/\ell)], \\ a_n &= \frac{1}{\ell}\int_{-\ell}^{\ell} f(x)\cos(n\pi x/\ell)\,dx, \\ b_n &= \frac{1}{\ell}\int_{-\ell}^{\ell} f(x)\sin(n\pi x/\ell)\,dx. \end{aligned} \tag{3.3}$$

Next, suppose we know $f(x)$ only on $0 < x < \ell$. Then its Fourier sine and cosine series are computable, but if we want to compute the *full* Fourier series of f, this requires knowledge of f on an interval of the form $-\ell < x < \ell$. How should we define $f(x)$ on $-\ell < x \leq 0$ to accomplish this?

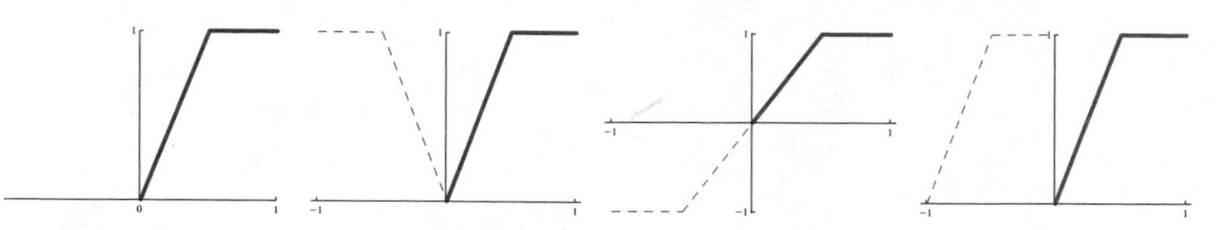

Figure 3.1: A given f, its even extension f_{even}, odd extension f_{odd}, and shift extension f_{shift}.

One answer is to define $f(x)$ on $-\ell < x \leq 0$ in a computationally helpful manner. Two natural ways to do this are with the *even extension* or *odd extension*, defined as

$$f_{\text{even}}(x) := \begin{cases} f(-x), & -\ell < x < 0, \\ f(0^+), & x = 0, \\ f(x), & 0 < x < \ell, \end{cases} \qquad f_{\text{odd}}(x) := \begin{cases} -f(-x), & -\ell < x < 0, \\ f(0^+), & x = 0, \\ f(x), & 0 < x < \ell. \end{cases}$$

Here $f(a^+) := \lim_{x \to a^+} f(x)$ and $f(a^-) := \lim_{x \to a^-} f(x)$ are just shorthand for the one-sided limits involved. Note that these extended versions of f have not altered the definition of f on $0 < x < \ell$, they have just extended f to the left half of a symmetric interval about the origin in such a way that the extension is either an even or odd function[1]. See Figure 3.1.

We can now apply the full Fourier series formulas (3.3) to $f_{\text{even}}(x)$ or $f_{\text{odd}}(x)$, since they are bona fide functions on $-\ell < x < \ell$. The advantage of using an even extension or odd extension is that by design the extension is an even or odd function, which results in the following useful theorem, the proof of which is outlined in Exercise 2.

Theorem 3.1.

(a) *The full Fourier series of* $f_{\text{even}}(x)$ *on* $-\ell < x < \ell$, *when restricted to* $0 < x < \ell$, *is equivalent to the Fourier cosine series of* $f(x)$ *on* $0 < x < \ell$.

(b) *The full Fourier series of* $f_{\text{odd}}(x)$ *on* $-\ell < x < \ell$, *when restricted to* $0 < x < \ell$, *is equivalent to the Fourier sine series of* $f(x)$ *on* $0 < x < \ell$.

This theorem is consistent with the fact that f_{even} is an even function, and thus its Fourier series requires only even basis functions, while f_{odd} is an odd function, and thus its Fourier series requires only odd basis functions. In fact, these last two observations were the motivation for extending $f(x)$ to $-\ell < x < 0$ the way we did (evenly or oddly) rather than by some random method because these extensions eliminate half the terms in the full Fourier series.

[1]Recall that f is an *even function*, provided $f(-x) = f(x)$ for all x in the domain of f, and f is an *odd function* provided $f(-x) = -f(x)$ for all x in the domain of f.

However, we could also extend f to $-\ell < x \leq 0$ any way we like. For example, we can define the *shift extension* via

$$f_{\text{shift}}(x) := \begin{cases} f(x+\ell), & -\ell < x < 0, \\ f(0^+), & x = 0, \\ f(x), & 0 < x < \ell. \end{cases}$$

See Figure 3.1. This is a perfectly valid extension, but there is no guarantee that the cosine or sine terms in the full Fourier series will vanish.

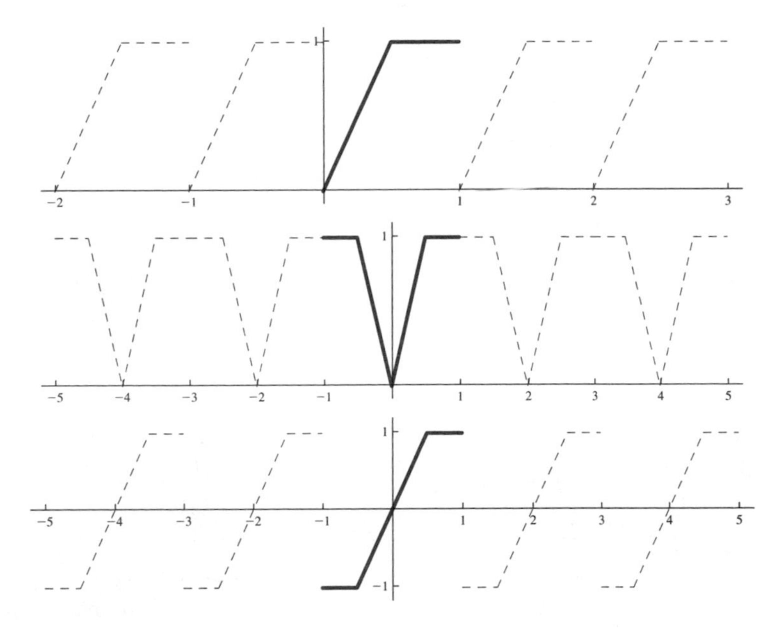

Figure 3.2: The periodic shift, periodic even, and periodic odd extensions of f from Figure 3.1.

Once we have extended f to all of $-\ell < x < \ell$ with either f_{even}, f_{odd}, or f_{shift}, we can periodically extend each of these to all of $\mathbb{R}$ by defining

$$\tilde{f}(x) := f(x + 2\ell).$$

Doing this, we obtain the *periodic even extension* $\tilde{f}_{\text{even}}$, the *periodic odd extension* $\tilde{f}_{\text{odd}}$, and the *periodic shift extension* $\tilde{f}_{\text{shift}}$ as shown in Figures 3.2 and 3.3.

In practice, when given $f(x)$ on $0 < x < \ell$, we are free to extend f to $-\ell < x \leq 0$ however we wish before computing its full Fourier series on $-\ell < x < \ell$. The resulting full Fourier series for f_{even}, f_{odd}, or f_{shift} will all converge to $f(x)$ on $0 < x < \ell$. However, these series very well may *not* converge to the same functions on $-\ell < x < \ell$. Also, the type of convergence for each may be different. We will explore this in detail in Section 3.3.

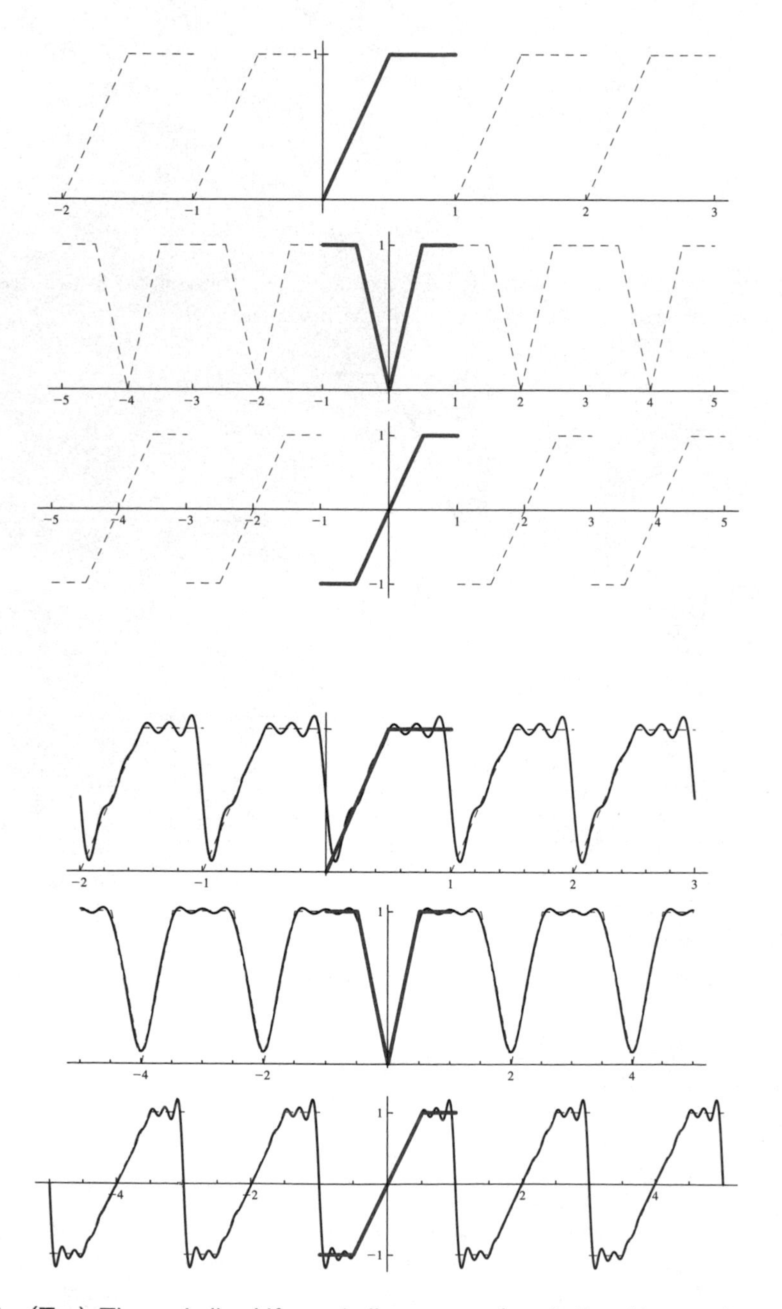

Figure 3.3: (Top) The periodic shift, periodic even, and periodic odd extensions of f from Figure 3.1. (Bottom) The $N = 10$ partial sum of the Fourier series for $\tilde{f}_{\text{shift}}$, the $N = 5$ partial sum of the Fourier series for $\tilde{f}_{\text{even}}$, and the $N = 10$ partial sum of the Fourier series for $\tilde{f}_{\text{odd}}$. All three are have period 2 and agree on $0 < x < 1$, but they are different functions on $-1 < x < 1$.

Exercises

1. (a) Show that if f is even and g is odd, then fg is odd.
   (b) Show that if f and g are even, then fg is even.
   (c) Show that if f and g are odd, then fg is even.

2. Let $f(x)$, $0 < x < \ell$ be given.
   (a) Use Exercise 1(a) to show that every sine coefficient, b_n, $n = 1, 2, \ldots$, in the full Fourier series of f_{even} must be zero. (Hint: Recall from calculus that if $h(x)$ is odd, then $\int_{-a}^{a} h(x)\,dx = 0$.)
   (b) Use Exercise 1(b) to show that the coefficients a_n for f_{even} also satisfy
   $$a_n = \frac{2}{\ell}\int_0^\ell f(x)\cos(n\pi x/\ell)\,dx.$$
   (Hint: Recall from calculus that if $h(x)$ is even, then $\int_{-a}^{a} h(x)\,dx = 2\int_0^a h(x)\,dx$.) Conclude that the full Fourier series of f_{even} is equivalent to the Fourier cosine series of f.
   (c) Use Exercise 1(a) to show that every cosine coefficient, a_n, $n = 0, 1, 2, \ldots$, in the full Fourier series of f_{odd} must be zero.
   (d) Use Exercise 1(c) to show that the coefficients b_n for f_{odd} also satisfy
   $$b_n = \frac{2}{\ell}\int_0^\ell f(x)\sin(n\pi x/\ell)\,dx.$$
   Conclude that the full Fourier series of f_{odd} is equivalent to the Fourier sine series of f.

3. *Without computing any Fourier series*, answer the following regarding the functions below.

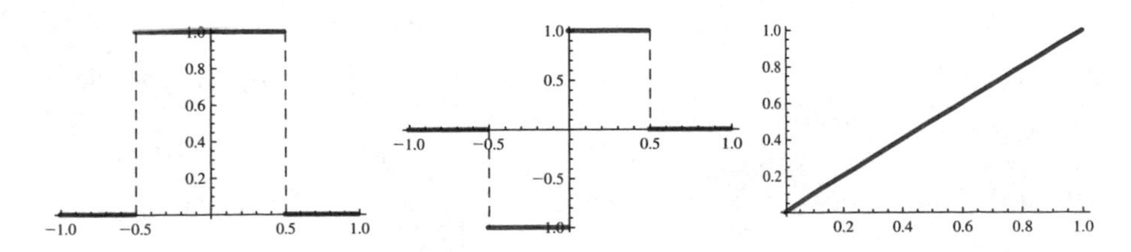

   (a) Sketch the graph of the full Fourier series of each function for $-3 < x < 3$.
   (b) For the first graph, find the value of b_{99}, the 99th Fourier sine coefficient.
   (c) For the second graph, find the value of a_{99}, the 99th Fourier cosine coefficient.
   (d) What is the value of the full Fourier series of each at $x = 0$? $x = 1$? $x = -1$?

✵ 4. Let $f(x) = x$ on $0 < x < 1$.

(a) Plot $f_{\text{even}}(x)$, $f_{\text{odd}}(x)$, and $f_{\text{shift}}(x)$ on $-1 < x < 1$.

(b) Plot three periods of $\tilde{f}_{\text{even}}(x)$, $\tilde{f}_{\text{odd}}(x)$, and $\tilde{f}_{\text{shift}}(x)$.

(c) Plot three periods of the $N = 10$ partial sum of the full Fourier series for each of $f_{\text{even}}(x)$, $f_{\text{odd}}(x)$, and $f_{\text{shift}}(x)$.

✵ 5. Let $f(x) = \sin(9.5x^2) + 5x$ on $0 < x < 1$.

(a) Plot $f_{\text{even}}(x)$, $f_{\text{odd}}(x)$, and $f_{\text{shift}}(x)$ on $-1 < x < 1$.

(b) Plot three periods of $\tilde{f}_{\text{even}}(x)$, $\tilde{f}_{\text{odd}}(x)$, and $\tilde{f}_{\text{shift}}(x)$.

(c) Plot three periods of the $N = 10$ partial sum of the full Fourier series for each of $f_{\text{even}}(x)$, $f_{\text{odd}}(x)$, and $f_{\text{shift}}(x)$.

✵ 6. Let $f(x) = 9x^3(1 - x)$ on $0 < x < 1$.

(a) Plot $f_{\text{even}}(x)$, $f_{\text{odd}}(x)$, and $f_{\text{shift}}(x)$ on $-1 < x < 1$.

(b) Plot three periods of $\tilde{f}_{\text{even}}(x)$, $\tilde{f}_{\text{odd}}(x)$, and $\tilde{f}_{\text{shift}}(x)$.

(c) Plot three periods of the $N = 10$ partial sum of the full Fourier series for each of $f_{\text{even}}(x)$, $f_{\text{odd}}(x)$, and $f_{\text{shift}}(x)$.

✵ 7. Let $f(x) = \sqrt{x}$ on $0 < x < 1$.

(a) Plot $f_{\text{even}}(x)$, $f_{\text{odd}}(x)$, and $f_{\text{shift}}(x)$ on $-1 < x < 1$.

(b) Plot three periods of $\tilde{f}_{\text{even}}(x)$, $\tilde{f}_{\text{odd}}(x)$, and $\tilde{f}_{\text{shift}}(x)$.

(c) Plot three periods of the $N = 10$ partial sum of the full Fourier series for each of $f_{\text{even}}(x)$, $f_{\text{odd}}(x)$, and $f_{\text{shift}}(x)$.

✵ 8. Let $f(x) = e^{-4x}$ on $0 < x < 1$.

(a) Plot $f_{\text{even}}(x)$, $f_{\text{odd}}(x)$, and $f_{\text{shift}}(x)$ on $-1 < x < 1$.

(b) Plot three periods of $\tilde{f}_{\text{even}}(x)$, $\tilde{f}_{\text{odd}}(x)$, and $\tilde{f}_{\text{shift}}(x)$.

(c) Plot three periods of the $N = 10$ partial sum of the full Fourier series for each of $f_{\text{even}}(x)$, $f_{\text{odd}}(x)$, and $f_{\text{shift}}(x)$.

9. (a) Find values of b_n so that $1 + 2x = \sum_{n=1}^{\infty} b_n \sin(n\pi x/3)$ holds for all $0 < x < 3$.

(b) Sketch the graph of the series in (a) for $-9 < x < 9$.

10. (a) Find values of a_n so that $1 + 2x = \frac{1}{2}a_0 + \sum_{n=1}^{\infty} a_n \cos(n\pi x/3)$ holds for all $0 < x < 3$.

(b) Sketch the graph of the series in (a) for $-9 < x < 9$.

11. Consider the partial sums of the Fourier series shown below.

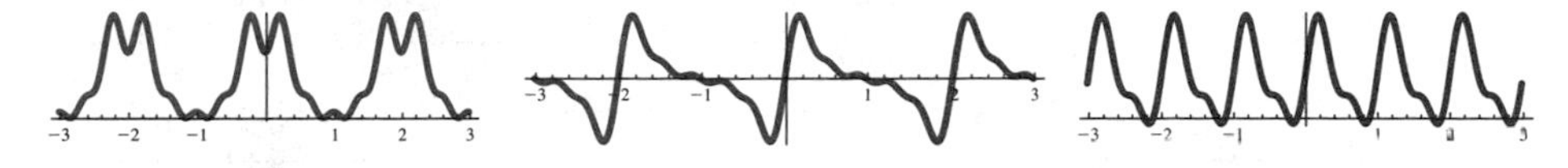

Which of these is a full Fourier series? A Fourier sine series? A Fourier cosine series? Explain your answer.

12. Suppose $f(x)$ is defined on $-\ell < x < \ell$. What conclusions can you make about the full Fourier series for $f(x)$ if you are told the following?

    (a) $\langle f, \sin(n\pi x/\ell)\rangle = 0$ for all $n = 1, 2, \ldots$.

    (b) $\langle f, \cos(n\pi x/\ell)\rangle = 0$ for all $n = 0, 1, 2, \ldots$.

    (c) $\langle f, \sin(n\pi x/\ell)\rangle = \langle f, \cos(n\pi x/\ell)\rangle = 0$ for all $n = 0, 1, 2, \ldots$.

13. Suppose $f : [0, \ell] \to \mathbb{R}$ is continuous.

    (a) Under what conditions is $\tilde{f}_{\text{even}}$ continuous on all of $\mathbb{R}$?

    (b) Under what conditions is $\tilde{f}_{\text{odd}}$ continuous on all of $\mathbb{R}$?

3.2 Fourier Series vs. Taylor Series: Global vs. Local Approximations

We have seen how Fourier series arise naturally from the method of separation of variables for solving PDEs. Alternatively, we could have presented Fourier series as an infinite series expansion of a given function, analogous to a Taylor series. The goal of this section is to compare and contrast these two different types of infinite series.

Recall from calculus that if $f(x)$ is infinitely differentiable in an open interval $a < x < b$ about $x = x_0$, then the *Taylor series* for f centered at $x = x_0$ is defined as

$$\sum_{n=0}^{\infty} c_n (x - x_0)^n, \quad c_n := \frac{f^{(n)}(x_0)}{n!}, \qquad a < x < b,$$

and the Nth degree *Taylor polynomial* $T_N(x)$ of f centered at $x = x_0$ is the Nth partial sum of the Taylor series,

$$T_N(x) := \sum_{n=0}^{N} c_n (x - x_0)^n, \quad a < x < b.$$

Intuitively, an approximation is called *local* if its underlying method emphasizes the accuracy of the approximation at or around some specified point, whereas an approximation is *global* if it emphasizes the accuracy of the approximation on the interval as a whole. The consistent use of the phrase "centered at $x = x_0$" hints that Taylor series/polynomials are local approximations.

Fourier Series	Taylor Series
$-\ell < x < \ell$	$x_0 - R < x < x_0 + R$
$\frac{1}{2}a_0 + \sum_{n=1}^{\infty} a_n \cos(n\pi x/\ell) + b_n \sin(n\pi x/\ell)$	$\sum_{n=0}^{\infty} c_n (x - x_0)^n$
$a_n = \frac{1}{\ell}\int_{-\ell}^{\ell} f(x)\cos(n\pi x/\ell)\,dx$ $b_n = \frac{1}{\ell}\int_{-\ell}^{\ell} f(x)\sin(n\pi x/\ell)\,dx$	$c_n = \frac{f^{(n)}(x_0)}{n!}$
global approximation of f	local approximation of f

In fact, recall how Taylor polynomials were developed in calculus: the Nth degree Taylor polynomial for $f(x)$ centered at $x = x_0$ is the best polynomial approximation of $f(x)$ at $x = x_0$ in the sense that the Taylor coefficient formulas are prescribed so that $T_N(x)$ and $f(x)$ match in all derivatives of order less than or equal to N at $x = x_0$; that is,

$$\left.\frac{d^k}{dx^k} T_N(x)\right|_{x=x_0} = \left.\frac{d^k}{dx^k} f(x)\right|_{x=x_0}, \quad k = 0, 1, \ldots, N. \tag{3.4}$$

This is why we say $T_1(x)$ is the best linear approximation for $f(x)$ at $x = x_0$, that $T_2(x)$ is the best quadratic approximation for $f(x)$ at $x = x_0$, etc. We need to be clear what we mean by "best" here, and (3.4) accomplishes this.

The local nature of the approximation is also revealed by the fact that a Taylor series converges on some interval $(x_0 - R, x_0 + R)$ around the point $x = x_0$ where the series expansion is anchored. The radius of convergence R is usually calculated via the Ratio Test.

On the other hand, Fourier series are *global approximations*: their aim is to approximate $f(x)$ not necessarily at some particular point, but *for all* $x \in (a, b)$. This is reflected in how the Fourier coefficients are calculated by integrating $f(x)$ across an interval of x values. Finally, there is no notion of a radius of convergence here.

It is important to distinguish between a partial Fourier or Taylor sum (as in Figure 3.4) and the infinite Fourier or Taylor series, where issues of convergence of the series must be discussed. We will deal with the convergence of these infinite series in the rest of this chapter.

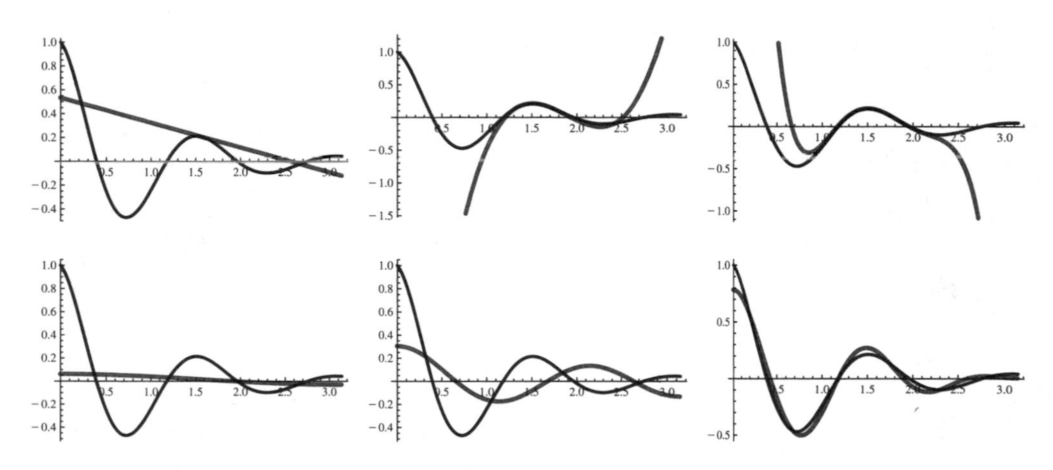

Figure 3.4: (Top) The Taylor polynomials $T_N(x)$ (in blue) centered at $x = \pi/2$ for $N = 1, 3, 5$, where $f(x) = e^{-x}\cos(4x)$ on $[0, \pi]$, in black. (Bottom) The $N = 1, 3, 5$ partial sums $F_N(x)$ (in blue) of the Fourier cosine series. The Taylor polynomials stress approximating f very near $x = x_0$, while the Fourier sums strive to approximate f over the whole interval without necessarily trying to match f at any one point.

Exercises

1. Let $f(x) = 5x + \sin(9.5x^2)$, $0 \le x \le 1$.

   (a) Compute and plot the Taylor polynomials $T_N(x)$ centered at $x = 1/2$ for $N = 1, 5, 10$.

   (b) Compute and plot the partial sums $F_N(x)$ of the Fourier sine series for $N = 1, 5, 10$.

   (c) Compute and plot the partial sums $F_N(x)$ of the Fourier cosine series for $N = 1, 5, 10$.

   (d) Compare (a)–(c) and discuss the local vs. global nature of the approximations. Explain the accuracy differences between the two Fourier series.

2. The *error function*[2], defined as

$$\operatorname{erf}(x) := \frac{2}{\sqrt{\pi}} \int_0^x e^{-t^2}\, dt,$$

   is used in applied mathematics, physics, engineering, as well as probability and statistics. Consider $\operatorname{erf}(x)$ for $0 \le x \le 3$.

   (a) Compute and plot the Taylor polynomials $T_N(x)$ centered at $x = 0$ for $N = 1, 5, 10$.

[2] In *Mathematica*, the command is `Erf[x]`.

(b) Compute and plot the partial sums $F_N(x)$ of the Fourier sine series for $N = 1, 5, 10$.

(c) Compute and plot the partial sums $F_N(x)$ of the Fourier cosine series for $N = 1, 5, 10$.

(d) Compare (a)–(c) and discuss the local vs. global nature of the approximations. Explain the accuracy differences between the two Fourier series.

3. Let $f(x) = \sqrt{1 - x^2}$, $0 \leq x \leq 1$.

(a) Compute and plot the Taylor polynomials $T_N(x)$ centered at $x = 1/2$ for $N = 1, 3, 5$.

(b) Compute and plot the partial sums $F_N(x)$ of the Fourier cosine series for $N = 1, 3, 5$.

(c) Compare (a), (b) and discuss the local vs. global nature of the approximations.

4. Let $f(x) = \sqrt{x}$, $0 \leq x \leq 2$.

(a) Compute and plot the Taylor polynomials $T_N(x)$ centered at $x = 1/2$ for $N = 1, 3, 5$.

(b) Compute and plot the partial sums $F_N(x)$ of the Fourier sine series for $N = 1, 3, 5$.

(c) Compare (a), (b) and discuss the local vs. global nature of the approximations.

(d) Compute and plot the Taylor polynomials $T_N(x)$ centered at $x = 1/2$ for several large values of N. Why does $T_N(x)$ fail to approximate $f(x)$ well for $x > 1$? Does this happen with $F_N(x)$? Discuss.

5. The *sine integral function*[3], defined as

$$\mathrm{Si}(x) := \int_0^x \frac{\sin t}{t}\, dt,$$

is widely used in applications. Consider $\mathrm{Si}(x)$ for $0 \leq x \leq \pi$.

(a) Compute and plot the Taylor polynomials $T_N(x)$ centered at $x = \pi/2$ for $N = 1, 3, 5$.

(b) Compute and plot the partial sums $F_N(x)$ of the Fourier sine series for $N = 1, 3, 5$.

(c) Compare (a), (b) and discuss the local vs. global nature of the approximations.

[3] In *Mathematica*, the command is `SinIntegral[x]`.

6. The difference between the Nth degree Taylor polynomial for $f(x)$ centered at $x = x_0$ and $f(x)$ is called the *Nth remainder* and is given by

$$R_N(x) := f(x) - T_N(x).$$

Taylor's Theorem states that if $f^{(N+1)}(x)$ exists and is continuous, then

$$R_N(x) = \frac{1}{N!}\int_{x_0}^{x}(x-t)^N f^{(N+1)}(t)\,dt.$$

When $R_N(x) \to 0$ as $N \to \infty$, we say the Taylor series converges to f. If we view $|R_N(x)| = |f(x) - T_N(x)|$ as the absolute error in the approximation $f(x) \approx T_N(x)$ for $x \approx x_0$, then Taylor's Theorem provides a useful way to quantify the absolute error for a given value of x.

(a) Plot $|R_N(x)|$ for Exercise 1(a) and discuss.
(b) Plot $|R_N(x)|$ for Exercise 2(a) and discuss.
(c) Plot $|R_N(x)|$ for Exercise 3(a) and discuss.
(d) Plot $|R_N(x)|$ for Exercise 4(a),(d) and discuss.
(e) Plot $|R_N(x)|$ for Exercise 5(a) and discuss.

7. Answer the following, referencing your calculus text as needed.

(a) Under what conditions can the Taylor series for $f(x)$ centered at $x = x_0$ be differentiated term-by-term? That is, if

$$f(x) = \sum_{n=0}^{\infty} c_n(x-x_0)^n = c_0 + c_1(x-x_0) + c_2(x-x_0)^2 + \cdots,$$

then when is it true that

$$f'(x) = \sum_{n=1}^{\infty} nc_n(x-x_0)^{n-1} = c_1 + 2c_2(x-x_0) + 3c_3(x-x_0)^2 + \cdots?$$

(b) Under what conditions can the Taylor series for $f(x)$ centered at $x = x_0$ be integrated term-by-term? That is, if

$$f(x) = \sum_{n=0}^{\infty} c_n(x-x_0)^n = c_0 + c_1(x-x_0) + c_2(x-x_0)^2 + \cdots,$$

then when is it true that

$$\int f(x)\,dx = C + \sum_{n=0}^{\infty} c_n\frac{(x-x_0)^{n+1}}{n+1}$$

$$= C + c_0(x-x_0) + c_1\frac{(x-x_0)^2}{2} + c_2\frac{(x-x_0)^3}{3} + \cdots,$$

where C is an arbitrary constant?

3.3 Error Analysis and Modes of Convergence

Error Analysis

The techniques of this chapter have produced infinite series representations for the solutions of initial-boundary value problems. However, in practice, we always have to truncate the infinite series to form a finite sum approximation of the solution. Thus, in order for the theory of this chapter to be useful, we must be able to quantify the error involved when we replace the infinite series with a finite sum.

Before accomplishing this in the context of truncating infinite series, let's first discuss what we mean by the error between two functions on an interval.

Definition 3.2. Given $f(x)$ and $g(x)$ on $a \leq x \leq b$, we define the (absolute) *pointwise error function* by

$$\boxed{\begin{array}{l}\text{pointwise error between}\\ f \text{ and } g \text{ on } a \leq x \leq b\end{array} = p(x) := |f(x) - g(x)|, \quad a \leq x \leq b.}$$

The *uniform error* between f and g is defined as the maximum of the pointwise error on the interval $a \leq x \leq b$; that is,

$$\boxed{\begin{array}{l}\text{uniform error between}\\ f \text{ and } g \text{ on } a \leq x \leq b\end{array} = \max_{a \leq x \leq b} p(x) = \max_{a \leq x \leq b} |f(x) - g(x)|.}$$

The L^2 *error* between f and g is defined as

$$\boxed{\begin{array}{l}L^2 \text{ error between } f\\ \text{and } g \text{ on } a \leq x \leq b\end{array} = \sqrt{\int_a^b p^2(x)\,dx} = \sqrt{\int_a^b |f(x) - g(x)|^2 dx}.}$$

The pointwise error function tracks the absolute value of the difference between $f(x)$ and $g(x)$ on a point-by-point basis throughout the interval $a \leq x \leq b$. Since it plays a key role in all three definitions above, we often analyze $p(x)$ as follows.

Example 3.3.1. Consider $f(x) = x^3 - 2x^2 + 10$ and $g(x) = 3x^2 + 4x - 10$ on $-2 \leq x \leq 2$. The pointwise error function is

$$p(x) = |f(x) - g(x)| = |x^3 - 5x^2 - 4x + 20|, \quad -2 \leq x \leq 2.$$

A little calculus reveals that the maximum value of $p(x)$ on $-2 \leq x \leq 2$ is $\frac{2}{27}(55 + 37\sqrt{37}) \approx 20.75$ and occurs at $x = \frac{1}{3}(5 - \sqrt{37}) \approx -0.361$. The graph of $p(x)$ confirms this as well. Therefore, the uniform error on $-2 \leq x \leq 2$ is approximately 20.75. Finally, the L^2 error on $-2 \leq x \leq 2$ is

$$\sqrt{\int_{-2}^{2} p^2(x)\,dx} = 16\sqrt{\frac{358}{105}} \approx 29.54.$$

See Figure 3.5 for the geometric interpretation of these quantities. ◊

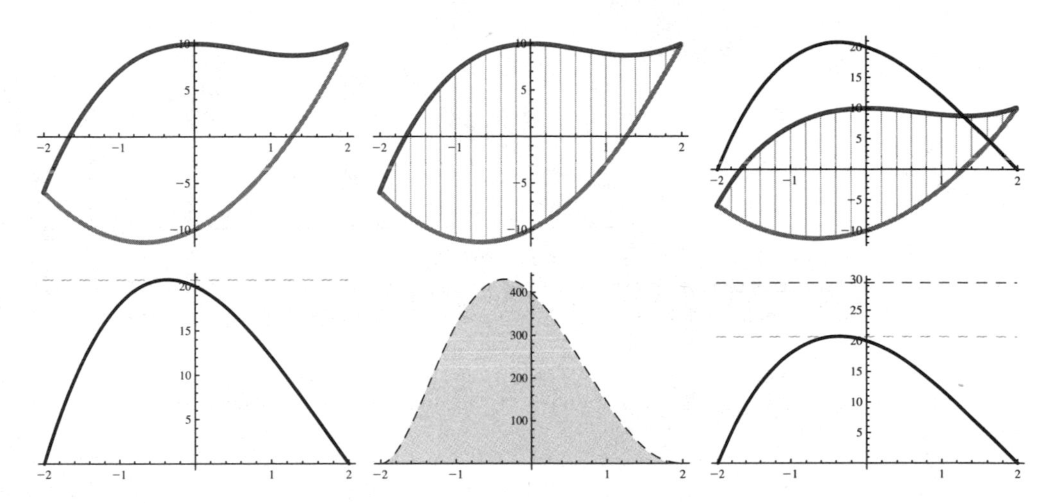

Figure 3.5: (Top, from left to right) The functions $f(x)$ (blue) and $g(x)$ (gray) from Example 3.3.1. The bars in the middle graph show the pointwise error at various points across the interval. On the right, the pointwise error function $p(x)$ (black) is shown. (Bottom, from left to right) The maximum of the pointwise error function on the interval (left, dashed) is the uniform error on the interval. For the L^2 error, we first square $p(x)$ and compute the area under the curve (center), before taking the square root. On the right, $p(x)$ is shown together with the computed values of the uniform error (lower dashed) and L^2 error (upper dashed).

The notions of uniform and L^2 error can be related back to the properties of a norm discussed in Section 2.1. More specifically,

$$\begin{array}{c}\text{uniform error between}\\ f \text{ and } g \text{ on } a \leq x \leq b\end{array} = \max_{a \leq x \leq b} p(x) = \max_{a \leq x \leq b} |f(x) - g(x)| = \|f - g\|,$$

where the norm here is the one in Exercise 2.1.11(b), called the *uniform norm* or *max norm*. On the other hand,

$$\begin{array}{c}L^2 \text{ error between } f\\ \text{and } g \text{ on } a \leq x \leq b\end{array} = \sqrt{\int_a^b p^2(x)\, dx} = \sqrt{\int_a^b |f(x) - g(x)|^2 dx} = \|f - g\|,$$

where the norm here is the familiar norm from Exercise 2.1.11(a), called the L^2 *norm*. Different concepts of error arise when different norms are applied to the pointwise error function.

Figure 3.5 shows that the uniform error and L^2 error "measure" two different characteristics of the pointwise error $p(x)$. The pointwise error might be "small" in one norm, but "large" in another, as Figure 3.6 illustrates. If so, then an approximate solution of a PDE might be good in one norm (since the error is "small" in that norm), but not so good in another norm (because the error is "large" in the other norm). Being able to say precisely in what sense an approximate solution is "close" to the exact solution is a vital part of applied mathematics.

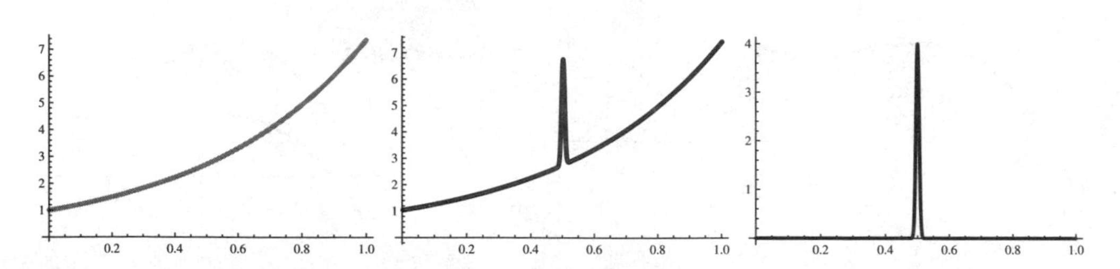

Figure 3.6: Two functions can be "close" in one norm but "far" in another norm. Here, f (gray), g (blue), and the pointwise error function p (black) are shown. Although the functions are identical except on the small interval where g spikes, the uniform error is 4 (large), whereas the L^2 error is approximately 0.38 (small). This is because a large deviation at even a single point results in a large uniform error, but a large L^2 error requires the accumulation of error over an interval since it involves the area under $p^2(x)$.

Modes of Convergence

In Section 2.4, we showed that the eigenfunctions of $X''+\lambda X = 0$ satisfying symmetric boundary conditions on $a < x < b$ form an orthogonal family on $a < x < b$. Moreover, a given function $f(x)$ can be expressed as an infinite sum of these orthogonal eigenfunctions, called an *eigenfunction expansion*[4]

$$f(x) = \sum_{n=1}^{\infty} c_n X_n(x) \qquad \text{where} \qquad c_n = \frac{\langle f, X_n \rangle}{\|X_n\|^2}. \tag{3.5}$$

In what sense is $f(x)$ "equal" to $\sum_{n=1}^{\infty} c_n X_n(x)$ on $a < x < b$? This is a question about the *convergence* of the infinite series (3.5), and the answer depends on the way the infinite series converges—called the *mode of convergence.* Next, we define three modes of convergence by relating them to the three notions of error we studied above.

Definition 3.3. Consider (3.5) and its Nth partial sum $S_N(x) := \sum_{n=1}^{N} c_n X_n(x)$.

- The infinite series (3.5) *converges pointwise* to $f(x)$ on $a \le x \le b$ if

$$|f(x) - S_N(x)| \to 0 \quad \text{as } N \to \infty, \text{ for every } x \in [a, b].$$

  That is, if the (absolute) pointwise error between $f(x)$ and $S_N(x)$ tends to zero as $N \to \infty$ for every x in $[a, b]$.

- The infinite series (3.5) *converges uniformly* to $f(x)$ on $a \le x \le b$ if

$$\max_{a \le x \le b} |f(x) - S_N(x)| \to 0 \quad \text{as } N \to \infty.$$

  That is, if the uniform error between $f(x)$ and $S_N(x)$ tends to zero as $N \to \infty$.

[4]Some texts refer to these as *generalized Fourier series.*

- The infinite series (3.5) *converges in the L^2 sense* to $f(x)$ on $a \le x \le b$ if

$$\sqrt{\int_a^b |f(x) - S_N(x)|^2\, dx} \to 0 \qquad \text{as } N \to \infty.$$

That is, if the L^2 error between $f(x)$ and $S_N(x)$ tends to zero as $N \to \infty$.

Example 3.3.2. Suppose the Nth partial sum of an infinite series is $S_N(x) = \frac{x+N}{N}$ and let $f(x) \equiv 1$, $0 \le x \le 1$. Let's examine whether the infinite series (3.5) converges pointwise, uniformly, or in the L^2 sense to f on $0 \le x \le 1$.

To answer this, we first need the pointwise error function,

$$p(x) = |f(x) - S_N(x)| = \left|1 - \frac{x+N}{N}\right| = \left|\frac{-x}{N}\right| = \frac{x}{N}, \quad 0 \le x \le 1.$$

Since $p(x_0) = \frac{x_0}{N} \to 0$ as $N \to \infty$ for each fixed x_0 in $[0, 1]$, the infinite series converges pointwise to f on $0 \le x \le 1$.

On the other hand,

$$\max_{0\le x\le 1} p(x) = \max_{0\le x\le 1} \frac{x}{N} = \frac{1}{N} \to 0 \qquad \text{as } N \to \infty,$$

so the infinite series does converge uniformly to f on $0 \le x \le 1$.

Finally,

$$\sqrt{\int_0^1 p^2(x)\, dx} = \sqrt{\int_0^1 \left(\frac{x}{N}\right)^2 dx} = \sqrt{\frac{1}{3N^2}} \to 0 \qquad \text{as } N \to \infty,$$

so the infinite series does converge in the L^2 sense to f on $0 \le x \le 1$. ◊

Example 3.3.3. Consider the infinite series

$$\sum_{n=0}^{\infty} (-1)^n x^{2n}.$$

We want to determine if this series converges pointwise, uniformly, or in the L^2 sense, and to what function and on what interval(s) it does so.

- **Pointwise convergence:** To get started, we need a candidate for the sum of the series, i.e., the $f(x)$ in Definition 3.3. Since this is a geometric series with geometric ratio $-x^2$, it sums[5] to

$$\frac{1}{1-(-x^2)} = \frac{1}{1+x^2}, \quad -1 < x < 1.$$

[5] From calculus, recall $\sum_{n=0}^{\infty} r^n = \frac{1}{1-r}$ for $|r| < 1$.

Therefore, the given series converges pointwise to $f(x) = \frac{1}{1+x^2}$ on the interval $(-1,1)$. This means that for any fixed number $a \in (-1,1)$, the numerical value of $\sum_{n=0}^{\infty}(-1)^n a^{2n}$ will equal $f(a) = \frac{1}{1+a^2}$.

- **Uniform convergence:** Does this series converge uniformly to $f(x) = \frac{1}{1+x^2}$ on $-1 \le x \le 1$? To answer this, let $S_N(x) := \sum_{n=0}^{N}(-1)^n x^{2n}$. We need to check whether

$$\max_{-1\le x\le 1}\left|\frac{1}{1+x^2} - S_N(x)\right| \to 0 \quad \text{as } N \to \infty. \tag{3.6}$$

The Nth partial sum of a geometric series is given by $\frac{a(1-r^N)}{1-r}$, where a is the first term in the series and r is the geometric ratio. Using this fact and working from (3.6),

$$\begin{aligned}\max_{-1\le x\le 1}\left|\frac{1}{1+x^2} - S_N(x)\right| &= \max_{-1\le x\le 1}\left|\frac{1}{1+x^2} - \frac{1-(-x^2)^N}{1-(-x^2)}\right| \\ &= \max_{-1\le x\le 1}\left|\frac{(-1)^N x^{2N}}{1+x^2}\right| \\ &= \max_{-1\le x\le 1}\frac{x^{2N}}{1+x^2} \\ &= 1/2,\end{aligned}$$

which of course does not tend to zero as $N \to \infty$. Therefore, this series does not converge uniformly to $f(x)$ on $-1 \le x \le 1$.

- **L^2 convergence:** Does this series converge in the L^2 sense to $f(x) = \frac{1}{1+x^2}$ on $-1 \le x \le 1$? To answer this, we need to check whether

$$\sqrt{\int_{-1}^{1}\left|\frac{1}{1+x^2} - S_N(x)\right|^2 dx} \to 0 \quad \text{as } N \to \infty. \tag{3.7}$$

Since (3.7) holds if and only if the integral (without the square root) tends to zero, it is enough to work with the latter. Again, computing the Nth partial sum, the integral in (3.7) becomes

$$\begin{aligned}\int_{-1}^{1}\left|\frac{(-1)^{N+1}x^{2N}}{1+x^2}\right|^2 dx = \int_{-1}^{1}\frac{x^{4N}}{(1+x^2)^2}\,dx &= \int_{-1}^{1}\frac{1}{(1+x^2)^2}\,x^{4N}\,dx \\ &\le \int_{-1}^{1} x^{4N}\,dx \\ &= \frac{2}{4N+1} \to 0 \quad \text{as } N \to \infty.\end{aligned}$$

Therefore, this series does converge in the L^2 sense to $f(x)$ on $-1 \le x \le 1$. ◊

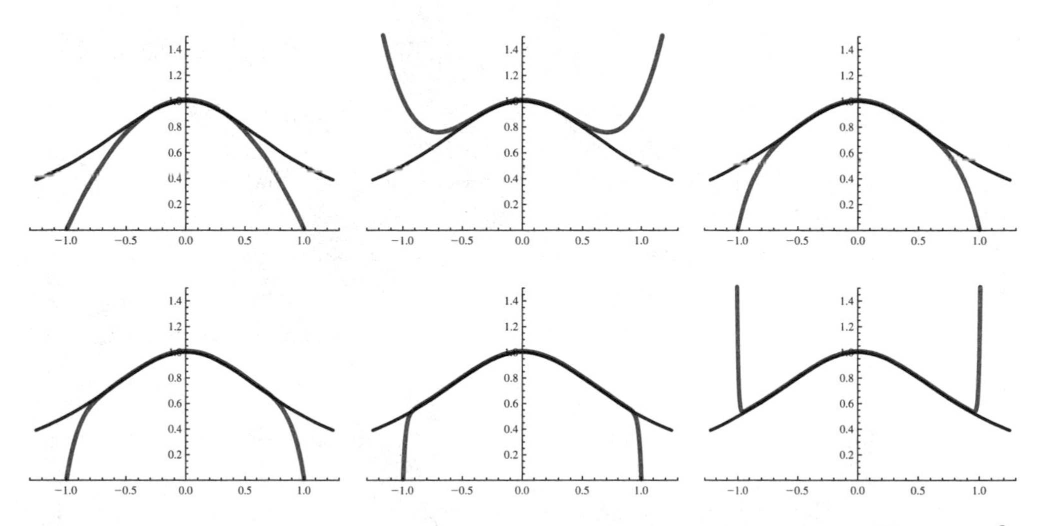

Figure 3.7: Pointwise convergence of the Nth partial sum of the geometric series $\sum_{n=0}^{\infty}(-1)^n x^{2n}$ (blue) to $\frac{1}{1+x^2}$ (black) for each $x \in (-1, 1)$. However, the convergence is *not* uniform since the maximum pointwise error over the closed interval $[-1, 1]$ (which occurs at $x = \pm 1$) does not tend to zero as $N \to \infty$. This series does converge in the L^2 sense on $[-1, 1]$.

Exercises

1. Suppose $f(x) = \sin(9.5x^2) + 5x$ on $0 \le x \le 1$ is approximated by the first three nonzero terms of its Fourier sine series.

   (a) Compute the (absolute) pointwise error in this approximation at $x = 0.25$, 0.5, and 0.75.

   (b) Compute the uniform error in this approximation.

   (c) Compute the L^2 error in this approximation.

   (d) Generate graphs for each part above analogous to those in Figure 3.5.

2. Suppose $f(x) = x$ on $0 \le x \le 1$ is approximated by the first five nonzero terms of its Fourier sine series.

   (a) Compute the (absolute) pointwise error in this approximation at $x = 0.25$, 0.5, and 0.75.

   (b) Compute the uniform error in this approximation.

   (c) Compute the L^2 error in this approximation.

   (d) Generate graphs for each part above analogous to those in Figure 3.5.

   (e) Repeat each part above using the first five nonzero terms of its Fourier cosine series.

✵ 3. Consider the function $\operatorname{erf}(x) = \frac{2}{\sqrt{\pi}} \int_0^x e^{-t^2}\, dt$, $0 \le x \le 3$ from Exercise 3.2.2.

(a) Plot the (absolute) pointwise error between $\operatorname{erf}(x)$ and its Taylor polynomials $T_N(x)$ centered at $x = 1/2$ for $N = 1, 5, 10$.

(b) Plot the (absolute) pointwise error between $\operatorname{erf}(x)$ and $F_N(x)$, its partial Fourier sine series for $N = 1, 5, 10$.

(c) Compute the uniform errors in (a) and (b).

(d) Compute the L^2 errors in (a) and (b).

(e) Discuss each of these in the context of local vs global approximations.

✵ 4. Consider the function $f(x) = \sqrt{1 - x^2}$, $0 \le x \le 1$ from Exercise 3.2.3.

(a) Plot the (absolute) pointwise error between $f(x)$ and its Taylor polynomials $T_N(x)$ centered at $x = 1/2$ for $N = 1, 5, 10$.

(b) Plot the (absolute) pointwise error between $f(x)$ and $F_N(x)$, its partial Fourier sine series for $N = 1, 5, 10$.

(c) Compute the uniform errors in (a) and (b).

(d) Compute the L^2 errors in (a) and (b).

(e) Discuss each of these in the context of local vs global approximations.

5. Consider the sequence $f_n(x) = (1 - x)x^{n-1}$, $0 \le x \le 1$, for $n = 1, 2, \ldots$.

(a) Show that $\sum_{n=1}^{N} f_n(x) = 1 - x^N$ by recognizing this as a telescoping sum. Use the definition of pointwise convergence to show that this series converges pointwise to $f(x) \equiv 1$ on $0 < x < 1$.

(b) Use the definition of uniform convergence to show that the convergence of this series is not uniform on $0 \le x \le 1$.

(c) Use the definition of L^2 convergence to show that this series converges to $f(x) \equiv 1$ in the L^2 sense on $0 \le x \le 1$.

6. Consider the sequence $f_n(x) = e^{-nx}$, $0 \le x \le 1$, for $n = 1, 2, \ldots$.

✵ (a) Plot several members of the sequence and use them to make a conjecture for the pointwise limit $f(x) := \lim_{n\to\infty} f_n(x)$.

(b) Using the plots to aid you, does $f_n(x)$ converge uniformly to $f(x)$ on $0 \le x \le 1$? Prove your assertion without the graphs.

(c) Use the definition of L^2 convergence to prove that $f_n(x)$ converges in the L^2 sense to $f(x)$ for $0 \le x \le 1$.

7. (a) Show that if a Fourier series fails to converge pointwise on (a, b), then it must fail to converge uniformly on $[a, b]$.

(b) Show that if a Fourier series converges uniformly on $[a, b]$, then it must converge pointwise on $[a, b]$ and it must converge in the L^2 sense on $[a, b]$.

3.4 Convergence Theorems

While issues of convergence might seem like unnecessarily abstract concepts, they are important for several reasons. In applications we are often concerned about the eigenfunction expansion converging to $f(x)$ in *some* sense—even in a weak sense—so that our problem has a solution. Therefore, it is beneficial to have more than one mode of convergence at our disposal. But there are deeper issues too:

(Q1) What conditions on $f(x)$ are required for each type of convergence?

(Q2) If $f(x)$ "equals" $\sum_{n=1}^{\infty} c_n X_n(x)$, does $f'(x)$ "equal" $\sum_{n=1}^{\infty} c_n X_n'(x)$?

(Q3) If $f(x)$ "equals" $\sum_{n=1}^{\infty} c_n X_n(x)$, does $\int_a^b f(x)\,dx$ "equal" $\sum_{n=1}^{\infty} c_n \int_a^b X_n(x)\,dx$?

The answers to these questions depend on which type of convergence (which brand of "equals") we have. Therefore, it behooves us to understand all the types of convergence at our disposal. Interestingly, Fourier and his successors wrestled greatly with these deep mathematical issues while solving PDEs, thereby laying the foundation of modern mathematical analysis.

We'd like to answer the first question posed earlier: Under what conditions on f can we guarantee that its Fourier series (eigenfunction expansion) will converge pointwise, uniformly, or in the L^2 sense? We will answer this question first in the context of *classical Fourier series*, i.e.,

$$f(x) = \frac{1}{2}a_0 + \sum_{n=1}^{\infty}[a_n \cos(n\pi x/\ell) + b_n \sin(n\pi x/\ell)], \quad -\ell < x < \ell. \tag{3.8}$$

After doing so, we will be able to give a very general answer concerning the convergence of the orthogonal expansion (3.5). We begin with some definitions.

Definition 3.4. A function f is *piecewise continuous on* (a, b) provided the one-sided limits $\lim_{x\to a^+} f(x)$ and $\lim_{x\to b^-} f(x)$ are finite and f is continuous on (a, b), except perhaps at finitely many points in (a, b), each having finite left- and right-hand limits. A function g is *piecewise smooth on* (a, b) if g and g' are piecewise continuous on (a, b).

Example 3.4.1. Consider the following functions on $0 < x < 1$.

1. The sawtooth function

$$f(x) = \begin{cases} x, & 0 < x < \frac{1}{2}, \\ 1 - x, & \frac{1}{2} < x < 1, \end{cases}$$

is piecewise continuous on $(0, 1)$, but $g(x) = 1/x$ is not piecewise continuous on $(0, 1)$.

2. The sawtooth function above is in fact piecewise smooth on $(0, 1)$.
3. The function $h(x) = x^{2/3}$ is piecewise continuous, but not piecewise smooth on $(0, 1)$. ◊

The first theorem gives conditions on the function f sufficient to guarantee the pointwise convergence of its classical Fourier series.

Theorem 3.2 (Pointwise Convergence of Fourier Series).
If f is piecewise smooth on $(-\ell, \ell)$, then the Fourier series of f given by (3.8) converges pointwise on $(-\ell, \ell)$ and

$$\frac{1}{2}a_0 + \sum_{n=1}^{\infty}[a_n \cos(n\pi x/\ell) + b_n \sin(n\pi x/\ell)] = \frac{f(x^+) + f(x^-)}{2}, \quad x \in (-\ell, \ell). \quad (3.9)$$

Here, $f(x^+) := \lim_{w \to x^+} f(w)$ and $f(x^-) := \lim_{w \to x^-} f(w)$.

Note that if $f(x)$ is continuous at x_0, then $f(x_0^+) = f(x_0^-) = f(x_0)$, so (3.9) says

$$\sum_{n=1}^{\infty} c_n X_n(x_0) = f(x_0), \quad x_0 \in (a, b),$$

which is in harmony with the definition of pointwise convergence. On the other hand, if $f(x)$ has a jump discontinuity at x_0, then the Fourier series converges to the average of the left- and right-hand limit values. See Figure 3.8.

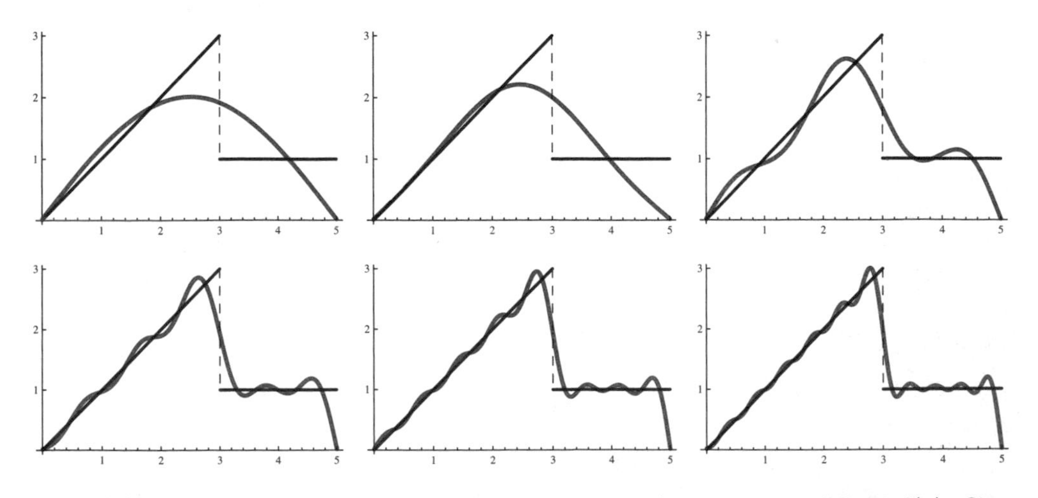

Figure 3.8: The Nth partial sums of the Fourier sine series for a piecewise defined $f(x)$. Since there is a jump discontinuity in f at $x = 3$, the Fourier series converges to the "average jump" at $x = 3$, which is 2.

Since uniform convergence is much stronger than pointwise convergence, one might suspect that more stringent requirements must be placed on f to get uniform convergence of its classical Fourier series. This is indeed the case.

Theorem 3.3 (Uniform Convergence of Fourier Series).

(a) *If f has a continuous periodic extension and f' is piecewise continuous on $(-\ell, \ell)$, then the Fourier series for f given by* (3.8) *converges uniformly to $f(x)$ on $[-\ell, \ell]$.*

(b) *If f does not have a continuous periodic extension, then the Fourier series for f given by* (3.8) *does not converge uniformly to $f(x)$ on $[-\ell, \ell]$.*

However, the condition needed to ensure L^2 convergence is very mild. This is one reason why L^2 convergence is so well studied—it is easily checkable and requires very little of the function f.

Theorem 3.4 (L^2 Convergence of Fourier Series).
The Fourier series for f given by (3.8) *converges in the L^2 sense to $f(x)$ on $[-\ell, \ell]$ if and only if $\int_{-\ell}^{\ell} |f(x)|^2\, dx < \infty$, i.e., $\|f\| < \infty$, where this is the L^2 norm.*

Example 3.4.2. Let $f(x) = x^2$, $0 \le x \le 1$ and consider the even and odd extensions of $f(x)$. Since $f_{\text{even}}(x)$ is continuous and $f'(x)$ is piecewise continuous on $(-1, 1)$, by Theorem 3.3(a), the full Fourier series of $f_{\text{even}}(x)$ converges uniformly on $[-1, 1]$. In contrast, $f_{\text{odd}}(x)$ does not have a continuous periodic extension, so by Theorem 3.3(b), the full Fourier series of $f_{\text{odd}}(x)$ does not converge uniformly on $[-1, 1]$.

On the other hand, both $\int_{-1}^{1} |f_{\text{even}}(x)|^2\, dx < \infty$ and $\int_{-1}^{1} |f_{\text{odd}}(x)|^2\, dx < \infty$, so by Theorem 3.4, the full Fourier series of both $f_{\text{even}}(x)$ and $f_{\text{odd}}(x)$ converge in the L^2 sense on $[-1, 1]$. See Figure 3.9. ◊

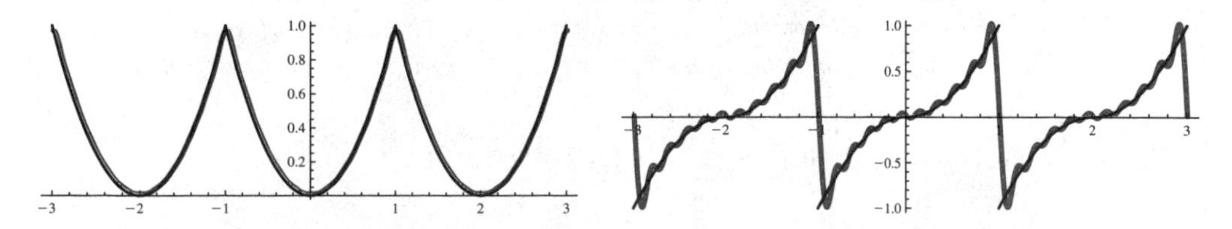

Figure 3.9: (Left) The Fourier series of $f_{\text{even}}(x)$ which converges uniformly and in the L^2 sense on $[-1, 1]$. (Right) The Fourier series of $f_{\text{odd}}(x)$ which does not converge uniformly on $[-1, 1]$, but does converge in the L^2 sense.

Differentiation/Antidifferentiation of Fourier Series

Properly equipped with these convergence criteria, we can address questions (Q2) and (Q3) above. We do so in the context of the classical Fourier series (3.8).

If $f(x)$ "equals" $\sum_{n=1}^{\infty} c_n X_n(x)$, does that mean $f'(x)$ "equals" $\sum_{n=1}^{\infty} c_n X_n'(x)$? The answer is that *it depends on the type of convergence.* For example, the Fourier sine series for $f(x) = x$, $0 < x < 1$, is given by

$$x = 2\sum_{n=1}^{\infty} \frac{(-1)^{n+1}}{n\pi} \sin(n\pi x), \qquad 0 < x < 1.$$

If we blindly differentiate both sides, the left-hand side is 1, while the right-hand side becomes $2\sum_{n=1}^{\infty}(-1)^{n+1}\cos(n\pi x)$, which diverges because the nth term does not tend to zero. Clearly, equality does not hold here even though the Fourier series converges pointwise (due to Theorem 3.2). However, the following theorem tells us when we can legitimately differentiate a Fourier series term-by-term.

Theorem 3.5 (Term-By-Term Differentiation of Fourier Series).
Suppose f satisfies the hypotheses of Theorem 3.3(a). Then, at every point $x \in (-\ell, \ell)$ where $f''(x)$ exists, the Fourier series for f given by (3.8) is differentiable and

$$f'(x) = \sum_{n=1}^{\infty} \frac{d}{dx}[a_n \cos(n\pi x/\ell) + b_n \sin(n\pi x/\ell)].$$

Here, equality is meant in the sense of pointwise convergence.

However, the conditions under which a Fourier series can be integrated term-by-term are much milder. In fact, the series is not even required to converge pointwise!

Theorem 3.6 (Term-By-Term Integration of Fourier Series).
Let f be a piecewise continuous function on $(-\ell, \ell)$. Then, regardless of whether the Fourier series of f given by (3.8) converges, the following equation holds:

$$\int_a^x f(s)\,ds = \int_a^x \frac{1}{2}a_0\,ds + \sum_{n=1}^{\infty} \int_a^x [a_n \cos(n\pi s/\ell) + b_n \sin(n\pi s/\ell)]\,ds.$$

Here, equality is meant in the sense of pointwise convergence.

Example 3.4.3. The function $f(x) = |x|$, $-1 < x < 1$, has Fourier series

$$|x| = \frac{1}{2}a_0 + \sum_{n=1}^{\infty} a_n \cos(n\pi x/\ell) + b_n \sin(n\pi x/\ell),$$

$$a_0 = \pi, \ a_n = \frac{2(-1+(-1)^n)}{\pi^2 n^2}, \ b_n = 0,$$

and satisfies the conditions of Theorem 3.5 for all $x \in (-1, 1)$ except $x = 0$ since $f''(0)$ does not exist. Thus, for $x \neq 0$, the Fourier series of the derivative is

$$f'(x) = \begin{cases} -1, & -1 < x < 0, \\ 1, & 0 < x < 1, \end{cases} = \sum_{n=1}^{\infty} -a_n n\pi \sin(n\pi x),$$

where equality is in the sense of pointwise convergence. Also, f satisfies the conditions of Theorem 3.6, so the Fourier series of the antiderivative is

$$\int_{-1}^{x} f(s)\,ds = \int_{-1}^{x} |s|\,ds = \begin{cases} \frac{1-x^2}{2}, & -1 < x < 0, \\ \frac{1+x^2}{2}, & 0 < x < 1, \end{cases} = \frac{1}{2}a_0(x+1) + \sum_{n=1}^{\infty} \frac{a_n}{n\pi} \sin(n\pi x).$$

See Figure 3.10. ◊

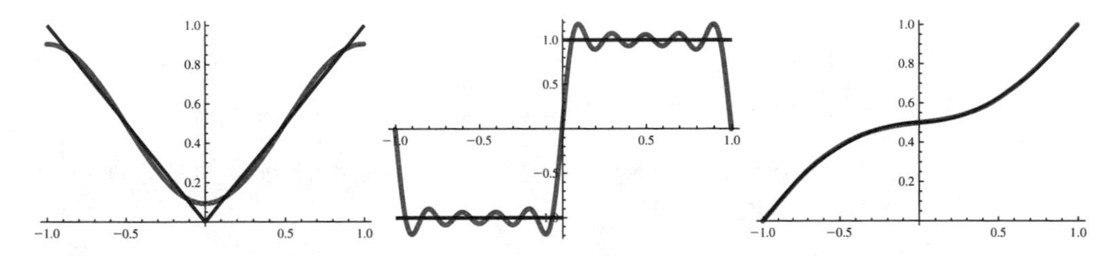

Figure 3.10: (Left) $f(x) = |x|$ (black) and its partial Fourier sum $F_2(x)$ (blue). (Center) $f'(x)$ (black) and the differentiated partial Fourier sum $F_{10}(x)$ (blue). (Right) $\int_{-1}^{x} f(s)\,ds$ (black) and the integrated partial Fourier sum $F_2(x)$ (blue).

Exercises

✵ 1. Let $f(x) = \frac{\sin(3x)}{x}$, $-\pi \leq x \leq \pi$, $x \neq 0$.

(a) Plot $f(x)$ and the first N terms of its full Fourier series on the same plane, where $N = 1, 2, 3, 4, 5$.

(b) Notice that $f(x)$ is undefined at $x = 0$, yet its Fourier series appears to be converging to 3 when $x = 0$. Explain this in light of the Pointwise Convergence Theorem.

(c) Using the theorems from this section, does the Fourier series for this function converge uniformly on $[-\pi, \pi]$?

(d) Using the theorems from this section, does the Fourier series for this function converge in the L^2 sense on $[-\pi, \pi]$?

2. Let $f(x) = \begin{cases} -1 - x, & -1 \leq x \leq 0, \\ 1 - x, & 0 < x \leq 1. \end{cases}$

(a) Compute the Fourier series for this function on $-1 \leq x \leq 1$ first by hand and then using *Mathematica*.

(b) Plot $f(x)$ and the N term Fourier series approximation on the same plane, where $N = 1, 3, 5, 10, 20$.

(c) Using the theorems from this section, does the Fourier series for this function converge pointwise on $-1 \leq x \leq 1$?

(d) Using the theorems from this section, does the Fourier series for this function converge uniformly on $-1 \leq x \leq 1$?

(e) Using the theorems from this section, does the Fourier series for this function converge in the L^2 sense on $-1 \leq x \leq 1$?

3. Consider $f(x) = x^2$, $0 < x < \ell$. *Without computing any series*, answer the following.

(a) What is the value of the Fourier cosine series of $f(x)$ at $x = \pm\ell$?

(b) What is the value of the Fourier sine series of $f(x)$ at $x = \pm\ell$?

(c) What is the value of the full Fourier series of $f(x)$ at $x = \pm\ell$?

4. Motivated by Theorem 3.4, we say that f is in $L^2[a, b]$ provided $\int_a^b f^2(x)\,dx < \infty$. Use this to answer the following.

(a) Find all values of p such that $\frac{1}{x^p}$ is in $L^2[0, 1]$.

(b) Find all values of p such that $\frac{1}{x^p}$ is in $L^2[1, \infty)$.

(c) Give an example of a function *not* of the form $\frac{1}{x^p}$ that *is* in $L^2[0, 1]$. Justify your answer.

(d) Give an example of a function *not* of the form $\frac{1}{x^p}$ that is *not* in $L^2[0, 1]$. Justify your answer.

5. Without computing any Fourier series, but instead using the theorems from this section, determine if we can conclude whether the full Fourier series for each function below converges pointwise, uniformly, or in the L^2 sense on the given interval and whether the Fourier series can be differentiated term-by-term on the open interval $-\ell < x < \ell$.

(a) $f(x) = x^3$ on $[0, 1]$, using an even extension to $[-1, 1]$

(b) $f(x) = x^3$ on $[0, 1]$, using an odd extension to $[-1, 1]$

(c) $f(x) = \sqrt{x}$ on $[0, 1]$, using an even extension to $[-1, 1]$

(d) $f(x) = \sqrt{x}$ on $[0, 1]$, using an odd extension to $[-1, 1]$

(e) $f(x) = |x|$ on $[-1, 1]$

(f) $f(x) = 1/x$ on $[-1, 1]$

(g) $f(x) = |\sin x|$ on $[-2\pi, 2\pi]$

6. (a) Show that the Fourier series for $f(x) = |x|$ on $-\pi \leq x \leq \pi$ is given by

$$|x| = \frac{\pi}{2} - \frac{4}{\pi}\left[\frac{\cos(x)}{1^2} + \frac{\cos(3x)}{3^2} + \frac{\cos(5x)}{5^2} + \cdots\right], \qquad -\pi \leq x \leq \pi.$$

(b) Verify that f satisfies the conditions of the Pointwise Convergence Theorem. Then use that theorem (with a judicious choice for x) and part (a) to show

$$\frac{1}{1^2} + \frac{1}{3^2} + \frac{1}{5^2} + \cdots = \frac{\pi^2}{8}.$$

(c) From (a), deduce the Fourier series for the *sign function* or *signum function* defined as

$$\operatorname{sgn}(x) := \begin{cases} -1, & -\pi < x < 0, \\ 1, & 0 < x < \pi. \end{cases}$$

Is this Fourier series for $\operatorname{sgn}(x)$ valid for all x? Explain.

7. Let $f(x) = e^x$ on $0 < x < 1$.

(a) Extend $f(x)$ to $-1 < x < 0$ in such a way that its Fourier series will converge pointwise at every point in $\mathbb{R}$. Clearly state the formula for your extension, $f_{\text{ext}}(x)$, and explain why the Fourier series of your extended f will converge pointwise on $\mathbb{R}$.

(b) What is the value of the Fourier series of $f_{\text{ext}}(x)$ from (a) at $x = 101$?

8. In signal processing, the *rectangle function* $\Pi(x)$ and *triangle function* $\Lambda(x)$ are defined as

$$\Pi(x) := \begin{cases} 0, & -1 \leq x < -\frac{1}{2}, \\ 1, & -\frac{1}{2} \leq x \leq \frac{1}{2}, \\ 0, & \frac{1}{2} < x \leq 1, \end{cases} \qquad \Lambda(x) := \begin{cases} 0, & -1 \leq x \leq -\frac{1}{2}, \\ 2x+1, & -\frac{1}{2} < x \leq 0, \\ 1-2x, & 0 < x \leq \frac{1}{2}, \\ 0, & \frac{1}{2} < x \leq 1. \end{cases}$$

Viewing these as functions on $-1 \leq x \leq 1$, answer the following.

(a) Discuss the pointwise, uniform, and L^2 convergence of the full Fourier series of $\Pi(x)$.

(b) If they exist, find the value of the full Fourier series for $\Pi(x)$ at $x = 100$ and $x = 101.5$.

(c) Discuss the pointwise, uniform, and L^2 convergence of the full Fourier series of $\Lambda(x)$.

(d) If they exist, find the value of the full Fourier series for $\Lambda(x)$ at $x = 100$ and $x = 101.5$.

(e) Can the Fourier series for $\Lambda(x)$ be differentiated term-by-term? If so, where? Discuss.

9. Consider the function

$$f(x) = \begin{cases} e^x, & -1 \leq x \leq 0, \\ mx + b, & 0 \leq x \leq 1. \end{cases}$$

Without computing any Fourier coefficients, answer the following.

(a) For what value(s) of m and b (if any) will the full Fourier series of $f(x)$ converge pointwise on $-1 < x < 1$? Justify your answer.

(b) For what value(s) of m and b (if any) will the full Fourier series of $f(x)$ converge uniformly on $-1 \leq x \leq 1$? Justify your answer.

(c) For what value(s) of m and b (if any) will the full Fourier series of $f(x)$ converge in the L^2 sense on $-1 \leq x \leq 1$? Justify your answer.

10. Let $f(x) = x$, $-1 < x < 1$. Plot the $N = 10$ partial Fourier sum for f, as well as the term-by-term antiderivative and term-by-term derivative. Explain your results in light of Theorems 3.5 and 3.6.

11. **(Weierstrass M-Test for Uniform Convergence)** Another way to prove uniform convergence of $\sum_{n=1}^{\infty} c_n X_n(x)$, $a \leq x \leq b$, is the following result.

If $\{M_n\}_{n=1}^{\infty}$ are positive constants such that $|c_n X_n(x)| \leq M_n$ for all x in $[a, b]$ and $\sum_{n=1}^{\infty} M_n < \infty$, then $\sum_{n=1}^{\infty} c_n X_n(x)$ converges uniformly on $[a, b]$.

Use this to answer the following.

(a) Use the Weierstrass M-Test to show that the Fourier series for $|x|$, $-\pi \leq x \leq \pi$, given in Exercise 6 converges uniformly on $-\pi \leq x \leq \pi$.

(b) More generally, use the Weierstrass M-Test to show that if a_n and b_n are the Fourier coefficients in (3.8) and $\sum_{n=1}^{\infty}(|a_n| + |b_n|) < \infty$, then (3.8) converges uniformly on $-\ell \leq x \leq \ell$.

(c) Suppose $f(x)$, $-\ell \leq x \leq \ell$, has Fourier coefficients

$$a_0 = \frac{2}{3}, \quad a_n = \frac{4(-1)^n}{\pi^2 n^2}, \quad b_n = 0, \quad n = 1, 2, \ldots.$$

Does the full Fourier series of $f(x)$ converge uniformly on $[-\ell, \ell]$? Explain.

3.5 Basic L^2 Theory

Our goal now is to focus on the geometric aspects of orthogonality. We begin with the defining properties of a general inner product which allows us to state the proper notion of the space $L^2[a, b]$.

Definition 3.5. Suppose u, v, w are in a vector space V and c is a real scalar. An *inner product* on V is a function that assigns each pair of vectors to a real number and satisfies:

(IP1) $\langle u, v\rangle = \langle v, u\rangle$ for all $u, v \in V$,

(IP2) $\langle u + v, w\rangle = \langle u, w\rangle + \langle v, w\rangle$ for all $u, v, w \in V$,

(IP3) $\langle cu, v\rangle = c\langle u, v\rangle$ for all $u, v \in V$ and scalars c,

(IP4) $\langle u, u\rangle \geq 0$ with equality if and only if $u = 0$.

A vector space together with an inner product is called an *inner product space.*

We have dealt with two inner products thus far: the usual dot product on $\mathbb{R}^n$ and the $L^2[a, b]$ inner product defined by

$$\langle f, g\rangle := \int_a^b f(x)g(x)\,dx.$$

We can quickly confirm that both of these satisfy (IP1)–(IP4) and are therefore inner products.

Thus far, we have worked with the space $C[a, b]$ equipped with the usual inner product. However, this space has a rather undesirable property: there are sequences of functions $\{f_n\}$ in $C[a, b]$ which converge in the mean-square or L^2 sense to a function f which is *not* in $C[a, b]$.

Ideally, we want any space X that we work with to have the following property:

> Whenever $\{f_n\}$ is a sequence of functions in X and f_n converges to f in the L^2 sense, then f is also in X. (P)

Unfortunately, $C[a, b]$ with the L^2 norm does *not* have this property for two reasons: the set of continuous functions is not "large enough," and the Riemann integral contained in the definition of the inner product does not capture "enough" square integrable functions. A rigorous explanation is beyond the scope of this text, but suffice it to say that the space $L^2[a, b]$ overcomes these obstacles; so property (P) is satisfied.[6]

[6]The letter L in $L^2[a, b]$ is in honor of Henri Lebesgue, because the Lebesgue integral must be used instead of the Riemann integral in the definition of the inner product. Fortunately for us, these two integrals coincide for all the functions that we will study.

Definition 3.6. The vector space of functions

$$L^2[a,b] := \left\{ f : [a,b] \to \mathbb{R} \;\middle|\; \int_a^b |f(x)|^2 dx < \infty \right\},$$

equipped with the inner product $\langle u, v\rangle := \int_a^b u(x)v(x)\,dx$, is called *space of L^2 functions*[7] *on* $[a,b]$.

This inner product naturally defines the familiar norm

$$\|u\| = \langle u, u\rangle^{1/2} = \sqrt{\int_a^b |u(x)|^2\,dx}. \tag{3.10}$$

We call this the L^2 *norm on* $[a,b]$. We quickly see that a function f is in $L^2[a,b]$ if and only if $\|f\| < \infty$. $L^2[a,b]$ is viewed properly as the completion of $C[a,b]$ with respect to the norm (3.10) in the sense we described above.

Best Approximation

The next theorem says the way we chose the Fourier coefficients in (3.5) turns out to be the "best" way to do it—in the sense that it minimizes the L^2 norm of the error between $f(x)$ and the Nth partial sum $S_N(x)$.

Theorem 3.7 (Best Approximation Theorem).
Suppose $f \in L^2[a,b]$ has orthogonal expansion

$$f(x) = \sum_{n=1}^{\infty} c_n X_n(x), \quad a \leq x \leq b,$$

and let N be fixed. Among all possible choices of the coefficients $c_1, c_2, \ldots, c_N$, the choice that minimizes the L^2 error between $f(x)$ and the Nth partial sum $S_N(x) := \sum_{n=1}^{N} c_n X_n(x)$, i.e. minimizes

$$\|f - S_N\|$$

(where this is the L^2 norm), is given by $c_n = \frac{\langle f, X_n\rangle}{\|X_n\|^2}$, $n = 1, \ldots, N$.

[7] Strictly speaking, the members of $L^2[a,b]$ are not functions, but rather equivalence classes of functions: two functions are equivalent if they differ only on a set of measure zero (e.g., finite number of points). Keep this in mind whenever we make a statement of equality involving members of $L^2[a,b]$.

Proof. Let E_N denote the error between f and the Nth partial sum of the orthogonal expansion in (3.5). Then

$$E_N^2 := \left\| \sum_{n=1}^N c_n X_n(x) - f(x) \right\|^2 = \int_a^b \left(\sum_{n=1}^N c_n X_n(x) - f(x) \right)^2 dx$$
$$= \int_a^b \left(\sum_{n=1}^N [c_n X_n(x)]^2 - 2c_n X_n(x) f(x) + f^2(x) \right) dx.$$

Rewriting this last expression in terms of L^2 inner products and then norms, we see

$$E_N^2 = \sum_{n=1}^N \left(\int_a^b [c_n X_n(x)]^2\, dx - 2\int_a^b c_n X_n(x) f(x)\, dx \right) + \int_a^b f^2(x)\, dx$$
$$= \sum_{n=1}^N \langle c_n X_n, c_n X_n \rangle - 2\langle c_n X_n, f \rangle + \langle f, f \rangle$$
$$= \sum_{n=1}^N \|c_n X_n\|^2 - 2\langle c_n X_n, f \rangle + \|f\|^2$$
$$= \sum_{n=1}^N \|X_n\|^2 \underbrace{\left(c_n^2 - 2\frac{c_n \langle X_n, f \rangle}{\|X_n\|^2} \right)}_{\text{complete the square}} + \|f\|^2$$
$$= \sum_{n=1}^N \|X_n\|^2 \left(c_n^2 - 2\frac{c_n \langle X_n, f \rangle}{\|X_n\|^2} + \frac{\langle X_n, f \rangle^2}{\|X_n\|^4} \right) + \|f\|^2 - \sum_{n=1}^N \frac{\langle X_n, f \rangle^2}{\|X_n\|^2}$$
$$= \sum_{n=1}^N \|X_n\|^2 \left(c_n - \frac{\langle X_n, f \rangle}{\|X_n\|^2} \right)^2 + \|f\|^2 - \sum_{n=1}^N \frac{\langle X_n, f \rangle^2}{\|X_n\|^2}.$$

However, this last expression is smallest when the first summation is zero, i.e., when $c_n = \frac{\langle X_n, f \rangle}{\|X_n\|^2}$, which is exactly how the c_n were chosen in (3.5). □

However, this completion of the square provides us with further insight. Notice that if we chose the coefficients c_n as in (3.5), then the last equation for the error says

$$0 \le E_N^2 = \|f\|^2 - \sum_{n=1}^N \frac{\langle X_n, f \rangle^2}{\|X_n\|^2} = \|f\|^2 - \sum_{n=1}^N c_n^2 \|X_n\|^2. \tag{3.11}$$

But since everything here is nonnegative, this yields

$$\sum_{n=1}^N c_n^2 \int_a^b |X_n(x)|^2\, dx \le \int_a^b |f(x)|^2\, dx = \|f\|^2 < \infty,$$

which is true for each N, and hence true as $N \to \infty$. Therefore, the choice of the Fourier coefficients in (3.5) leads us to another discovery known as *Bessel's Inequality*:

$$\sum_{n=1}^{\infty} c_n^2 \|X_n\|^2 \leq \|f\|^2 < \infty, \tag{3.12}$$

that is, the infinite (numerical) series on the left must converge whenever $f \in L^2[a,b]$.

Complete Orthogonal Families

Next, we want to establish the concept of an orthogonal basis for $L^2[a,b]$.

Definition 3.7. Suppose $\{X_n\}_{n=1}^{\infty}$ is an orthogonal family in $L^2[a,b]$. We say $\{X_n\}_{n=1}^{\infty}$ is *complete* if

$$f(x) = \sum_{n=1}^{\infty} c_n X_n(x), \quad c_n := \frac{\langle f, X_n \rangle}{\|X_n\|^2}, \tag{3.13}$$

holds (in the sense of L^2 convergence) for all $f \in L^2[a,b]$. That is, $\{X_n\}_{n=1}^{\infty}$ is complete if, for all $f \in L^2[a,b]$, the orthogonal expansion of f converges to f in the L^2 sense.

Hopefully the motivation for this concept is clear: If the orthogonal expansion (in terms of the X_n) of some $f \in L^2[a,b]$ converges in the L^2 sense to something other than f (or fails to converge at all), that would indicate a serious flaw in the orthogonal family $\{X_n\}_{n=1}^{\infty}$. Reflecting on our work in Chapter 2, we certainly needed the family of orthogonal eigenfunctions to be complete, otherwise we couldn't be sure that the final solution would converge appropriately.

Proving that a given orthogonal family is complete is often quite difficult. The next theorem provides two alternate characterizations of completeness.

Theorem 3.8 (Complete Orthogonal Families).
Suppose $\{X_n\}_{n=1}^{\infty}$ is an orthogonal family in $L^2[a,b]$. The following are equivalent:

(a) *$\{X_n\}_{n=1}^{\infty}$ is complete.*

(b) *The only function orthogonal to every member of the orthogonal family is the zero function. That is, if $\langle f, X_n \rangle = 0$ for all $n = 1, 2, \ldots,$ then $f \equiv 0$.*

(c) *Parseval's Equality holds for all $f \in L^2[a,b]$:*

$$\|f\|^2 = \sum_{n=1}^{\infty} c_n^2 \|X_n\|^2.$$

A complete orthogonal family is also called an orthogonal basis.

Proof. (a)$\Longleftrightarrow$(b): Suppose (a) holds. Then $f(x) = \sum_{n=1}^{\infty} c_n X_n(x)$ in the sense of L^2 convergence, where $c_n := \frac{\langle f, X_n \rangle}{\|X_n\|^2}$. If $\langle f, X_n \rangle = 0$ for all n, then $c_n = 0$ for all n, and $f \equiv 0$.

Conversely, suppose (b) holds and consider the function

$$g(x) := \sum_{n=1}^{\infty} c_n X_n(x), \quad c_n := \frac{\langle f, X_n \rangle}{\|X_n\|^2}. \tag{3.14}$$

It can be shown that $g \in L^2[a,b]$. For any fixed n,

$$\begin{aligned}
\langle f - g, X_n \rangle &= \langle f, X_n \rangle - \langle g, X_n \rangle \\
&= \langle f, X_n \rangle - \left\langle \sum_{m=1}^{\infty} \frac{\langle f, X_m \rangle}{\|X_m\|^2} X_m(x), X_n \right\rangle \\
&= \langle f, X_n \rangle - \sum_{m=1}^{\infty} \frac{\langle f, X_m \rangle}{\|X_m\|^2} \langle X_m, X_n \rangle \\
&= \langle f, X_n \rangle - \frac{\langle f, X_n \rangle}{\|X_n\|^2} \langle X_n, X_n \rangle \\
&= 0,
\end{aligned}$$

so $f - g$ is orthogonal to every member of the orthogonal family, and thus $f - g \equiv 0$. Since $f \equiv g$, (3.14) and (3.13) together yield (a).

(a)$\Longleftrightarrow$(c): Let $f \in L^2[a,b]$ so that the error expression (3.11) is valid:

$$E_N^2 = \|f\|^2 - \sum_{n=1}^{N} c_n^2 \|X_n\|^2.$$

If (a) holds, then $E_N^2 \to 0$ as $N \to \infty$ (from the definition of L^2 convergence), forcing $\|f\|^2 - \sum_{n=1}^{N} c_n^2 \|X_n\|^2 \to 0$ as $N \to \infty$ as well; so (c) holds. Conversely, if (c) holds, then the right-hand side tends to zero as $N \to \infty$, forcing the left-hand side to zero, so (a) holds. □

Although $L^2[a,b]$ is an infinite dimensional space, we can see why a complete orthogonal family is called an orthogonal *basis* for $L^2[a,b]$: part (b) implies $\{X_n\}_{n=1}^{\infty}$ is linearly independent, while (3.13) says that any $f \in L^2[a,b]$ is in the span of $\{X_n\}_{n=1}^{\infty}$, i.e., f can be written uniquely as a (infinite!) linear combination of the X_n.

Example 3.5.1. The family $\{\sin(nx)\}_{n=2}^{\infty}$ is an orthogonal family in $L^2[0,\pi]$, but it is not a complete orthogonal family since $\langle \sin x, \sin(nx) \rangle = 0$ for all $n = 2, 3, \ldots$, but of course $\sin x \not\equiv 0$. Thus, $\{\sin(nx)\}_{n=2}^{\infty}$ is not an orthogonal basis for $L^2[0,\pi]$. ◊

That example is reminiscent of the finite dimensional setting: Remove even one member of the basis and the result is no longer a basis. This is why, in Chapter 2 it was crucial for us to find every possible eigenfunction. If we omitted even one, the resulting family of orthogonal eigenfunctions would not be an orthogonal basis for the solution space.

Once we have an orthogonal basis, the following useful theorem is available to us.

Theorem 3.9 (Riesz-Fischer Theorem).
Suppose $\{X_n\}_{n=1}^{\infty}$ *is a complete orthogonal family in* $L^2[a,b]$ *and*

$$f(x) = \sum_{n=1}^{\infty} c_n X_n(x), \quad c_n := \frac{\langle f, X_n \rangle}{\|X_n\|^2}.$$

The following are equivalent:

(a) $f \in L^2[a,b]$.

(b) *The Fourier series of* f *converges to* f *in the* L^2 *sense on* $[a,b]$.

(c) *Parseval's Equality holds:* $\|f\|^2 = \sum_{n=1}^{\infty} c_n^2 \|X_n\|^2$.

The Riesz-Fischer Theorem justifies the L^2 Convergence Theorem in Section 3.4 and validates the convergence of all of the Fourier series solutions we obtained in Chapter 2 under rather mild conditions.

Note that Bessel's Inequality (3.12) does not require the completeness of the orthogonal family. However, when the orthogonal family is complete, Bessel's Inequality becomes the stronger statement known as Parseval's Equality, which can be used to sum certain infinite series.

Example 3.5.2. Consider $f(x) = x$ on $[0, \pi]$. Its Fourier sine series has coefficients

$$b_n = \frac{2}{\pi} \int_0^{\pi} x \sin(nx)\, dx = \frac{(-1)^{n+1} 2}{n}, \quad n = 1, 2, \ldots.$$

The family $\{\sin(nx)\}_{n=1}^{\infty}$ is orthogonal on $[0, \pi]$ (from Section 2.4), complete (we will show this in Chapter 4), and $f \in L^2[0, \pi]$, so computing the ingredients of Theorem 3.9(c) we see

$$\frac{\pi^3}{3} = \sum_{n=1}^{\infty} \frac{4}{n^2} \cdot \frac{\pi}{2} = 2\pi \sum_{n=1}^{\infty} \frac{1}{n^2},$$

and thus deduce the rather interesting identity

$$\sum_{n=1}^{\infty} \frac{1}{n^2} = \frac{\pi^2}{6}.$$

Other curious identities like this are explored in the exercises. ◊

Exercises

1. (a) Let $V = \mathbb{R}^n$ with the inner product $\langle \mathbf{u}, \mathbf{v} \rangle := \mathbf{u} \cdot \mathbf{v}$ (the usual dot product). Show that (IP1)–(IP4) all hold.

   (b) Let $V = L^2[a, b]$ with the inner product $\langle u, v \rangle := \int_a^b u(x)v(x)\,dx$. Show that (IP1)–(IP4) all hold.

   (c) Let $V = C[0, 1]$ with $\langle u, v \rangle := \max\limits_{0 \leq x \leq 1} |u(x)v(x)|$. Is this an inner product? Why or why not?

2. In calculus, we learned that the Taylor series of $f(x)$ at $x = a$ is given by

$$f(x) = \sum_{n=0}^{\infty} c_n (x - x_0)^n, \quad c_n = \frac{f^{(n)}(x_0)}{n!}.$$

   The formula for the coefficients above is different than the one in the Best Approximation Theorem. Were we using a "poor" choice of coefficients for Taylor series? Explain.

3. Which of the following functions are in $L^2[0, 1]$? Justify your answer.

   (a) $\ln x$

   (b) $x^{-1/2}$

   (c) $\csc x$

   (d) $\frac{\sin x}{x}$

4. Suppose we want to make the approximation

$$x \approx \frac{1}{2}a_0 + a_1 \cos x + a_2 \cos(2x) + a_3 \cos(3x), \qquad 0 \leq x \leq \pi,$$

   in such a way that the L^2 norm of the error on $0 \leq x \leq \pi$ is minimized.

   (a) What values of a_0, a_1, a_2, a_3 accomplish this?

   (b) What is this minimum L^2 error? (Note: L^2 error calculations normally require numerical integration[8], because symbolic integration is very difficult and therefore slow.)

5. Consider the orthogonal family of Legendre polynomials on $-1 \leq x \leq 1$ discussed in Exercise 2.4.6.

   (a) Let $f(x) = |x|$. Using the three term orthogonal series $\sum_{n=0}^{2} c_n P_n(x)$ to approximate $f(x)$ on $-1 \leq x \leq 1$, what formulas for the coefficients will minimize the L^2 error on this interval?

[8] In *Mathematica*, this is accomplished via `NIntegrate`.

(b) Plot $f(x)$ and the orthogonal expansion on the same coordinate plane and compute this minimum error in (a).

6. Pick your favorite function $f(x)$ defined on $0 < x < \ell$.

   (a) Find the constant function that best approximates your function on $0 < x < \ell$, where "best" is in the sense of minimizing the mean-square error over $0 < x < \ell$.

   (b) Show that this constant is the average value (in the calculus sense) of $f(x)$ over the interval $0 < x < \ell$.

   (c) Plot your function $f(x)$ and this average value on the same coordinate plane for $0 < x < \ell$ to verify (b).

7. There are two ways to compute the error in a partial sum approximation: using the definition of L^2 error,

$$E_N := \left\| \sum_{n=1}^{N} c_n X_n(x) - f(x) \right\| = \left(\int_a^b \left| \sum_{n=1}^{N} c_n X_n(x) - f(x) \right|^2 dx \right)^{1/2}, \qquad (*)$$

or the equivalent expression computed in (3.11),

$$E_N := \left(\|f\|^2 - \sum_{n=1}^{N} c_n^2 \|X_n\|^2 \right)^{1/2}. \qquad (**)$$

Note that E_N^2 rather than E_N was dealt with throughout the proof of Theorem 3.7, but this was just to avoid carrying square roots in all of the calculations. Engineers refer to the errors above as the *root mean-square* (RMS) error.

For $f(x) = x^3$, $0 \le x \le 1$, answer the following using these two different expressions for E_N.

   (a) Compute the Fourier sine series for f on $0 \le x \le 1$. Make a table of numerical values of E_N using each of $(*)$ and $(**)$. Was either method significantly faster to compute? How many terms are needed to ensure the error is less than 0.1?

   (b) Compute the Fourier cosine series for f on $0 \le x \le 1$. Make a table of numerical values of E_N using each of $(*)$ and $(**)$. Was either method significantly faster to compute? How many terms are needed to ensure the error is less than 0.1?

   (c) Compare and contrast the results from (a) and (b).

8. This is a continuation of Exercise 7 where we outline an analytic approach for finding N based on Parseval's Equality. Let $f(x) = x^3$, $0 \le x \le 1$.

❀ (a) Show that the coefficients in the Fourier sine series of f are given by

$$c_n = \frac{2(-1)^{n+1}(\pi^2 n^2 - 6)}{\pi^3 n^3}.$$

(b) Use Parseval's Equality to show

$$E_N^2 = \sum_{n=N+1}^{\infty} \frac{2(\pi^2 n^2 - 6)^2}{\pi^6 n^6}.$$

Then use the Integral and Comparison Tests to show $E_N \leq \sqrt{\frac{2}{\pi^2(N+1)}}$ for sufficiently large N.

(c) Find the smallest N for which $E_N < 0.1$. Compare with Exercise 7(a).

9. (a) For the classical Fourier sine series (3.1), show that Parseval's Equality is

$$\|f\|^2 = \frac{\ell}{2} \sum_{n=1}^{\infty} b_n^2.$$

(b) For the classical Fourier cosine series (3.2), show that Parseval's Equality is

$$\|f\|^2 = \frac{\ell}{2} \left(\frac{1}{2} a_0^2 + \sum_{n=1}^{\infty} a_n^2 \right).$$

(c) For the classical full Fourier series (3.3), show that Parseval's Equality is

$$\|f\|^2 = \ell \left(\frac{1}{2} a_0^2 + \sum_{n=1}^{\infty} (a_n^2 + b_n^2) \right).$$

10. (a) Using Exercise 9, show that in the Fourier series of $f \in L^2$, the Fourier coefficients must tend to zero as $n \to \infty$. (This result is a version of the celebrated *Riemann-Lebesgue Lemma.*)

(b) Using (a), show that $\int_0^\pi \ln x \sin(nx)\, dx \to 0$ as $n \to \infty$.

(c) Using (a), show that $\displaystyle\int_{-\pi}^{\pi} \frac{\cos(nx)}{\sqrt[3]{|x|}}\, dx \to 0$ as $n \to \infty$.

11. Consider $f(x) = x^2$ on $0 < x < \ell$.

❀ (a) Compute the Fourier cosine series to show

$$x^2 = \frac{\ell^2}{3} + \sum_{n=1}^{\infty} \frac{(-1)^n 4\ell^2}{n^2\pi^2} \cos(n\pi x/\ell), \quad 0 < x < \ell.$$

(b) Verify that f satisfies the conditions of Theorem 3.9. Then use that theorem and part (a) to show

$$\sum_{n=1}^{\infty} \frac{1}{n^4} = \frac{\pi^4}{90}.$$

12. Use Theorem 3.9 and the Fourier sine series of $f(x) \equiv 1$, $0 < x < \ell$, to find the sum of the infinite series

$$1 + \frac{1}{9} + \frac{1}{25} + \frac{1}{49} + \cdots.$$

13. Is $\displaystyle\sum_{n=1}^{\infty} \frac{1}{\sqrt{n}} \sin(n\pi x/\ell)$ the Fourier series of *any* function in $L^2[-\ell, \ell]$? Explain.

14. In signal processing, the *energy* of a signal $f(t)$ is defined as $E = \int_{-\infty}^{\infty} |f(t)|^2 dt$, which is just $\|f\|^2$, where this is the L^2 norm on $(-\infty, \infty)$. Signals with finite energy are called *energy signals.* Since not all signals have finite energy, another useful concept is the *power* of a signal, defined as

$$P = \lim_{L \to \infty} \frac{1}{2L} \int_{-L}^{L} |f(t)|^2 dt.$$

Signals with nonzero but finite power are called *power signals.*

(a) Compute the energy and power of the signal $f(t) = \sin t$.

(b) Compute the energy and power of the signal $f(t) = \begin{cases} e^t, & |t| \le 1, \\ 0, & |t| > 1. \end{cases}$

(c) Give an example of an energy signal which is not a power signal; a power signal which is not an energy signal; a signal which is neither an energy signal nor a power signal.

15. The definition given for an inner product space is sometimes referred to as a *real inner product space*, since the vectors and scalars were assumed to be real-valued. A *complex inner product space* is one in which the vectors and scalars involved are complex-valued. In this case, (IP1) is replaced with

(IP1′) $\langle u, v \rangle = \overline{\langle v, u \rangle}$ for all $u, v \in V$,

while (IP2)–(IP4) remain the same.

(a) Let $V = \mathbb{C}^2$, i.e., two-dimensional vectors whose entries are complex numbers, and $\langle \mathbf{u}, \mathbf{v} \rangle := \mathbf{u} \cdot \mathbf{v}$ is the usual dot product. Show that this is a complex inner product space.

(b) Let $V = \{f : [0,1] \to \mathbb{C} \mid f \text{ is continuous}\}$ with $\langle u, v \rangle := \int_0^1 u(x)\overline{v(x)}\, dx$. Show that this is a complex inner product space.

3.6 The Gibbs Phenomenon

If $f(x)$ has a jump discontinuity at $x = a$, we know from Section 3.3 that its Fourier series cannot converge uniformly in any interval about $x = a$, but will converge pointwise to the average of the left- and right-hand limits at $x = a$. In practical applications, since we cannot sum infinitely many terms in a Fourier series, instead we truncate the infinite series to a finite, partial Fourier sum. This truncation results in a peculiar behavior around a jump discontinuity in f. We explore this with a specific example.

Consider the function

$$f(x) = \begin{cases} -\frac{1}{2}, & -\pi < x < 0, \\ \frac{1}{2}, & 0 < x < \pi, \end{cases} \tag{3.15}$$

with Fourier series given by

$$\begin{aligned} f(x) &= \sum_{n=1}^{\infty} b_n \sin(nx), \quad b_n = \frac{1}{\pi}\int_{-\pi}^{\pi} f(x)\sin(nx)\,dx = \frac{1+(-1)^{n+1}}{n\pi}, \\ &= \frac{2}{\pi}\left(\sin x + \frac{\sin 3x}{3} + \frac{\sin 5x}{5} + \cdots\right). \end{aligned}$$

Since only the odd terms don't vanish, we will denote the partial sum by

$$F_{2N+1}(x) := \frac{2}{\pi}\left(\sin x + \frac{\sin(3x)}{3} + \frac{\sin(5x)}{5} + \cdots + \frac{\sin((2N+1)x)}{2N+1}\right).$$

See Figure 3.11.

To quantify the overshoot in $F_{2N+1}(x)$ to the right of $x = 0$, we differentiate and then sum the cosines (see Exercise 2):

$$\begin{aligned} \frac{d}{dx}F_{2N+1}(x) &= \frac{2}{\pi}\left(\cos x + \cos(3x) + \cos(5x) + \cdots + \cos((2N+1)x)\right) \\ &= \frac{2}{\pi}\left(\frac{\frac{1}{2}\sin(2(N+1)x)}{\sin x}\right), \end{aligned} \tag{3.16}$$

so the critical numbers are at $x = \frac{k\pi}{2(N+1)}$ and $x = k\pi$, $k = 0, \pm 1, \pm 2, \ldots$. The one that interests us is the first positive critical number, $x^* = \frac{\pi}{2(N+1)}$. Using either the First or Second Derivative Test, we see that indeed a local maximum occurs here, and is given by

$$F_{2N+1}(x^*) = \frac{2}{\pi}\left(\sin(x^*) + \frac{\sin(3x^*)}{3} + \cdots + \frac{\sin((2N+1)x^*)}{2N+1}\right).$$

The key to computing this quantity is to view it as an appropriate Riemann sum by multiplying and dividing by x^* to obtain

$$F_{2N+1}(x^*) = \frac{2x^*}{\pi}\left(\frac{\sin(x^*)}{x^*} + \frac{\sin(3x^*)}{3x^*} + \cdots + \frac{\sin((2N+1)x^*)}{(2N+1)x^*}\right).$$

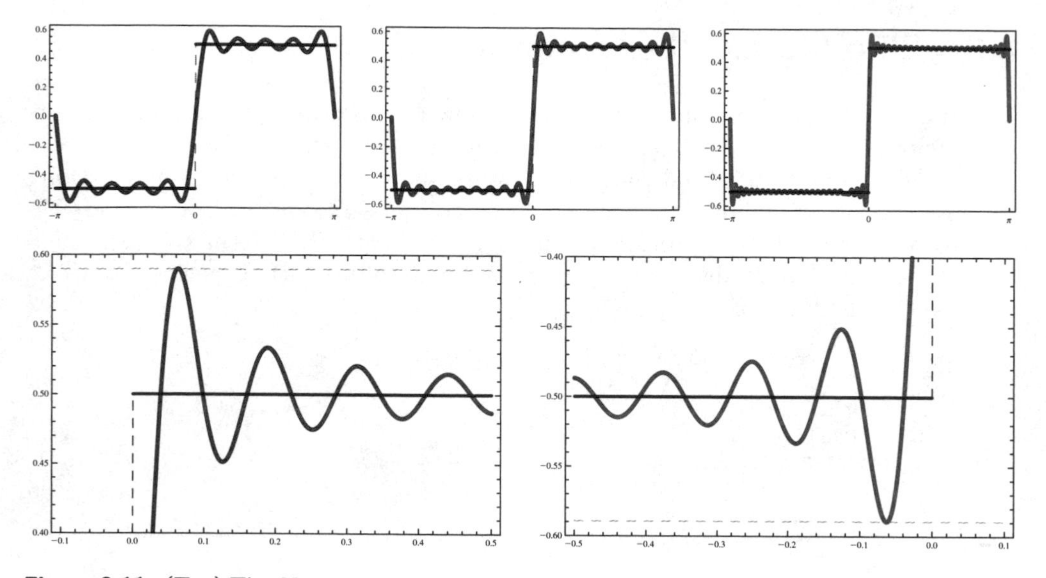

Figure 3.11: (Top) The $N = 10, 20, 40$ partial sums of $f(x)$ are shown in blue with $f(x)$ in black. (Bottom) The overshoots calculated in (3.18) are the dashed horizontal lines. A closer look (note the change in scale) to the right and left of the jump discontinuity. Regardless of how large a partial sum we take, the overshoot remains.

This is a Riemann sum for $\frac{\sin x}{x}$ on $0 < x < \pi$, where the partition is $\{x^*, 3x^*, \cdots, (2N+1)x^*\}$, and so the width is $\Delta x = 2x^*$. As $N \to \infty$, $\Delta x \to 0$, and

$$\begin{aligned}\lim_{N\to\infty} F_{2N+1}(x^*) &= \frac{1}{\pi}\lim_{N\to\infty}\left(\frac{\sin(x^*)}{x^*} + \frac{\sin(3x^*)}{3x^*} + \cdots + \frac{\sin((2N+1)x^*)}{(2N+1)x^*}\right)\Delta x \\ &= \frac{1}{\pi}\int_0^{\pi}\frac{\sin x}{x}dx.\end{aligned}$$

Although N remains finite in practice, this limit nonetheless accomplishes our goal of quantifying the local maximum,

$$F_{2N+1}(x^*) \approx \frac{1}{\pi}\int_0^{\pi}\frac{\sin x}{x}dx \quad \text{for large } N.$$

This is often expressed in terms of the *sine integral function*,[9] $\mathrm{Si}(z) := \int_0^z \frac{\sin x}{x}dx$ as

$$F_{2N+1}(x^*) \approx \frac{\mathrm{Si}(\pi)}{\pi} = 0.58949\ldots \quad \text{for large } N. \tag{3.17}$$

Therefore, the overshoot to the right of $x = 0$ is

$$F_{2N+1}(x^*) - f(0^+) \approx \frac{\mathrm{Si}(\pi)}{\pi} - \frac{1}{2} = 0.08949\ldots \quad \text{for large } N,$$

[9]In *Mathematica*, the command is `SinIntegral[z]`.

and arguing from symmetry a similar statement is true to the left of $x = 0$. Thus, we have shown that regardless of how large N is—as long as it is finite—there is a *constant* overshoot of the original function on either side of the jump discontinuity. This is called the *Gibbs phenomenon*. See Figure 3.12.

In fact, we can make a more general statement characterizing the overshoot in the Gibbs phenomenon. Suppose $f(x)$ has a jump discontinuity at $x = a$ and $f(a^+) > f(a^-)$. Then to the right of $x = a$,

$$\text{overshoot} = A + \frac{\text{Si}(\pi)}{\pi} J - f(a^+) \quad \text{for large } N, \tag{3.18}$$

where $J := f(a^+) - f(a^-)$ is the distance of the jump discontinuity and $A := \frac{f(a^+)+f(a^-)}{2}$ is the average of the left- and right-hand limits at $x = a$. The Gibbs phenomenon is sometimes called the "Gibbs 9% overshoot" because

$$A + \frac{\text{Si}(\pi)}{\pi} J - f(a^+) = \left(\frac{\text{Si}(\pi)}{\pi} - \frac{1}{2} \right) J \approx 0.08949J, \tag{3.19}$$

so the overshoot is about 9% of the distance of the jump discontinuity J.

It is important here to distinguish between the *finite* Fourier partial sum of f—which, for a given N, will be the best possible $L^2[-\pi, \pi]$ approximation of f by Theorem 3.7—and the *infinite* Fourier series representing $f(x)$ on $-\pi < x < \pi$. The Gibbs phenomenon will *always* occur in the finite partial sum and the overshoot is quantified in (3.18). However, the Gibbs phenomenon can *never* occur in the infinite Fourier series due to the fact that the Fourier series converges pointwise on $-\pi < x < \pi$, provided the conditions of Theorem 3.2 are satisfied. This is a curious instance in which the *infinite series* behaves in some sense "better" than the *finite* series.

Mitigating the Gibbs Phenomenon

There are various ways to mitigate the Gibbs phenomenon so that the overshoot (also called *ringing* in electrical engineering and signal processing) does not introduce unwanted error in a finite Fourier series approximation. One popular method that we outline here is due to Cornelius Lanczos.

Lanczos proposed multiplying the nonconstant terms in the Nth partial Fourier sum by a factor[10] of the form

$$\sigma(n, N) := \text{sinc}(n\pi/N) \quad \text{where} \quad \text{sinc}(z) := \begin{cases} \frac{\sin z}{z}, & z \neq 0, \\ 1, & z = 0. \end{cases}$$

Denote this new partial sum for $f(x)$, $-\pi < x < \pi$, by

$$S_N(x) := \frac{1}{2} a_0 + \sum_{n=1}^{N} \sigma(n, N) \left[a_n \cos(nx) + b_n \sin(nx) \right].$$

[10] Called a *Lanczos sigma factor*, *sigma factor*, *Lanczos smoother*, or *sigma smoother*, depending on the source.

Figure 3.12: Josiah Willard Gibbs (1839–1903) was an American mathematician, physicist, and engineer who worked primarily on problems in thermodynamics and electromagnetic theory. When he was awarded a doctorate from Yale in 1863, it was the first doctorate of engineering to be conferred in the United States. In 1899 Gibbs published a paper describing the overshoot phenomenon that now bears his name. However, Henry Wilbraham was the first to document the phenomenon in an 1848 paper that went unnoticed by the mathematical community. Because of this, the Gibbs phenomenon is sometimes called the *Wilbraham-Gibbs phenomenon* .

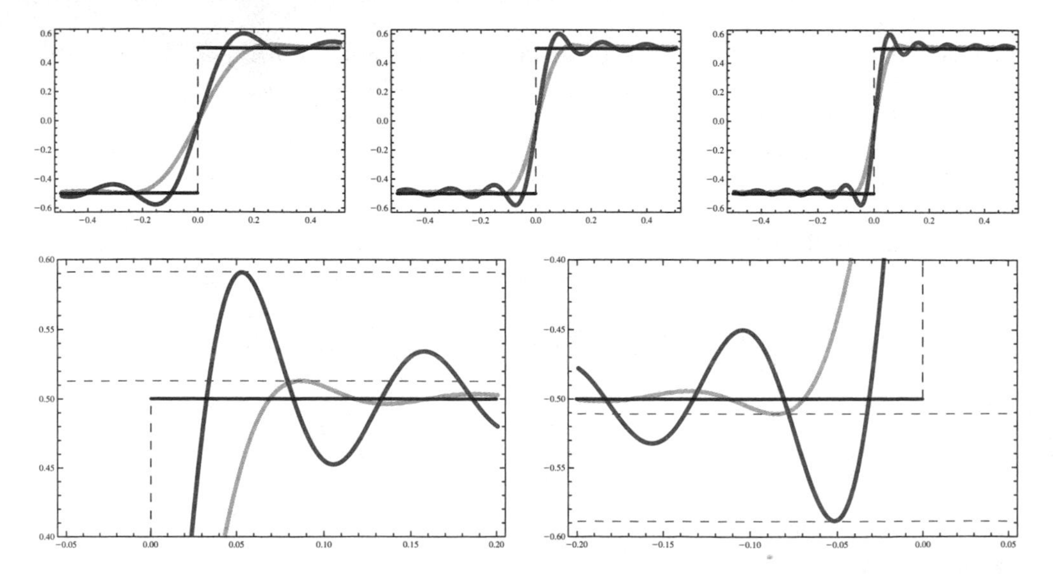

Figure 3.13: (Top) The $N = 20, 40, 60$ partial sums (standard in dark blue, Lanczos in light blue) along with the original $f(x)$ in black. (Bottom) A closer look to the right and left of the jump discontinuity. Note how much smaller the overshoots are for the Lanczos sums, but they also have larger L^2 error.

Although we will not prove it here, the Lanczos sigma factor has the effect of greatly smoothing out the Gibbs phenomenon, while still preserving the pointwise convergence as $N \to \infty$. This improvement does come at a cost though: notice that near the jump discontinuity the new partial sum converges more slowly than the usual partial sum, resulting in a larger L^2 error. See Figure 3.13 and Table 3.1.

	Standard Sums, $F_N(x)$		**Lanczos Sums, $S_N(x)$**	
N	Overshoot	L^2 error	Overshoot	L^2 error
10	0.091164	0.251898	0.011368	0.310176
20	0.089907	0.178338	0.011748	0.219505
40	0.089594	0.126143	0.011840	0.155245
60	0.089536	0.103002	0.011857	0.126762
80	0.089516	0.089204	0.011863	0.109781

Table 3.1: Comparing the standard partial Fourier sum vs. the Lanczos sums. For $N = 80$, Lanczos sums decrease the overshoot by about 87%, but increase the L^2 error by about 23%.

Exercises

1. Determine if the following statements are true or false and justify your answer.
   (a) The Gibbs phenomenon refers to any type of overshoot in a partial Fourier sum.
   (b) The Gibbs phenomenon never occurs in a Fourier series.
   (c) Suppose f has a uniformly convergent Fourier series. The Nth partial Fourier sum for f might display the Gibbs phenomenon.
   (d) Lanczos sigma factors completely remove the Gibbs phenomenon.
   (e) The Lanczos sums $S_N(x)$ in Figure 3.13 converge uniformly to $f(x)$ on $[-\pi, \pi]$.

2. A crucial step in obtaining (3.16) was summing the series

$$\cos x + \cos(3x) + \cos(5x) + \cdots + \cos((2N+1)x) = \frac{\sin(2(N+1)x)}{2\sin x}. \qquad (*)$$

   We will prove this as follows.

   (a) Multiply the left-hand side of $(*)$ by $2\sin x$, then use the identity $2\cos u \sin v = \sin(u+v) - \sin(u-v)$ to show

$$\sum_{k=0}^{N} 2\cos((2k+1)x)\sin x = \sum_{k=0}^{N} \left[\sin((2k+2)x) - \sin(2kx)\right].$$

(b) Write out several terms of the right-hand side above to confirm that the series "telescopes" (meaning most of the terms cancel out) and sums to $\sin(2(N+1)x)$.

(c) Use (a) and (b) to deduce (*).

3. Show that (3.19) holds.

✵ 4. Consider the function

$$f(x) = \begin{cases} -\frac{1}{2}, & -\pi < x < 0, \\ 1, & 0 < x < \pi. \end{cases}$$

(a) Numerically verify (3.18) by examining appropriate partial Fourier sums.

(b) Generate plots similar to Figures 3.11 and 3.13 for this function.

(c) Compile a table similar to Table 3.1 for this function. Discuss your results.

(d) By what percentage did the Lanczos sums reduce the Gibbs phenomenon for the values of N used in (c)? What was the percentage difference of L^2 error between the Lanczos and standard sums?

✵ 5. Consider the function

$$f(x) = \begin{cases} -\frac{1}{\pi}x - 1, & -\pi < x < 0, \\ \frac{1}{\pi}x + 1, & 0 < x < \pi. \end{cases}$$

(a) Numerically verify (3.18) by examining appropriate partial Fourier sums.

(b) Generate plots similar to Figures 3.11 and 3.13 for this function.

(c) Compile a table similar to Table 3.1 for this function. Discuss your results.

(d) By what percentage did the Lanczos sums reduce the Gibbs phenomenon for the values of N used in (c)? What was the percentage difference of L^2 error between the Lanczos and standard sums?

✵ 6. In certain applications, engineers are interested in the *rise time* of a partial Fourier sum, meaning the time it takes the partial sum to first reach $f(a^+)$, where the jump discontinuity is at $x = a$. Figures 3.11 and 3.13 illustrate that as rise time decreases, overshoot increases.

(a) Compute the rise times for $F_N(x)$ vs. $S_N(x)$ for $f(x)$ in (3.15) with $N = 20, 40, 60$. Confirm your results graphically.

(b) Compute the rise times for $F_N(x)$ vs. $S_N(x)$ for $f(x)$ from Exercise 4 with $N = 20, 40, 60$. Confirm your results graphically.

(c) Compute the rise times for $F_N(x)$ vs. $S_N(x)$ for $f(x)$ from Exercise 5 with $N = 20, 40, 60$. Confirm your results graphically.

7. In this exercise, we compute $\mathrm{Si}(\pi)$ in (3.17) using calculus.

   (a) Let $g(x) := \int_0^x \frac{\sin t}{t}\,dt$. Use the Taylor series for $\sin t$ about $t = 0$ and term-by-term integration to show

   $$g(x) = x - \frac{x^3}{3\cdot 3!} + \frac{x^5}{5\cdot 5!} - \cdots = \sum_{n=0}^{\infty} \frac{(-1)^n x^{2n+1}}{(2n+1)(2n+1)!}.$$

   (b) Verify that the series for $g(\pi)$ satisfies the conditions needed for the Alternating Series Test. Deduce that the truncation error for $g(\pi)$ for the first N terms is less than $\frac{\pi^{2N+3}}{(2N+3)(2N+3)!}$.

   ✲ (c) Using (b), find the smallest N for which the truncation error is less than 10^{-6}.

   (d) Use this N and the series from (a) to arrive at (3.17).

   (e) In this argument, we used the Taylor series about the origin to compute an integral on $[0, \pi]$, i.e., for values "far" from the origin. Why was this justified?

8. **(Dirichlet Kernel)** In this exercise, we develop an alternative representation for the Nth partial Fourier sum that arises in more advanced theory of Fourier analysis. Consider $f(x)$ on $-\pi < x < \pi$ and denote its Nth partial Fourier sum by $F_N(x)$.

   (a) Using the standard formulas for the Fourier coefficients, show

   $$F_N(x) = \frac{1}{2\pi}\int_{-\pi}^{\pi}\left(1 + 2\sum_{n=1}^{N} \cos(ny)\cos(nx) + \sin(ny)\sin(nx)\right) f(y)\,dy.$$

   Note that the usage of a dummy variable of integration is important here.

   (b) Next, use the identity $\cos u \cos v + \sin u \sin v = \cos(u - v)$ to show

   $$F_N(x) = \frac{1}{2\pi}\int_{-\pi}^{\pi}\left(1 + 2\sum_{n=1}^{N} \cos(n(y-x))\right) f(y)\,dy.$$

   This is usually written as

   $$F_N(x) = \frac{1}{2\pi}\int_{-\pi}^{\pi} D_N(y-x) f(y)\,dy,$$

   where $D_N(\theta) := 1 + 2\sum_{n=1}^{N} \cos(n\theta)$ is called the *Dirichlet kernel.*

(c) Use the identity $2\cos u \sin v = \sin(u+v) - \sin(u-v)$ and a telescoping argument (as in Exercise 2) to sum the Dirichlet kernel and obtain

$$D_N(\theta) = \frac{\sin((N+1/2)\theta)}{\sin(\theta/2)}.$$

Conclude that alternative representation for the Nth partial Fourier sum is

$$\begin{aligned} F_N(x) &= \frac{1}{2\pi}\int_{-\pi}^{\pi} D_N(y-x) f(y)\,dy \\ &= \frac{1}{2\pi}\int_{-\pi}^{\pi} \frac{\sin((N+1/2)(y-x))}{\sin((y-x)/2)} f(y)\,dy. \end{aligned} \tag{**}$$

9. (a) Prove that $D_N(\theta)$ is an even function.
   (b) Plot $D_N(\theta)$, $-\pi < \theta < \pi$ for $N = 1, 2, \ldots, 5$.

10. (a) Consider $f(x)$ from (3.15). Use (**) to compute $F_N(x)$ for $N = 5, 10, 20$.
    (b) Repeat (b) for $f(x)$ from Exercise 4.
    (c) Repeat (b) for $f(x)$ from Exercise 5.

11. The Dirichlet kernel formulation of Exercise 8 can be used to give a different proof of the Gibbs phenomenon. We outline the steps below.

    (a) Consider $f(x)$ in (3.15). Use (**) to show

$$F_N(x) = \frac{1}{4\pi}\left(\int_0^{\pi} D_N(y-x)\,dy - \int_{-\pi}^{0} D_N(y-x)\,dy\right).$$

    (b) Let $s = y - x$ in the first integral and $s = x - y$ in the second to deduce

$$F_N(x) = \frac{1}{4\pi}\left(\int_{-x}^{\pi-x} D_N(s)\,ds - \int_{x}^{x+\pi} D_N(s)\,ds\right).$$

    Argue that these integrals cancel on $x < s < \pi - x$ (ordering the various limits of integration on a number line might help), leaving

$$F_N(x) = \frac{1}{4\pi}\left(\int_{-x}^{x} D_N(s)\,ds - \int_{\pi-x}^{\pi+x} D_N(s)\,ds\right).$$

    (c) Since we are interested in the first positive overshoot in the Gibbs phenomenon, argue that we only need to consider the first integral, and since it involves the integral of an even function over a symmetric interval,

$$F_N(x) \approx \frac{1}{2\pi}\int_0^{x} D_N(s)\,ds.$$

(d) Show that the first positive critical number for $F_N(x)$ is $x^* = \frac{\pi}{N+\frac{1}{2}}$ and indeed a maximum occurs there.

(e) Use (c) and an appropriate change of variables to show

$$\begin{aligned} F_N(x^*) &\approx \frac{1}{2\pi} \int_0^{\frac{\pi}{N+\frac{1}{2}}} D_N(s)\, ds \\ &= \frac{1}{2\pi} \int_0^{\pi} \frac{\sin t}{\sin\left(\frac{t}{2(N+\frac{1}{2})}\right)} \cdot \frac{1}{N+\frac{1}{2}}\, dt. \end{aligned}$$

The benefit of this last expression is that N only appears in the integrand.

(f) Although we are interested in large, finite N for the Gibbs phenomenon, we can approximate this by considering the limit as $N \to \infty$. Motivated by the denominator in (e), use L'Hôpital's Rule to show

$$\lim_{N\to\infty} \left(N + \frac{1}{2}\right) \sin\left(\frac{t}{2(N+\frac{1}{2})}\right) = \frac{t}{2}.$$

(g) Use (f) to conclude

$$F_N(x^*) \approx \frac{1}{2\pi} \int_0^{\pi} \frac{\sin t}{t/2}\, dt = \frac{\mathrm{Si}(\pi)}{\pi}, \qquad \text{for large } N,$$

just as in (3.17). From there, the argument to quantify the Gibbs phenomenon proceeds as before.

Chapter 4

General Orthogonal Series Expansions

4.1 Regular and Periodic Sturm-Liouville Theory

In this chapter, we will explore the technique of separation of variables and eigenfunction expansions for problems beyond those in rectangular coordinates, namely, polar, cylindrical, and spherical coordinates. In doing so, we need a suitably generalized analogue of Theorem 2.1.

Thus far, the eigenvalue problem has always taken the form $X'' + \lambda X = 0$ together with relevant boundary conditions. Our goal now is to generalize this to eigenvalue problems of the form

$$\begin{aligned} a_2(x)y'' + a_1(x)y' + a_0(x)y + \lambda y = 0, &\quad a < x < b, \\ \alpha_1 y(a) + \alpha_2 y'(a) = 0, & \\ \beta_1 y(b) + \beta_2 y'(b) = 0, & \end{aligned}$$

where we assume the coefficient functions a_2, a_1, a_0 are continuous and $a_2(x) > 0$ on $a < x < b$. The first step is to transform the operator

$$Ly := a_2(x)y'' + a_1(x)y' + a_0(x)y$$

into its equivalent *Sturm-Liouville operator*, named in honor of Charles François Sturm and Joseph Liouville (see Figure 4.1) which has the form

$$\mathcal{S}y := \frac{1}{w(x)}[(p(x)y')' + q(x)y], \quad a < x < b.$$

Figure 4.1: Joseph Liouville (1809–1882) of France was strongly influenced by Cauchy and Poisson, but Dirichlet was his close friend and primary mathematical collaborator. Sturm and Liouville published their systematic theory for eigenvalue problems arising from differential equations and in a sequence of papers between 1836 and 1837. Liouville was prolific in many areas of mathematics, publishing over 400 papers.

To determine p, q, w, we expand $\mathcal{S}y$ and set it equal to Ly:

$$\begin{aligned}\mathcal{S}y &:= \frac{1}{w(x)}[(p(x)y')' + q(x)y] \\ &= \frac{1}{w(x)}[p(x)y'' + p'(x)y' + q(x)y] \\ &= Ly := a_2(x)y'' + a_1(x)y' + a_0(x)y.\end{aligned}$$

Equating coefficients yields

$$p(x) = a_2(x)w(x), \quad p'(x) = a_1(x)w(x), \quad q(x) = a_0(x)w(x). \tag{4.1}$$

Differentiating $p(x)$ and applying the second equation, we obtain the first order ODE in w,

$$w' = \left(\frac{a_1(x) - a_2'(x)}{a_2(x)}\right) w,$$

which has solution

$$w(x) = \exp\left(\int \left(\frac{a_1(x) - a_2'(x)}{a_2(x)}\right) dx\right). \tag{4.2}$$

Knowing $w(x)$, all three parts of (4.1) are determined. Note that $p(x), w(x) > 0$ on $a < x < b$, provided $a_2(x) > 0$ on $a < x < b$.

Definition 4.1. Consider the Sturm-Liouville problem

$$\frac{1}{w(x)}[(p(x)y')' + q(x)y] + \lambda y = 0, \quad a < x < b, \tag{4.3a}$$

$$\alpha_1 y(a) + \alpha_2 y'(a) = 0, \tag{4.3b}$$

$$\beta_1 y(b) + \beta_2 y'(b) = 0. \tag{4.3c}$$

Here, λ is an *eigenvalue* of (4.3), provided there is a nontrivial solution $y(x)$ to (4.3) corresponding to that value of λ. We call $y(x)$ an *eigenfunction* corresponding to λ.

If each of the following holds,

(1) p, q, w, p' are continuous on $a \le x \le b$,

(2) $p(x), w(x) > 0$ on $a \le x \le b$,

(3) $\alpha_1^2 + \alpha_2^2 \ne 0$ and $\beta_1^2 + \beta_2^2 \ne 0$,

we call (4.3) a *regular Sturm-Liouville problem*, and we call $\mathcal{S}$ a *regular Sturm-Liouville operator.*

If (4.3a) satisfies (1), (2) and $p(a) = p(b)$, but with the periodic boundary conditions

$$y(a) = y(b), \; y'(a) = y'(b), \tag{4.4}$$

then we call (4.3a), (4.4) a *periodic Sturm-Liouville problem* and we call $\mathcal{S}$ a *periodic Sturm-Liouville operator.*

Example 4.1.1. The familar eigenvalue problem

$$X'' + \lambda X = 0, \quad 0 < x < \ell,$$
$$X(0) = 0, \; X(\ell) = 0,$$

is already in Sturm-Liouville form with $p(x) \equiv 1$, $q(x) \equiv 0$, $w(x) \equiv 1$, and $\alpha_1 = \beta_1 = 1$, $\alpha_2 = \beta_2 = 0$. Since all of the conditions in the definition hold, this is a regular Sturm-Liouville problem.

The (also familiar) eigenvalue problem

$$\Theta'' + \lambda\Theta = 0, \quad -\pi < \theta < \pi,$$
$$\Theta(-\pi) = \Theta(\pi),$$
$$\Theta'(-\pi) = \Theta'(\pi),$$

is a periodic Sturm-Liouville problem. ◇

Example 4.1.2. Consider the eigenvalue problem

$$2y'' + xy' + x^2 y + \lambda y = 0, \quad 0 < x < 1,$$
$$y'(0) = 0, \; y(1) = 0.$$

Although the given operator, $Ly := 2y'' + xy' + x^2y$, is not in Sturm-Liouville form, using (4.2), (4.1) we compute $w(x) = e^{\int x/2\,dx} = e^{x^2/4}$, $p(x) = 2e^{x^2/4}$, $q(x) = x^2e^{x^2/4}$, and thus the equivalent Sturm-Liouville form of the problem is

$$\frac{1}{e^{x^2/4}}\left[\left(2e^{x^2/4}y'\right)' + x^2e^{x^2/4}y\right] + \lambda y = 0, \quad 0 < x < 1,$$
$$y'(0) = 0, \; y(1) = 0.$$

Since conditions (1)–(3) in the definition hold, the new problem is in fact a regular Sturm-Liouville problem. ◊

Theorem 4.1 (Lagrange's Identity and Green's Identity).
Let $\mathcal{S}$ be a Sturm-Liouville operator with $p \in C^1[a,b]$ and $u, v \in C^2[a,b]$. Then

$$u\mathcal{S}v - v\mathcal{S}u = \frac{1}{w}\frac{d}{dx}\left[p(uv' - u'v)\right] \tag{4.5}$$

is called Lagrange's Identity. The weighted integration of (4.5) *is*

$$\int_a^b \left[u(x)\mathcal{S}v(x) - v(x)\mathcal{S}u(x)\right] w(x)\,dx = p(x)\left[u(x)v'(x) - u'(x)v(x)\right]\Big|_a^b, \tag{4.6}$$

which can be written more compactly as

$$\langle u, \mathcal{S}v\rangle_w - \langle \mathcal{S}u, v\rangle_w = p(x)\left[u(x)v'(x) - u'(x)v(x)\right]\Big|_a^b. \tag{4.7}$$

Both (4.6) *and* (4.7) *are called Green's Identity.*

Lagrange's Identity is named in honor of Joseph-Louis Lagrange (see Figure 4.2). The proof of Lagrange's Identity is computational and outlined in Exercise 1. Green's Identity is obtained by simply multiplying (4.5) by the weight function $w(x)$ and integrating over $[a,b]$. Theorem 4.1 allows us to prove parts of the next theorem, which outlines several key properties of regular Sturm-Liouville problems.

Theorem 4.2 (Regular Sturm-Liouville Operators are Symmetric).
Let $\mathcal{S}$ be a regular Sturm-Liouville operator and $u, v \in C^2[a,b]$ satisfy the boundary conditions (4.3b), (4.3c). *Then*

$$\langle u, \mathcal{S}v\rangle_w = \langle \mathcal{S}u, v\rangle_w, \tag{4.8}$$

and we say that $\mathcal{S}$ is symmetric with respect to the weighted inner product.

Figure 4.2: Joseph-Louis Lagrange (1736–1813) was born in Italy, but spent most of his life in France, making tremendous contributions to analysis, number theory, and mechanics. He corresponded with and had great respect for Euler. Upon the recommendation of d'Alembert and Euler, Lagrange succeeded Euler as the director of the Prussian Academy of Science in Berlin. Lagrange famously quipped that if he had been rich, he probably would not have devoted himself to mathematics.

Proof. Since the left-hand side of Green's Identity (4.6) is $\langle u, \mathcal{S}v\rangle_w - \langle \mathcal{S}u, v\rangle_w$, we only need to show that the right-hand side of (4.6) vanishes in order to prove (4.8). To accomplish this, consider the evaluation at $x = b$ in (4.6). Since β_1 and β_2 are not both zero in the boundary condition (4.3c), suppose $\beta_1 \neq 0$. Then

$$u(b) = -\frac{\beta_2}{\beta_1}u'(b), \quad v(b) = -\frac{\beta_2}{\beta_1}v'(b),$$

so that

$$u(b)v'(b) - v(b)u'(b) = -\frac{\beta_2}{\beta_1}u'(b)v'(b) + \frac{\beta_2}{\beta_1}u'(b)v'(b) = 0.$$

Arguing similarly, if instead we suppose $\beta_2 \neq 0$,

$$u'(b) = -\frac{\beta_1}{\beta_2}u(b), \quad v'(b) = -\frac{\beta_1}{\beta_2}v(b),$$

so that

$$u(b)v'(b) - v(b)u'(b) = -\frac{\beta_1}{\beta_2}u(b)v(b) + \frac{\beta_1}{\beta_2}u(b)v(b) = 0.$$

Therefore, the part of the (4.6) evaluated at $x = b$ vanishes. The same type of argument shows that the part evaluated at $x = a$ vanishes. Hence, the right-hand side of (4.6) is zero; therefore, (4.8) holds. □

The fact that a regular Sturm-Liouville operator is symmetric is crucial in developing generalizations of the theory from Section 2.4 for general Sturm-Liouville problems.

Theorem 4.3 (Properties of Regular Sturm-Liouville Problems). *Consider the regular Sturm-Liouville problem (4.3).*

(a) *The eigenvalues are real and can be arranged into an increasing sequence*

$$\lambda_1 < \lambda_2 < \cdots < \lambda_n < \lambda_{n+1} < \cdots,$$

where $\lambda_n \to \infty$ as $n \to \infty$.

(b) *The sequence of eigenfunctions $\{y_n(x)\}_{n=1}^{\infty}$ forms a complete orthogonal family on $a < x < b$ with respect to the weight function $w(x)$; that is, if λ_n and λ_m are distinct eigenvalues with corresponding eigenfunctions $y_n(x)$ and $y_m(x)$, then*

$$\langle y_n, y_m \rangle_w := \int_a^b y_n(x) y_m(x) w(x)\, dx = 0, \quad n \neq m.$$

(c) *The eigenfunction $y_n(x)$ corresponding to the eigenvalue λ_n is unique up to a constant multiple.*

(d) *If $f \in L_w^2[a,b]$ is expanded in an infinite series of these eigenfunctions,*

$$f(x) = \sum_{n=1}^{\infty} c_n y_n(x), \quad a < x < b, \tag{4.9}$$

then the coefficients are given by

$$c_n = \frac{\langle f, y_n \rangle_w}{\langle y_n, y_n \rangle_w} = \frac{\int_a^b f(x) y_n(x) w(x)\, dx}{\int_a^b y_n^2(x) w(x)\, dx}. \tag{4.10}$$

Here, equality is meant in the sense of L^2 convergence weighted by $w(x)$. We denote the weighted L^2 space by $L_w^2[a,b]$, where

$$L_w^2[a,b] := \left\{ f : [a,b] \to \mathbb{R} \;\middle|\; \int_a^b |f(x)|^2 w(x) dx < \infty \right\}.$$

Proving that the eigenvalues form an increasing sequence and that the eigenfunctions are complete is beyond the scope of this text. The completeness of the eigenfunctions (i.e., that they form a basis for $L^2[a,b]$) is especially important; in retrospect, all of our earlier series solution methods relied on this crucial result. In particular, we used orthogonality to compute the coefficients in our series solutions.

Proof. (a) To see that the eigenvalues must all be real, suppose $\lambda = \alpha + \beta i$ is an eigenvalue with corresponding eigenfunction $y(x) = u(x) + iv(x)$. Then $\overline{\lambda}$ is also an eigenvalue with eigenfunction $\overline{y(x)} = u(x) - iv(x)$ (see Exercise 13):

$$\mathcal{S}y = -\lambda y, \quad \mathcal{S}\overline{y} = -\overline{\lambda}\,y. \tag{4.11}$$

Calculating $\langle y, \mathcal{S}\overline{y}\rangle_w - \langle \mathcal{S}y, \overline{y}\rangle_w$ using (4.11), yields

$$\int_a^b \left[y(x)\mathcal{S}\overline{y(x)} - \overline{y(x)}\mathcal{S}y(x)\right] w(x)\,dx = (\lambda - \overline{\lambda}) \int_a^b y(x)\overline{y(x)}w(x)\,dx.$$

Since $\mathcal{S}$ is symmetric and $y, \overline{y}$ both satisfy the boundary conditions (4.3b), (4.3c), the left-hand side is zero by Theorem 4.2. Therefore,

$$\begin{aligned}(\lambda - \overline{\lambda}) \int_a^b y(x)\overline{y(x)}w(x)\,dx &= 0\\ (\lambda - \overline{\lambda}) \int_a^b (u(x) + iv(x))(u(x) - iv(x))w(x)\,dx &= 0\\ (\lambda - \overline{\lambda}) \int_a^b (u^2(x) + v^2(x))w(x)\,dx &= 0.\end{aligned}$$

Since this is a regular Sturm-Liouville problem, $w(x) > 0$. Furthermore, $u^2(x)+v^2(x) > 0$ on (a, b) since $u(x)$ and $v(x)$ cannot both be zero, else $y(x)$ is not an eigenfunction. Therefore, the integral is positive, which forces $\lambda - \overline{\lambda} = 0$; that is, λ is real.

(b) To show that the eigenfunctions form an orthogonal family on $a < x < b$ with respect to the weight function $w(x)$, let λ_n and λ_m be any two distinct eigenvalues of (4.3) with corresponding eigenfunctions $y_n(x)$ and $y_m(x)$. That is,

$$\mathcal{S}y_n = -\lambda_n y_n, \quad \mathcal{S}y_m = -\lambda_m y_m, \quad n \neq m. \tag{4.12}$$

A direct calculation of $\langle y_n, \mathcal{S}y_m\rangle_w - \langle \mathcal{S}y_n, y_m\rangle_w$ using (4.12), yields

$$\int_a^b [y_n(x)\mathcal{S}y_m(x) - y_m(x)\mathcal{S}y_n(x)]\,w(x)\,dx = (\lambda_n - \lambda_m) \int_a^b y_n(x)y_m(x)w(x)\,dx,$$

and the left-hand side vanishes by Theorem 4.2, since $\mathcal{S}$ is symmetric. Hence,

$$(\lambda_n - \lambda_m) \int_a^b y_n(x)y_m(x)w(x)\,dx = 0.$$

But since $n \neq m$, we conclude

$$\int_a^b y_n(x)y_m(x)w(x)\,dx = 0,$$

which we could write as $\langle y_n, y_m\rangle_w = 0$. Therefore, y_n and y_m are orthogonal with respect to the weight function w.

(c) Suppose λ is an eigenvalue with eigenfunctions $y_1(x)$ and $y_2(x)$; that is,

$$\mathcal{S}y_1 = -\lambda y_1, \quad \mathcal{S}y_2 = -\lambda y_2. \tag{4.13}$$

Using the substitution (4.13) and Lagrange's Identity, then integrating, we see

$$\begin{aligned}
y_1\mathcal{S}y_2 - y_2\mathcal{S}y_1 &= -\lambda y_2 y_1 + \lambda y_1 y_2 \\
\frac{1}{w}\frac{d}{dx}\left[p(y_1 y_2' - y_1' y_2)\right] &= 0 \\
\int_a^x \frac{1}{w(x)}\frac{d}{dx}\left[p(x)(y_1(x)y_2'(x) - y_1'(x)y_2(x))\right] w(x)\,dx &= 0 \\
\left[p(x)(y_1(x)y_2'(x) - y_1'(x)y_2(x))\right]\Big|_a^x &= 0 \\
p(x)\left[y_1(x)y_2'(x) - y_1'(x)y_2(x)\right] - p(a)\left[y_1(a)y_2'(a) - y_1'(a)y_2(a)\right] &= 0.
\end{aligned}$$

We already showed in the proof of Theorem 4.2 that the second term vanishes, so the first term must vanish also. However, since this is a regular Sturm-Liouville problem, $p(x) > 0$ on (a, b), which forces

$$y_1(x)y_2'(x) - y_1'(x)y_2(x) = 0, \quad a < x < b.$$

But the left-hand side is the Wronskian of y_1 and y_2, which vanishes if and only if y_1 and y_2 are linearly dependent on $a < x < b$.

(d) The formula for the coefficients follows from the orthogonality of the eigenfunctions established in (b). Convergence of the series in the L^2 sense is guaranteed by the completeness of the eigenfunctions and Theorem 3.9. □

The importance of Theorem 4.3 cannot be overstated; we used all four parts in each of the separation of variables problems in Chapter 2, as the next example illustrates.

Example 4.1.3. Consider the regular Sturm-Liouville eigenvalue problem from the first example in this section, which arises in separation of variables for the heat equation and wave equation with Dirichlet-Dirichlet boundary conditions:

$$\begin{aligned}
X'' + \lambda X = 0, &\quad 0 < x < \ell, \\
X(0) = 0,\ X(\ell) = 0. &
\end{aligned}$$

Theorem 4.3(a) guarantees that the eigenvalues are all real and form an unbounded, increasing sequence. This agrees with our computation that $\lambda_n = (n\pi/\ell)^2$, $n = 1, 2, \ldots$.

Part (b) guarantees that the eigenfunctions form a complete orthogonal family on $0 < x < \ell$ with respect to the weight function $w(x) \equiv 1$. Although we could not prove completeness before, we showed that indeed the family of eigenfunctions $\{\sin(n\pi x/\ell)\}_{n=1}^{\infty}$ was orthogonal with respect to $w(x) \equiv 1$:

$$\langle \sin(n\pi x/\ell), \sin(m\pi x/\ell)\rangle = \int_0^\ell \sin(n\pi x/\ell)\sin(m\pi x/\ell)\,dx = 0, \quad n \neq m.$$

Part (c) guarantees that when we found the nth eigenfunction above, we didn't "miss" any others, except for constant multiples. This was illustrated in our computations when we claimed the eigenfunctions were $X_n(x) = C\sin(n\pi x/\ell)$, where C was an arbitrary constant.

Part (d) guarantees that the inner product formulas we used to determine the coefficients were the correct ones—not only in the sense that they generated *some* set of numerical coefficients, but that they generate a set of coefficients for which the resulting infinite series is guaranteed to converge in the L^2 sense to the function f. This allowed us to compute the Fourier sine series for $f(x)$ as

$$f(x) = \sum_{n=1}^{\infty} b_n \sin(n\pi x/\ell)$$

$$b_n = \frac{\langle f, \sin(n\pi x/\ell)\rangle}{\langle \sin(n\pi x/\ell), \sin(n\pi x/\ell)\rangle} = \frac{2}{\ell}\int_0^\ell f(x)\sin(n\pi x/\ell)\,dx, \qquad n = 1, 2, \ldots,$$

and be assured that the Fourier series converges to f in the L^2 sense on $[a, b]$. ◇

Example 4.1.4. Consider the eigenvalue problem

$$y'' + 4y' + y + \lambda y = 0, \quad 0 < x < 1, \tag{4.14a}$$
$$y'(0) = 0, \ y(1) = 0. \tag{4.14b}$$

Although the given operator, $Ly := y'' + 4y' + y$, is not in Sturm-Liouville form, using (4.2), (4.1) we compute $w(x) = e^{\int 4\,dx} = e^{4x}$, $p(x) = e^{4x}$, $q(x) = e^{4x}$, and thus the equivalent Sturm-Liouville form of the problem is

$$\frac{1}{e^{4x}}\left[\left(e^{4x}y'\right)' + e^{4x}y\right] + \lambda y = 0, \quad 0 < x < 1,$$
$$y(0) = 0, \ y(1) = 0.$$

We quickly verify that this is a regular Sturm-Liouville problem.

Theorem 4.3(a) guarantees that the eigenvalues are all real and form an unbounded, increasing sequence. We can independently confirm this by computing the eigenvalues and eigenfunctions from (4.14) as follows. The characteristic equation here is

$$r^2 + 4r + (1 + \lambda) = 0,$$

which has roots

$$r = -2 \pm \frac{1}{2}\sqrt{12 - \lambda}. \tag{4.15}$$

As in Section 1.4, we consider three cases based on the sign of the discriminant.

- Case 1: $12 - \lambda = 0$. Then (4.15) yields the double root $r = -2$, so the general solution to (4.14a) is $y(x) = c_1 e^{-2x} + c_2 x e^{-2x}$. The first boundary condition forces $c_1 = 0$ while the second forces $c_2 = 0$. Thus, $y(x)$ is the trivial solution, and this case yields no eigenvalues.

- CASE 2: $12-\lambda > 0$. Then (4.15) yields two distinct roots, $r_{1,2} = -2 \pm \frac{1}{2}\sqrt{12-\lambda}$, so the general solution to (4.14a) is $y(x) = c_1 e^{r_1 x} + c_2 e^{r_2 x}$. From the first boundary condition, $y(0) = c_1 + c_2 = 0$, and from the second boundary condition, $y(1) = c_1 e^{r_1} + c_2 e^{r_2} = 0$. Together, these imply $c_1 = c_2 = 0$, so $y(x)$ is the trivial solution, and this case yields no eigenvalues.

- CASE 3: $12 - \lambda < 0$. Then (4.15) becomes

$$r = -2 \pm \frac{1}{2}\sqrt{\lambda - 12}\, i,$$

so that we have a complex conjugate pair of roots $r = \alpha \pm \beta i$ where $\alpha = -2$ and $\beta = \frac{1}{2}\sqrt{\lambda - 12}$. The general solution of (4.14a) in this case is $y(x) = c_1 e^{\alpha x}\cos(\beta x) + c_2 e^{\alpha x}\sin(\beta x)$. From the first boundary condition, $y(0) = c_1 = 0$, and from the second boundary condition, $y(1) = c_2 e^{\alpha}\sin\beta = 0$. Thus, $\beta = n\pi$, $n = \pm 1, \pm 2, \ldots$, so that

$$\begin{aligned} \frac{1}{2}\sqrt{\lambda - 12} &= n\pi, & n &= \pm 1, \pm 2, \ldots \\ \lambda_n &= 12 + (2n\pi)^2, & n &= 1, 2, \ldots, \end{aligned}$$

with associated eigenfunctions $y_n(x) = e^{-2x}\sin(n\pi x)$, $n = 1, 2, \ldots$.

Part (b) says that the eigenfunctions form a complete orthogonal family on $0 < x < 1$ with respect to the weight function $w(x) = e^{4x}$. The orthogonality relation is

$$\begin{aligned} \langle e^{-2x}\sin(n\pi x), e^{-2x}\sin(m\pi x)\rangle_w &= \int_0^1 e^{-2x}\sin(n\pi x)e^{-2x}\sin(m\pi x)e^{4x}\,dx \\ &= 0, \quad n \neq m. \end{aligned}$$

Part (c) is demonstrated in our computations since each eigenvalue λ_n has one linearly independent eigenfunction $y_n(x)$.

Part (d) guarantees that given a function f in the weighted L^2 space $L^2_w[0,1]$, the eigenfunction expansion of f can be computed as

$$\begin{aligned} f(x) &= \sum_{n=1}^{\infty} c_n e^{-2x}\sin(n\pi x) \\ c_n &= \frac{\langle f(x), e^{-2x}\sin(n\pi x)\rangle_w}{\langle e^{-2x}\sin(n\pi x), e^{-2x}\sin(n\pi x)\rangle_w} \\ &= \frac{\int_0^1 f(x)e^{-2x}\sin(n\pi x)e^{4x}\,dx}{\int_0^1 (e^{-2x}\sin(n\pi x))^2\, e^{4x}\,dx}, \qquad n = 1, 2, \ldots. \end{aligned}$$

Moreover, we are assured that this eigenfunction expansion converges to f in the L^2_w sense on $[0,1]$. See Figure 4.3. ◇

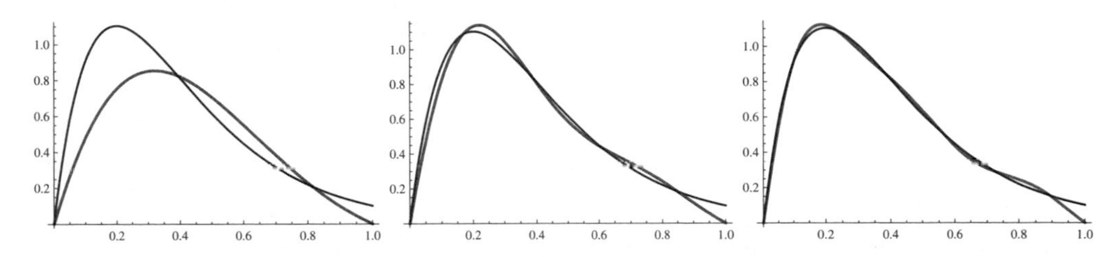

Figure 4.3: The $N = 1, 3, 5$ partial sums of the eigenfunction expansion of $f(x) = 15xe^{-5x}$ on $0 < x < 1$ using the eigenfunctions from Example 4.1.4.

Periodic Sturm-Liouville Theory

Periodic Sturm-Liouville problems enjoy many of the same properties that regular Sturm-Liouville problems have. As we saw earlier, if the underlying differential operator is symmetric, we can prove several other key results.

Theorem 4.4 (Periodic Sturm-Liouville Operators are Symmetric). *Let $\mathcal{S}$ be a periodic Sturm-Liouville operator and $u, v \in C^2[a, b]$ satisfy the periodic boundary conditions* (4.4). *Then $\mathcal{S}$ is symmetric with respect to the weighted inner product in the sense that $\langle u, \mathcal{S}v\rangle_w = \langle \mathcal{S}u, v\rangle_w$.*

The proof is very similar to Theorem 4.2; see Exercise 9. This in turn allows us to establish an analogue of Theorem 4.3 for periodic Sturm-Liouville problems.

Theorem 4.5 (Properties of Periodic Sturm-Liouville Problems). *Consider the periodic Sturm-Liouville problem* (4.3a), (4.4). *The statements in Theorem 4.3 still hold, with* (a) *and* (c) *replaced by*

(a′) *The eigenvalues are real and can be arranged into an increasing sequence*

$$\lambda_1 < \lambda_2 \leq \lambda_3 < \lambda_4 \leq \lambda_5 < \cdots,$$

where $\lambda_n \to \infty$ as $n \to \infty$.

(c′) *An unrepeated eigenvalue has exactly one linearly independent eigenfunction. A repeated eigenvalue has exactly two linearly independent eigenfunctions.*

The conclusions (a′) and (c′) show the subtle differences between regular Sturm-Liouville problems and their periodic counterparts: in periodic problems, an eigenvalue can have a two dimensional eigenspace.

Example 4.1.5. Consider the familiar periodic Sturm-Liouville problem

$$\begin{aligned} \Theta'' + \lambda\Theta &= 0, \quad -\pi < \theta < \pi, \\ \Theta(-\pi) &= \Theta(\pi), \\ \Theta'(-\pi) &= \Theta'(-\pi). \end{aligned}$$

The eigenvalues and eigenfunctions are

$$\begin{aligned} &\lambda_0 = 0, \quad \Theta_0(\theta) = 1, \\ &\lambda_n = n^2, \quad \Theta_n(\theta) = a_n \cos(n\theta) + b_n \sin(n\theta), \quad n = 1, 2, \ldots. \end{aligned}$$

Thus λ_0 is the only unrepeated eigenvalue and it has only one linearly independent eigenfunction (a one dimensional eigenspace), while each λ_n, $n = 1, 2, \ldots$ is repeated and has two linearly independent eigenfunctions (a two dimensional eigenspace). We can see this by listing the eigenvalues and linearly independent eigenfunctions as

$$\begin{gathered} 0 < 1^2 = 1^2 < 2^2 = 2^2 < 3^2 = 3^2 < \cdots \\ \{1, \cos x, \sin x, \cos 2x, \sin 2x, \cos 3x, \sin 3x, \ldots\}. \end{gathered}$$

Since Theorem 4.3(b), (d) still hold, these eigenfunctions form a complete orthogonal family with respect to $w(x) \equiv 1$ on $-\pi < x < \pi$, and the coefficient formulas (4.10) yield a full Fourier series which converges in the L^2 sense. ◊

Exercises

1. We will prove Lagrange's Identity two different ways.

   (a) By direct calculation: compute the left-hand side of (4.5) and simplify the resulting expression to obtain the right-hand side of (4.5).

   (b) By integration by parts: integrate the left-hand side of (4.5) by parts, then differentiate the result and simplify to obtain the right-hand side of (4.5).

   (c) Where did we use the assumption that u, v were twice differentiable?

2. Without solving the differential equation, state an orthogonality relation like the one in Theorem 4.3(b) that distinct eigenfunctions must satisfy.

   (a) $y'' + 2y' + 3y + \lambda y = 0$, $a < x < b$

   (b) $(1 + x^2)y'' + 4xy' + \lambda y = 0$, $0 < x < 1$

3. Consider the eigenvalue problem

$$\begin{gathered} y'' + y' + \lambda y = 0, \quad 0 < x < 1, \\ y(0) = 0, \ y(1) = 0. \end{gathered}$$

   (a) Write the problem in Sturm-Liouville form, identifying p, q, and w.

(b) Is the problem regular? Explain.

(c) Is the operator $\mathcal{S}$ symmetric? Explain.

(d) Find all eigenvalues and eigenfunctions by considering three cases based on the sign of the discriminant in the associated characteristic equation. Discuss your answer in light of Theorem 4.3.

(e) Find the orthogonal expansion of $f(x) = x$, $0 < x < 1$, in terms of these eigenfunctions.

(f) Find the smallest N such that the weighted L^2 error between the Nth partial sum in (e) and $f(x)$ is less than 10^{-1}. Plot f and $F_N(x)$ on the same coordinate plane for this value of N.

4. Consider the eigenvalue problem

$$x^2y'' + xy' + \lambda y = 0, \quad 1 < x < 2,$$
$$y(1) = 0, \; y(2) = 0.$$

(a) Write the problem in Sturm-Liouville form, identifying p, q, and w.

(b) Is the problem regular? Explain.

(c) Is the operator $\mathcal{S}$ symmetric? Explain.

(d) Find all eigenvalues and eigenfunctions. Discuss in light of Theorem 4.3.

(e) Find the orthogonal expansion of $f(x) = x^5(2-x)^3$, $1 < x < 2$, in terms of these eigenfunctions.

(f) Find the smallest N such that the weighted L^2 error between the Nth partial sum in (e) and $f(x)$ is less than 10^{-1}. Plot f and $F_N(x)$ on the same coordinate plane for this value of N.

5. Rework Exercise 4 using the boundary conditions $y(1) = 0$, $y'(2) = 0$.

6. Consider the eigenvalue problem

$$x^2y'' + xy' + 3y + \lambda y = 0, \quad 1 < x < 2,$$
$$y(1) = 0, \; y(2) = 0.$$

(a) Write the problem in Sturm-Liouville form, identifying p, q, and w.

(b) Is the problem regular? Explain.

(c) Is the operator $\mathcal{S}$ symmetric? Explain.

(d) Find all eigenvalues and eigenfunctions. Discuss in light of Theorem 4.3.

(e) Find the orthogonal expansion of $f(x) = \ln x$, $1 < x < 2$, in terms of these eigenfunctions.

(f) Find the smallest N such that the weighted L^2 error between the Nth partial sum in (e) and $f(x)$ is less than 10^{-1}. Plot f and $F_N(x)$ on the same coordinate plane for this value of N.

7. Consider the eigenvalue problem

$$y'' - 2y + \lambda y = 0, \quad 0 < x < \pi,$$
$$y(0) = y(\pi), \; y'(0) = y'(\pi).$$

   (a) Write the problem in Sturm-Liouville form, identifying p, q, and w.
   (b) Is the problem regular? Explain.
   (c) Is the problem periodic? Explain.
   (d) Is the operator $\mathcal{S}$ symmetric? Explain.
   (e) Find all eigenvalues and eigenfunctions. Discuss in light of Theorem 4.5.
   (f) Find the orthogonal expansion of

$$f(x) = \begin{cases} 0, & 0 < x < \frac{\pi}{2}, \\ x - \frac{\pi}{2}, & \frac{\pi}{2} < x < \pi, \end{cases}$$

   in terms of these eigenfunctions.
   (g) Find the smallest N such that the weighted L^2 error between the Nth partial sum in (f) and $f(x)$ is less than 10^{-1}. Plot f and $F_N(x)$ on the same coordinate plane for this value of N.

8. Verify that all of the eigenvalue problems in Section 2.5 are regular Sturm-Liouville problems.

9. Adapt the proof of Theorem 4.2 to the periodic boundary conditions setting to prove Theorem 4.4.

10. Give an example of an operator which is not symmetric, i.e., $\langle u, \mathcal{S}v\rangle_w \neq \langle \mathcal{S}u, v\rangle_w$.

11. Find all eigenvalues of the problem

$$y'' + \lambda y = 0, \quad 0 < x < 1,$$
$$2y(0) = y(1), \; 2y'(0) = -y'(1).$$

    Does your answer contradict Theorem 4.5? Discuss.

12. In Theorem 4.5, why can't an eigenvalue have *more than* two linearly independent eigenfunctions?

13. Let $\mathcal{S}$ be a regular Sturm-Liouville operator. Show that if $\mathcal{S}y = -\lambda y$, then $\mathcal{S}\overline{y} = -\overline{\lambda}\,\overline{y}$.

14. Compare the results of this section with the results of Section 2.4.

15. There are connections between the theory of symmetric operators in differential equations and $n \times n$ real symmetric matrices (meaning $A = A^T$, where A^T denotes the transpose of A) in linear algebra. Discuss each part of Theorem 4.3 in the matrix setting.

4.2 Singular Sturm-Liouville Theory

Unfortunately, not all Sturm-Liouville type problems that we will encounter in the method of separation of variables (especially in nonrectangular coordinate systems) will satisfy all of the requirements for the problem to be regular. In these cases, we would still like to have a sequence of real eigenvalues and for the eigenfunctions to form a complete orthogonal family in an appropriate weighted L^2 space. The aim of this section is to modify the regular Sturm-Liouville theory to achieve this.

Definition 4.2. Consider the Sturm-Liouville equation

$$\frac{1}{w(x)}[(p(x)y')' + q(x)y] + \lambda y = 0, \quad a < x < b, \tag{4.16a}$$

$$\alpha_1 y(a) + \alpha_2 y'(a) = 0, \tag{4.16b}$$

$$\beta_1 y(b) + \beta_2 y'(b) = 0. \tag{4.16c}$$

If each of the following holds:

(1) p, q, w, p' are continuous on $a < x < b$, and

(2) $p(x), w(x) > 0$ on $a < x < b$,

but we also have at least one of:

(a) $p(x)$ or $w(x)$ is zero at an endpoint,

(b) p, q, or w becomes infinite at an endpoint,

(c) $a = -\infty$ or $b = \infty$,

then we call (4.16) a *singular Sturm-Liouville problem* and the operator $\mathcal{S}$ is called a *singular Sturm-Liouville operator.* An endpoint where (a), (b), or (c) occurs is called *singular.*

Example 4.2.1. Consider the eigenvalue problem

$$x^2 y'' + 3xy' + e^x y + \lambda y = 0, \quad 0 < x < 1,$$
$$y'(0) = 0, \; y(1) = 0.$$

Using (4.2), (4.1), we compute $w(x) = e^{\int 1/x\, dx} = x$, $p(x) = x^2 \cdot x = x^3$, $q(x) = e^x x$, and thus the equivalent Sturm-Liouville form of the problem is

$$\frac{1}{x}\left[\left(x^3 y'\right)' + e^x xy\right] + \lambda y = 0, \quad 0 < x < 1,$$
$$y'(0) = 0, \; y(1) = 0.$$

Although $p(x), w(x) > 0$ on $0 < x < 1$, because $p(0) = 0$ and $w(0) = 0$, this is a singular Sturm-Liouville problem and $x = 0$ is the singular endpoint. $\Diamond$

We will see in Chapter 6 that the method of separation of variables in other coordinate systems often leads to singular Sturm-Liouville problems. Most importantly, we need to know that the eigenfunctions form a complete orthogonal family so that series solutions involving eigenfunctions of singular Sturm-Liouville problems are meaningful. To accomplish this, we need an analogue of Theorem 4.3 for singular Sturm-Liouville problems, but a prerequisite for that is the symmetry of the operator.

The proof of Theorem 4.2 shows that the key to adapting that same argument to the singular case is to ensure that the right-hand side of Green's Identity is zero:

$$\langle u, \mathcal{S}v\rangle_w - \langle \mathcal{S}u, v\rangle_w = p(x)\left[u(x)v'(x) - u'(x)v(x)\right]\Big|_a^b . \tag{4.17}$$

If, for example, $p = 0$ at an endpoint, it could still happen that a solution of (4.16a) (or its derivative) blows up at the endpoint; in fact, this is generally how solutions behave near a singular point. This would result in a $0 \cdot \infty$ indeterminate form that is not necessarily zero. To avoid this scenario, we will modify the boundary condition at the singular endpoint and impose conditions that ensure that the right-hand side of Green's Identity is zero, forcing the singular Sturm-Liouville operator to be symmetric. It is beyond the scope of this text to give a general procedure for determining what these modified boundary conditions should be. However, for the problems we will encounter, one of the following sets of modified boundary conditions generally works:

y, y' remain bounded as x approaches the singular finite endpoint,

$\sqrt{p}y$, $\sqrt{p}y'$ remain bounded as x approaches the singular infinite endpoint.

In practice, the correct choice of modified boundary conditions is influenced by the physics of the problem at hand.

Theorem 4.6 (Singular Sturm-Liouville Operators are Symmetric). *Let $\mathcal{S}$ be a singular Sturm-Liouville operator and $u, v \in C^2[a,b]$. If*

$$\lim_{x\to b^-} p(x)\left[u'(x)v(x) - u(x)v'(x)\right] = \lim_{x\to a^+} p(x)\left[u'(x)v(x) - u(x)v'(x)\right] \tag{4.18}$$

holds for all u, v that satisfy the (appropriately modified) boundary conditions, then

$$\langle u, \mathcal{S}v\rangle_w = \langle \mathcal{S}u, v\rangle_w,$$

and we say that $\mathcal{S}$ is symmetric with respect to the weighted inner product.

We expressed the condition in (4.18) in terms of limits since the ingredient functions may not be defined at the endpoints, or the endpoints may be themselves infinite. Note, however, that at a regular endpoint, the limit condition is the same as the evaluation in (4.17).

Example 4.2.2. Here are several famous singular Sturm-Liouville equations that we will deal with in Chapter 6.

- *Legendre's equation* is
$$(1-x^2)y'' - 2xy' + \lambda y = 0, \quad -1 < x < 1,$$
and its Sturm-Liouville form is
$$\left[(1-x^2)y'\right]' + \lambda y = 0, \quad -1 < x < 1.$$
This equation is singular at both endpoints since $p(x) = 1 - x^2$ vanishes at $x = \pm 1$. Therefore, the appropriate boundary conditions are
$$y, y' \text{ bounded as } x \to -1^+ \text{ and as } x \to 1^-.$$
Direct calculations show that the eigenvalues are $\lambda_n = n(n+1)$, $n = 0, 1, \ldots$, and the eigenfunctions are the Legendre polynomials $P_n(x)$, $n = 0, 1, 2, \ldots$. The functions satisfy (4.18), and hence the operator $\mathcal{S}y := \left[(1-x^2)y'\right]'$ is symmetric.

- *Chebyshev's equation* is
$$(1-x^2)y'' - xy' + \lambda y = 0, \quad -1 < x < 1,$$
and its Sturm-Liouville form is
$$\sqrt{1-x^2}\left[\sqrt{1-x^2}\,y'\right]' + \lambda y = 0, \quad -1 < x < 1.$$
This equation is singular at $x = \pm 1$ since $p(\pm 1) = 0$, but also because $w \to \infty$ as $x \to \pm 1$. Therefore, the appropriate boundary conditions are
$$y, y' \text{ bounded as } x \to -1^+ \text{ and as } x \to 1^-.$$
Direct calculations show that the eigenvalues are $\lambda_n = n^2$, $n = 0, 1, \ldots$, and the eigenfunctions are the Chebyshev polynomials $T_n(x)$, $n = 0, 1, 2, \ldots$. The functions satisfy (4.18), and hence the operator $\mathcal{S}y := \sqrt{1-x^2}\left[\sqrt{1-x^2}\,y'\right]'$ is symmetric. See Figure 4.4.

- *Laguerre's equation* is
$$xy'' + (1-x)y' + \lambda y = 0, \quad 0 < x < \infty,$$
and its Sturm-Liouville form is
$$e^x\left[xe^{-x}y'\right]' + \lambda y = 0, \quad 0 < x < \infty.$$
This equation is singular at $x = 0$ since $p(x) = xe^{-x}$ vanishes there, but is also singular at $x = \infty$. Therefore, the appropriate boundary conditions are
$$y, y' \text{ bounded as } x \to 0^+ \text{ and } \sqrt{xe^x}\,y,\ \sqrt{xe^x}\,y' \to 0 \text{ as } x \to \infty.$$
Direct calculations show that the eigenvalues are $\lambda_n = n$, $n = 0, 1, \ldots$, and the eigenfunctions are the Laguerre polynomials $L_n(x)$, $n = 0, 1, 2, \ldots$. The functions satisfy (4.18), and hence the operator $\mathcal{S}y := e^x\left[xe^{-x}y'\right]'$ is symmetric.

- *Hermite's equation* is

$$y'' - 2xy' + \lambda y = 0, \quad -\infty < x < \infty,$$

and its Sturm-Liouville form is

$$e^{x^2}\left[e^{-x^2}y'\right]' + \lambda y = 0, \quad -\infty < x < \infty.$$

This equation is singular at $x = \pm\infty$, so the appropriate boundary conditions are

$$e^{-x^2/2}y,\ e^{-x^2/2}y' \to 0 \text{ as } x \to \pm\infty.$$

Direct calculations show that the eigenvalues are $\lambda_n = 2n$, $n = 0, 1, \ldots$, and the eigenfunctions are the Hermite polynomials $H_n(x)$, $n = 0, 1, 2, \ldots$. The functions satisfy (4.18), and hence the operator $\mathcal{S}y := e^{x^2}\left[e^{-x^2}y'\right]'$ is symmetric.

- *Bessel's equation* (of order n) is

$$x^2y'' + xy' + (\lambda x^2 - n^2)y = 0, \quad 0 < x < 1,$$

and its Sturm-Liouville form is

$$\frac{1}{x}\left[(xy')' - \frac{n^2}{x}y\right] + \lambda y = 0, \quad 0 < x < 1.$$

Here, n is a parameter. This equation is singular at $x = 0$ since $p(x) = x$ and $w(x) = x$ vanish there, but also because $q \to \infty$ as $x \to 0^+$. The appropriate modified boundary condition at $x = 0$ is

$$y, y' \text{ bounded as } x \to 0^+.$$

Whatever boundary condition was given at $x = 1$ would remain unchanged. Direct calculations show that the eigenvalues are $\lambda_{nm} = z_{nm}^2$, $m = 1, 2, \ldots$, where z_{nm} denotes the mth positive root of $J_n(x)$, the Bessel function of order n. The corresponding eigenfunctions are $J_n(\sqrt{\lambda_{nm}}\, x)$, $m = 1, 2, \ldots$ and satisfy (4.18). Hence, the operator $\mathcal{S}y := \frac{1}{x}\left[(xy')' - \frac{n^2}{x}y\right]$ is symmetric. ◊

Once we have the symmetry of the singular Sturm-Liouville operator, we can prove that its eigenvalues are real and corresponding eigenfunctions are orthogonal with respect to the weight function w. However, conditions under which general singular Sturm-Liouville problems enjoy other properties listed in Theorem 4.3 are delicate and beyond the scope of this text. The good news is that the eigenfunction families in Example 4.2.2—Legendre, Chebyshev, Laguerre, and Hermite polynomials, as well as Bessel functions—all are complete in $L_w^2[a, b]$.

Figure 4.4: Pafnuty Chebyshev (1821–1894) was a Russian mathematician who made contributions to probability theory, statistics, and number theory. He published in Liouville's journal and said he especially enjoyed Dirichlet's lectures. His name is sometimes spelled Tchebycheff, and this is why we often denote the Chebyshev polynomials by T_n.

Exercises

1. Put the following problems in Sturm-Liouville form and state p, q, w. If the problem is singular, explain why. State appropriate modified boundary conditions where needed. *Do not solve the boundary value problem.*

   (a) $xy'' + 2y' + \lambda y = 0,\ 0 < x < 1; \quad y(0) = 0,\ y'(1) = 0$

   (b) $(1-x)y'' + y' - y + \lambda y = 0,\ 0 < x < 1; \quad y(0) = 0,\ y(1) = 0$

   (c) $(1-x^2)y'' + y' + y + \lambda y = 0,\ -1 < x < 1; \quad y(-1) = 0,\ y'(1) = 0$

   (d) $y'' + \cot x\, y' + \csc^2 x\, y + \lambda y = 0,\ 0 < x < \pi; \quad y(0) = 0,\ y(\pi) = 0$

2. State the appropriate modified boundary conditions for these infinite domain problems. *Do not solve the boundary value problem.*

   (a) $\sqrt{x}\, y'' + (1+\lambda)y = 0,\ 1 < x < \infty; \quad y(1) = 0$

   (b) $y'' + e^x y + \lambda y = 0,\ -\infty < x < \infty$

3. Find all eigenvalues and eigenfunctions for

$$y'' + \lambda y = 0, \quad 0 < x < \infty,$$
$$y(0) = 0, \quad y, y' \text{ bounded as } x \to \infty.$$

4. Consider the problem

$$r^2\frac{d^2R}{dr^2}+r\frac{dR}{dr}-n^2R=0, \quad 0<r<\rho,$$
$$R(0)=0, \; R(\rho)=0,$$

where $n=0,1,2,\ldots$, and $\rho>0$.

   (a) Put the problem in Sturm-Liouville form and explain the nature of any singular points.
   (b) State the appropriate modified boundary conditions.
   (c) Show that $n=0$ is an eigenvalue with eigenfunction $R_0(r)\equiv 1$.
   (d) Show that n^2, $n=1,2,\ldots$ are eigenvalues with eigenfunctions $R_n(r)=r^n$.

5. Consider the problem

$$x^2y''+xy'+\lambda y=0, \quad 0<x<1,$$
$$y(0)=0, \; y(1)=0.$$

   (a) Put the problem in Sturm-Liouville form and explain the nature of any singular points.
   (b) State the appropriate modified boundary conditions.
   (c) Find all eigenvalues and eigenfunctions for the modified problem.

6. (a) Explain why the Legendre, Chebyshev, Laguerre, and Hermite polynomials form orthogonal families (with respect to their weight functions). Do the same for the Bessel functions.
   (b) For what weighted L^2 space does each family of special functions in (a) form an orthogonal basis?

7. What conditions on α and β will result in

$$y''+\lambda y=0, \quad 0<x<1,$$
$$y(0)+\alpha y'(1)=0,$$
$$y'(0)+\beta y(1)=0,$$

having a symmetric Sturm-Liouville operator $\mathcal{S}$? Justify your answer.

8. Consider the Sturm-Liouville problem

$$(py')'+qy+\lambda y=0, \quad a<x<b,$$
$$y(a)=0, \; y(b)=0.$$

Show that if $p(x)\geq 0$ and $q(x)\leq M$, then $\lambda\geq -M$.

4.3 Orthogonal Expansions: Special Functions

The families of complete, orthogonal eigenfunctions—called the *special functions* of mathematical physics—that we studied in the last section can be used to form infinite series expansions of other L^2 functions.

The theory of Sections 4.1 and 4.2 showed that for regular, periodic, or singular Sturm-Liouville equations,

$$\frac{1}{w(x)}[(p(x)y')' + q(x)y] + \lambda y = 0, \quad a < x < b,$$

the eigenfunctions $\{y_n(x)\}_{n=1}^{\infty}$ form an orthogonal basis for $L^2[a,b]$. Therefore, as in Theorem 4.3(d), for any $f \in L^2_w[a,b]$, we can expand f in an infinite series of these basis functions,

$$f(x) = \sum_{n=1}^{\infty} c_n y_n(x), \quad a < x < b, \tag{4.19}$$

with coefficients given by

$$c_n = \frac{\langle f, y_n \rangle_w}{\langle y_n, y_n \rangle_w} = \frac{\int_a^b f(x)y_n(x)w(x)\,dx}{\int_a^b y_n^2(x)w(x)\,dx}, \quad n = 1, 2, \ldots. \tag{4.20}$$

The infinite series (4.19), (4.20) is called an *orthogonal expansion*, *eigenfunction expansion*, or *generalized Fourier series*. Note that Fourier series (sine, cosine, and full) are just special cases of (4.19), (4.20), where the eigenfunctions are sines, cosines, or sines and cosines.

Legendre Series

The method of separation of variables in spherical coordinates with rotational symmetry leads to the eigenvalue problem

$$(1 - x^2)y'' - 2xy' + \lambda y = 0, \quad -1 < x < 1,$$

known as *Legendre's equation*. The associated Sturm-Liouville operator is

$$\mathcal{S}y := \left[(1 - x^2)y'\right]'$$

which is singular at $x = \pm 1$, so the modified boundary conditions become

$$y, y' \text{ bounded as } x \to -1^+ \text{ and as } x \to 1^-. \tag{4.21}$$

Using the method of power series[1] from ODE theory, we can compute a power series solution about the ordinary point $x = 0$. Doing so shows that the only values of λ

[1]For a brief summary of power series methods, see the Appendix

which yield nontrivial solutions are $\lambda_n = n(n+1)$, $n = 0, 1, 2\ldots$, and in those cases, the associated solutions take the form

$$y_n(x) = c_1 P_n(x) + c_2 Q_n(x), \quad -1 < x < 1.$$

The $P_n(x)$ are called *Legendre functions of the first kind*, while the $Q_n(x)$ are called *Legendre functions of the second kind*, and these two families of functions are linearly independent. $P_n(x)$ is a terminating power series (i.e., a polynomial, and for this reason Legendre functions of the first kind are also called *Legendre polynomials*; see Figure 4.5) and $Q_n(x)$ is a nonterminating power series. Moreover, $Q_n(x)$ becomes unbounded as $x \to \pm 1$, violating the modified boundary conditions (4.21) and forcing $c_2 = 0$. Therefore, the eigenfunctions of the Legendre boundary value problem are (up to a constant multiple) given by

$$y_n(x) = P_n(x) = \frac{1}{2^n} \sum_{k=0}^{\lfloor n/2 \rfloor} \frac{(-1)^k (2n-2k)!}{k!(n-k)!(n-2k)!} x^{n-2k}, \tag{4.22}$$

where $\lfloor n/2 \rfloor$ denotes the greatest integer[2] function. The first few Legendre polynomials are listed below and shown in Figure 4.5.

$$\begin{aligned}
P_0(x) &\equiv 1, \\
P_1(x) &= x, \\
P_2(x) &= \frac{1}{2}(3x^2 - 1), \\
P_3(x) &= \frac{1}{2}(5x^3 - 3x), \\
P_4(x) &= \frac{1}{8}(35x^4 - 30x^2 + 3), \\
&\vdots
\end{aligned}$$

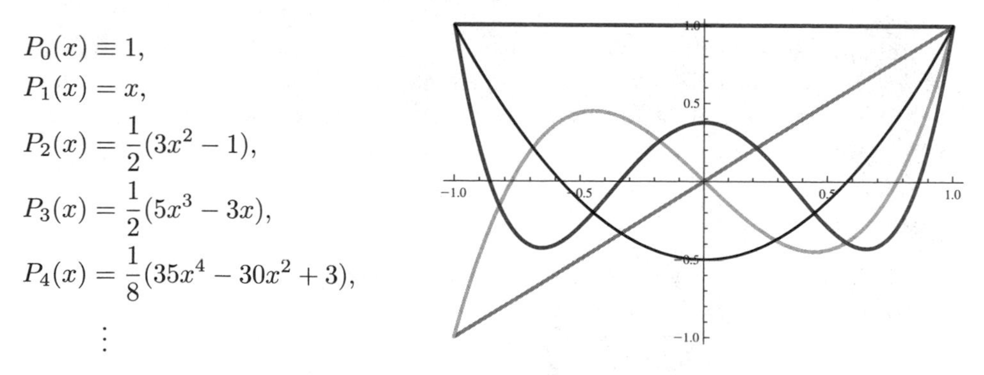

Figure 4.5: The first five Legendre polynomials.

There are several ways to generate the Legendre polynomials besides (4.22). A particularly useful one is *Rodrigues' Formula*:

$$P_n(x) = \frac{1}{2^n n!} \frac{d^n}{dx^n} (x^2 - 1)^n, \quad n = 0, 1, 2, \ldots, \tag{4.23}$$

which can then be used to prove *Bonnet's Recursion Formula*:

$$(n+1)P_{n+1}(x) = (2n+1)xP_n(x) - nP_{n-1}(x), \quad n = 1, 2, \ldots. \tag{4.24}$$

[2] Also called the *floor function*, $\lfloor x \rfloor$ is the greatest integer less than or equal to x.

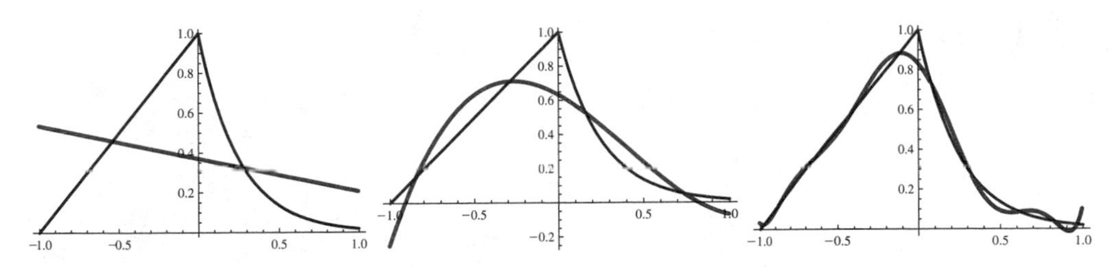

Figure 4.6: The function $f(x)$ from Example 4.3.1 in black and the Nth partial Legendre sum in blue for $N = 1, 3, 8$.

In addition to these three, virtually all modern mathematical software packages have the Legendre polynomials conveniently built into them.[3]

Now that we know how to compute the Legendre polynomials fairly efficiently, we can generate orthogonal expansions in terms of the Legendre polynomials using (4.19), (4.20). When computing the coefficients, it is helpful to use the fact that

$$\|P_n\|_w^2 = \int_{-1}^{1} P_n^2(x)\,dx = \frac{2}{2n+1}, \quad n = 0, 1, 2, \ldots, \tag{4.25}$$

which can be proved from Bonnet's Recursion Formula. Note that is $w(x) \equiv 1$ from the Sturm-Liouville form of the operator.

Example 4.3.1. Let's compute the orthogonal expansion of

$$f(x) = \begin{cases} x+1, & -1 < x < 0, \\ e^{-4x}, & 0 \le x < 1, \end{cases}$$

in terms of the Legendre polynomials. From (4.19) and (4.20),

$$f(x) = \sum_{n=0}^{\infty} c_n P_n(x), \quad -1 < x < 1,$$

with coefficients (recall that $w(x) \equiv 1$ here) given by

$$c_n = \frac{\langle f, P_n \rangle_w}{\langle P_n, P_n \rangle_w} = \frac{\int_{-1}^{1} f(x) P_n(x)\,dx}{\int_{-1}^{1} P_n^2(x)\,dx} = \frac{2n+1}{2} \int_{-1}^{1} f(x) P_n(x)\,dx,$$

where in the last equality we have used (4.25). Since $f \in L_w^2[-1,1]$, we are assured convergence of the series in the L_w^2 sense on $[-1,1]$. See Figure 4.6. ◊

[3]In *Mathematica*, the command is `LegendreP[n,x]`.

Chebyshev Series

On the other hand, *Chebyshev's equation*,

$$(1-x^2)y'' - xy' + \lambda y = 0, \quad -1 < x < 1,$$

has the associated Sturm-Liouville operator

$$\mathcal{S}y := \sqrt{1-x^2}\left[\sqrt{1-x^2}y'\right]',$$

which is singular at $x = \pm 1$, so the modified boundary conditions become

$$y, y' \text{ bounded as } x \to -1^+ \text{ and as } x \to 1^-. \tag{4.26}$$

Using the method of power series from ODE theory, we can compute a power series solution about the ordinary point $x = 0$. Doing so shows that the eigenvalues are $\lambda_n = n^2$, $n = 0, 1, 2\ldots$, and in those cases, the solution can be written as a linear combination of a terminating power series (i.e., a polynomial) and a nonterminating power series:

$$y_n(x) = c_1 T_n(x) + c_2 U_n(x), \quad -1 < x < 1.$$

The $T_n(x)$ are called *Chebyshev functions of the first kind*, while the $U_n(x)$ are called *Chebyshev functions of the second kind*, and these two families of functions are linearly independent. Since $T_n(x)$ terminates, these are also called *Chebyshev polynomials*, but $U_n(x)$ is a nonterminating power series. Moreover, $U_n'(x)$ becomes unbounded as $x \to \pm 1$, violating the modified boundary conditions (4.26) and forcing $c_2 = 0$. Therefore, the eigenfunctions of the Chebyshev boundary value problem are (up to a constant multiple) given by $y_0(x) = T_0(x) \equiv 1$ and

$$y_n(x) = T_n(x) = \frac{n}{2} \sum_{k=0}^{\lfloor n/2 \rfloor} \frac{(-1)^k}{n-k} \binom{n-k}{k} (2x)^{n-2k}, \quad n > 0. \tag{4.27}$$

Here, $\binom{n-k}{k}$ denotes the binomial coefficient.[4] The first few Chebyshev polynomials are listed below and shown in Figure 4.7.

There are other ways to generate the Chebyshev polynomials, such as a Rodrigues' Formula and a type of Bonnet Recursion Formula, but we defer these to the exercises, since we will take Chebyshev polynomials directly from mathematical software packages[5] rather than computing them from scratch.

---

[4] $\binom{p}{q}$ is read "p choose q" and computed via factorials by $\binom{p}{q} = \frac{p!}{q!(p-q)!}$, where $0 \le q \le p$.

[5] In *Mathematica*, the command is `ChebyshevT[n,x]`.

$$
\begin{aligned}
T_0(x) &\equiv 1, \\
T_1(x) &= x, \\
T_2(x) &= 2x^2 - 1, \\
T_3(x) &= 4x^3 - 3x, \\
T_4(x) &= 8x^4 - 8x^2 + 1, \\
&\vdots
\end{aligned}
$$

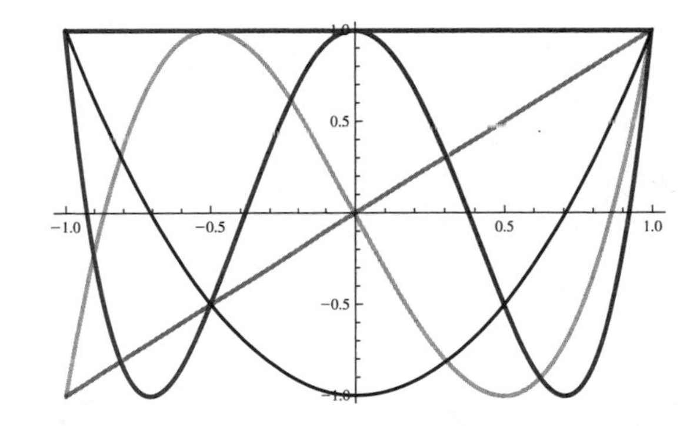

Figure 4.7: The first five Chebyshev polynomials.

Example 4.3.2. Let's compute the orthogonal expansion of

$$f(x) = \arctan(10x), \quad -1 < x < 1,$$

in terms of the Chebyshev polynomials. From (4.19) and (4.20),

$$f(x) = \sum_{n=0}^{\infty} c_n T_n(x), \quad -1 < x < 1,$$

with coefficients (recall that $w(x) = (1 - x^2)^{-1/2}$ here) given by

$$c_n = \frac{\langle f, T_n \rangle_w}{\langle T_n, T_n \rangle_w} = \frac{\int_{-1}^{1} f(x) T_n(x)(1 - x^2)^{-1/2}\, dx}{\int_{-1}^{1} T_n^2(x)(1 - x^2)^{-1/2}\, dx}.$$

Since $f \in L_w^2[-1, 1]$, the series converges in the L_w^2 sense on $[-1, 1]$. See Figure 4.8. ◊

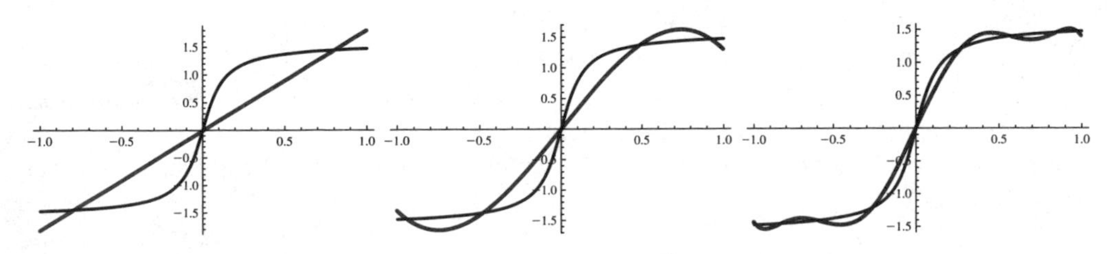

Figure 4.8: The function $f(x) = \arctan(10x)$ in black and the Nth partial Chebyshev sum in blue for $N = 1, 3, 8$.

Bessel Series

Finally, *Bessel's equation* arises from problems in polar and cylindrical coordinates and is given by

$$x^2y'' + xy' + (\lambda x^2 - n^2)y = 0, \quad 0 < x < \ell, \tag{4.28}$$

with Sturm-Liouville form

$$\frac{1}{x}\left[(xy')' - \frac{n^2}{x}y\right] + \lambda y = 0, \quad 0 < x < \ell.$$

Since the problem is singular at $x = 0$, the boundary condition there would be modified to take the form

$$y, y' \text{ bounded as } x \to 0^+. \tag{4.29}$$

Unlike Legendre's and Chebyshev's equation, here $x = 0$ is a singular point of the equation, so we must use the Method of Frobenius to compute the two linearly independent power series solutions. We will briefly outline the results here, but the computations are done in full detail in Section 4.4. Doing so yields the eigenvalues $\lambda_{nm} = (z_{nm}/\ell)^2$, $n = 0, 1, \ldots,$ $m = 1, 2, \ldots,$ where z_{nm} denotes the mth positive root of $J_n(x)$, the Bessel function of order n, with nontrivial solutions

$$y_{nm}(x) = c_1 J_n(\sqrt{\lambda_{nm}}\, x) + c_2 Y_n(\sqrt{\lambda_{nm}}\, x), \quad 0 < x < \ell.$$

Here, $J_n(x)$ is called a *Bessel function of the first kind of order* n and $Y_n(x)$ is called a *Bessel function of the second kind of order* n. Both are nonterminating power series, and $J_n(x)$ satisfies (4.29), but $Y_n(x)$ blows up as $x \to 0^+$, forcing $c_2 = 0$. Therefore, the corresponding eigenfunctions are (up to a constant multiple) given by

$$\begin{aligned} y_{nm}(x) &= J_n(\sqrt{\lambda_{nm}}\, x) \\ &= \sum_{k=0}^{\infty} \frac{(-1)^k}{k!\,\Gamma(k+n+1)} \left(\frac{1}{2}\sqrt{\lambda_{nm}}\, x\right)^{2k+n}, \quad n = 0, 1, \ldots, \quad m = 1, 2, \ldots, \end{aligned}$$

where Γ denotes the gamma function.[6] See Figure 4.9.

Unlike the previous two examples, these Bessel functions are *not* polynomials, but rather convergent power series. These functions are used so often that they are built into software packages,[7] as are the required numerical roots z_{nm} of the transcendental equation $J_n(x) = 0$.

---

[6] The gamma function is related to the factorial function by $\Gamma(n) = (n-1)!$, but also extends the factorial function to noninteger arguments.

[7] In *Mathematica*, the Bessel functions are given by `BesselJ[n,x]`. The mth positive zero of $J_n(x)$ is obtained by `BesselJZero[n,m]`.

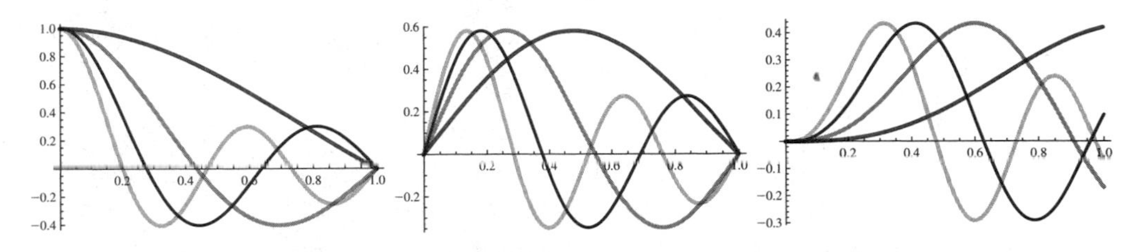

Figure 4.9: (Left) The first four Bessel functions of order zero, $J_0(\sqrt{\lambda_{0m}}\,x)$, $m = 1, \ldots, 4$. (Center) Bessel functions of order one, $J_1(\sqrt{\lambda_{1m}}\,x)$, $m = 1, \ldots, 4$. (Right) Bessel functions of order two, $J_2(\sqrt{\lambda_{2m}}\,x)$, $m = 1, \ldots, 4$. In all three cases we take $\ell = 1$.

Example 4.3.3. Let's compute the orthogonal expansion of $f(x) = x\sin(10x)$, $0 < x < 1$, in terms of Bessel functions of order zero. From (4.19) and (4.20),

$$f(x) = \sum_{m=1}^{\infty} c_m J_0(\sqrt{\lambda_{0m}}\,x), \quad 0 < x < 1,$$

with coefficients (recall that $w(x) = x$ here) given by

$$c_m = \frac{\langle f, J_0(\sqrt{\lambda_{0m}}\,x)\rangle_w}{\langle J_0(\sqrt{\lambda_{0m}}\,x), J_0(\sqrt{\lambda_{0m}}\,x)\rangle_w} = \frac{\int_0^1 f(x) J_0(\sqrt{\lambda_{0m}}\,x) x\,dx}{\int_0^1 J_0^2(\sqrt{\lambda_{0m}}\,x) x\,dx}.$$

Since $f \in L_w^2[0,1]$, the series converges in the L_w^2 sense on $[0,1]$. See Figure 4.10. ◊

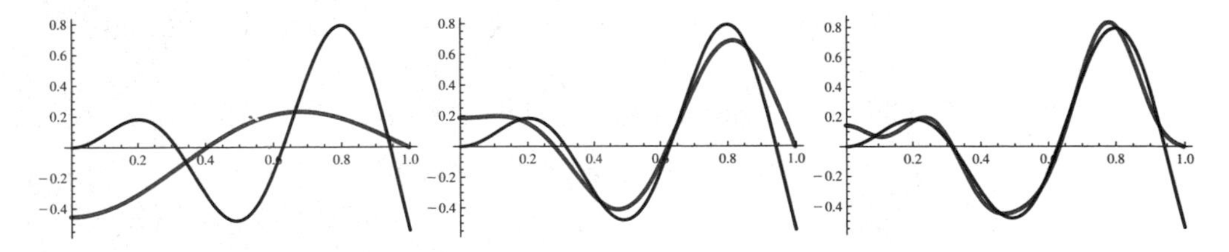

Figure 4.10: The function $f(x) = x\sin(10x)$ in black and the Nth partial Bessel sum in blue for $N = 2, 4, 8$.

Exercises

1. Compute the first 4 Legendre polynomials first by using each of the formulations in (4.22), (4.23), (4.24), and then by simply calling (not computing) them from a software package.

2. In this exercise, we outline the steps for finding power series solutions of Legendre's equation when $\lambda = n(n+1)$, $n = 0, 1, 2, \ldots$:

$$(1 - x^2)y'' - 2xy' + n(n+1)y = 0, \quad -1 < x < 1. \qquad (*)$$

(a) Verify that $x = 0$ is an ordinary point of $(*)$ (see the Appendix). Looking for a power series solution about $x = 0$, i.e., of the form

$$y(x) = \sum_{k=0}^{\infty} a_k x^k,$$

find the recurrence formula for the coefficients.

(b) Using (a), determine two linearly independent power series solutions of $(*)$—one that terminates and one that does not—by separately considering the cases when n is even and when n is odd.

(c) Show that the nonterminating power series has a radius of convergence equal to 1. Show that the series diverges at $x = \pm 1$, and hence does not satisfy (4.21).

(d) Verify that the polynomial solution found in (b) can be scaled to agree with (4.22).

(e) In (b), show that when n is even, the polynomial solution contains only even powers and when n is odd, it contains only odd powers.

3. (a) Compute the Legendre series for $f(x) = x^3 + |x|$, $-1 < x < 1$.

(b) Plot f and the Nth partial Legendre sum $F_N(x)$ on the same coordinate plane for $N = 1, 3, 6$ as in Figure 4.6.

(c) Plot the pointwise error function in (b).

(d) Compute (numerically, not analytically) the uniform error and weighted L^2 error in (b).

(e) Find the smallest N such that the weighted L^2 error in (b) is less than 10^{-1}.

(f) Discuss the pointwise, uniform, and L^2 convergence of this series.

4. Repeat Exercise 3 with

$$f(x) = \begin{cases} x+1, & -1 < x \le 0, \\ 1/2, & 0 < x < 1. \end{cases}$$

5. Repeat Exercise 3 with $f(x) = e^{-x}\sin(2\pi x)$.

6. In this exercise, we outline a proof of Rodrigues' Formula (4.23) for the Legendre polynomials.

(a) Show that

$$\frac{d^n}{dx^n} x^{2n-2k} = \begin{cases} \frac{(2n-2k)!}{(n-2k)!} x^{n-2k}, & \text{for } 0 \le k \le \lfloor n/2 \rfloor, \\ 0, & \text{for } \lfloor n/2 \rfloor < k \le n. \end{cases}$$

(b) Use (a) to rewrite (4.22) as

$$P_n(x) = \frac{1}{2^n n!} \sum_{k=0}^{n} (-1)^k \frac{n!}{k!(n-k)!} \frac{d^n}{dx^n} x^{2n-2k}.$$

(c) Finally, use the Binomial Formula,

$$(a+b)^n = \sum_{k=0}^{n} \frac{n!}{k!(n-k)!} a^{n-k} b^k,$$

with $a = x^2$ and $b = 1$ to rewrite (b) as (4.23).

7. In this exercise, we outline the steps for finding power series solutions of Chebyshev's equation when $\lambda = n^2$, $n = 0, 1, 2, \ldots$:

$$(1 - x^2)y'' - xy' + n^2 y = 0, \quad -1 < x < 1. \qquad (**)$$

(a) Verify that $x = 0$ is an ordinary point of $(**)$. Looking for a power series solution about $x = 0$, i.e., of the form

$$y(x) = \sum_{k=0}^{\infty} a_k x^k,$$

find the recurrence formula for the coefficients.

(b) Using (a), determine two linearly independent power series solutions of $(*)$—one that terminates and one that does not—by separately considering the cases when n is even and when n is odd.

(c) Show that the nonterminating power series has a radius of convergence equal to 1. Show that the series diverges at $x = \pm 1$, and hence does not satisfy (4.26).

(d) Verify that the polynomial solution found in (b) can be scaled to agree with (4.27).

(e) In (b), show that when n is even, the polynomial solution contains only even powers and when n is odd, it contains only odd powers.

8. The Chebyshev polynomials also have a Rodrigues' Formula,

$$T_n(x) = \frac{\Gamma(1/2)\sqrt{1-x^2}}{(-2)^n \Gamma(n+1/2)} \frac{d^n}{dx^n} (1-x^2)^{n-1/2},$$

as well as a Bonnet-type Recurrence Formula

$$T_{n+1}(x) = 2xT_n(x) - T_{n-1}(x), \quad n = 1, 2, \ldots$$

for computing Chebyshev polynomials. (You are not asked to prove these two formulas.) Compute the first 4 Chebyshev polynomials using (4.27), as well as the two methods above. Finally, call them using a software package.

9. (a) Compute the Chebyshev series for $f(x) = \sqrt{x+1}$, $-1 < x < 1$.
   (b) Plot f and the Nth partial Chebyshev sum $F_N(x)$ on the same coordinate plane for $N = 1, 3, 6$ as in Figure 4.8.
   (c) Plot the pointwise error function in (b).
   (d) Compute (numerically, not analytically) the uniform error and weighted L^2 error in (b).
   (e) Find the smallest N such that the weighted L^2 error in (b) is less than 10^{-1}.
   (f) Discuss the pointwise, uniform, and L^2 convergence of this series.

10. Repeat Exercise 9 with

$$f(x) = \begin{cases} 1, & -1 < x \leq 0, \\ \cos(2\pi x), & 0 < x < 1. \end{cases}$$

11. Repeat Exercise 9 with $f(x) = \sqrt{1 - x^2}$.

12. (a) Compute the Bessel series of order zero for $f(x) = xe^{-5x}$, $0 < x < 1$.
    (b) Plot f and the Nth partial Bessel sum $F_N(x)$ on the same coordinate plane for $N = 2, 4, 8$ as in Figure 4.10.
    (c) Plot the pointwise error function in (b).
    (d) Compute (numerically, not analytically) the uniform error and weighted L^2 error in (b).
    (e) Find the smallest N such that the weighted L^2 error in (b) is less than 10^{-1}.
    (f) Discuss the pointwise, uniform, and L^2 convergence of this series.

13. Repeat Exercise 12 using Bessel functions of order one.

14. Repeat Exercise 12 with $f(x) = x^{1/3}$.

15. In future applications, we will need to solve Bessel's equation on the interval $0 < x < \ell$ instead of $0 < x < 1$, as in this section. Show that the substitution $\xi := \ell x$ transforms the original Bessel's equation,

$$x^2 y''(x) + xy'(x) + (\lambda x^2 - n^2)y(x) = 0, \quad 0 < x < 1,$$

to the slightly more general form

$$\xi^2 y''(\xi) + \xi y'(\xi) + \left(\tilde{\lambda}\xi^2 - n^2\right) y(\xi) = 0, \quad 0 < \xi < \ell.$$

From this, conclude that the eigenvalues of the second equation are $\tilde{\lambda}_{nm} = (z_{nm}/\ell)^2$ with eigenfunctions

$$y_{nm}(\xi) = J_n(\sqrt{\tilde{\lambda}_{nm}}\,\xi) = J_n(z_{nm}\xi/\ell), \quad n = 0, 1, \ldots, \; m = 1, 2, \ldots,$$

where z_{nm} is the mth positive zero of $J_n(x)$.

4.4 Computing Bessel Functions: The Method of Frobenius

The eigenvalue equation (4.28) is one of various similar forms of parameterized ODEs referred to as Bessel's equation. In this section, we consider *Bessel's equation of order* p given by

$$x^2y''(x) + xy'(x) + (x^2 - p^2)y(x) = 0, \tag{4.30}$$

where $p \geq 0$ is a parameter. This equation is named in honor of Wilhelm Bessel; see Figure 4.11.

However, since $x = 0$ is not an ordinary point, but rather a (regular) singular point[8] of (4.30), the standard power series method as outlined for Legendre's equation in Exercise 4.3.2 and for Chebyshev's equation in Exercise 4.3.7, will not yield a solution. However, the Method of Frobenius explained next will produce the two linearly independent power series solutions that form the general solution of (4.30).

To begin, we look for a solution of the form

$$y(x) = x^m \sum_{n=0}^{\infty} a_n x^n = \sum_{n=0}^{\infty} a_n x^{n+m}, \tag{4.31}$$

for some value of the constant m which will be determined momentarily. Exploiting the fact that a power series can be differentiated term-by-term, we compute

$$\begin{aligned} y'(x) &= \sum_{n=0}^{\infty} a_n(n+m)x^{n+m-1}, \\ y''(x) &= \sum_{n=0}^{\infty} a_n(n+m)(n+m-1)x^{n+m-2}. \end{aligned}$$

Substituting these into (4.30) and simplifying,

$$\begin{aligned} \sum_{n=0}^{\infty} a_n(n+m)(n+m-1)x^{n+m} &+ \sum_{n=0}^{\infty} a_n(n+m)x^{n+m} \\ &+ \sum_{n=0}^{\infty} a_n x^{n+m+2} - \sum_{n=0}^{\infty} a_n p^2 x^{n+m} = 0. \end{aligned}$$

The next to last term is the only one with exponents $n + m + 2$ instead of $n + m$, so we shift the index there to obtain

$$\begin{aligned} \sum_{n=0}^{\infty} a_n(n+m)(n+m-1)x^{n+m} &+ \sum_{n=0}^{\infty} a_n(n+m)x^{n+m} \\ &+ \sum_{n=2}^{\infty} a_{n-2} x^{n+m} - \sum_{n=0}^{\infty} a_n p^2 x^{n+m} = 0. \end{aligned}$$

Figure 4.11: Wilhelm Bessel (1784–1846) of Germany made tremendous contributions to mathematics, physics, and especially astronomy despite leaving school at age 14 to work in the family business. His work on Bessel functions first appeared in an infinite series expansion used to solve a problem of Kepler for calculating the motion of three bodies moving under mutual gravitation. He was later granted an honorary doctorate based on the recommendation of Gauss.

Peeling off the $n = 0$ and $n = 1$ terms of these sums and equating the coefficients,

$$
\begin{aligned}
n = 0: \quad & a_0 m(m-1) + a_0 m - a_0 p^2 = 0 \\
& a_0[m^2 - p^2] = 0 \\
& m = \pm p \quad \text{(since } a_0 \text{ is the first } \textit{nonzero} \text{ coefficient)} && (4.32) \\
n = 1: \quad & a_1(1+m)m + a_1(1+m) - a_1 p^2 = 0 \\
& a_1[m^2 + 2m + 1 - p^2] = 0 \\
& a_1[(m+1)^2 - p^2] = 0 && (4.33) \\
n \geq 2: \quad & a_n(n+m)(n+m-1) + a_n(n+m) + a_{n-2} - a_n p^2 = 0 \\
& a_n[(m+n)^2 - p^2] = -a_{n-2} \\
& a_n = \frac{-a_{n-2}}{[(m+n)^2 - p^2]}. && (4.34)
\end{aligned}
$$

Substituting $m = p$ from (4.32) into (4.33), we conclude $a_1(2p+1) = 0$, i.e., $a_1 = 0$ or $p = -1/2$, but the latter is impossible since $p \geq 0$ in the statement of (4.30). Hence, $a_1 = 0$. On the other hand, substituting $m = p$ into (4.34) and simplifying, we get

$$a_n = \frac{-1}{n(2p+n)} a_{n-2}, \qquad n = 2, 3, \ldots. \tag{4.35}$$

First, note that (4.35) (together with $a_1 = 0$) forces

$$a_3 = a_5 = a_7 = \cdots = 0.$$

[8]See the Appendix for a discussion of singular points and power series solutions in general.

As such, we only need to consider the coefficients a_n when n is an even number, i.e., $n = 2k$ for $k = 1, 2, \ldots$ so that (4.35) becomes

$$a_{2k} = \frac{-1}{2k(2p+2k)} a_{2k-2}, \qquad k = 1, 2, 3, \ldots. \tag{4.36}$$

Putting together (4.36), $m = p$ from (4.32), and (4.31), we get a solution to (4.30):

$$y(x) = a_0 x^p + a_0 \sum_{k=1}^{\infty} \frac{(-1)^k p!}{2^{2k} k!(k+p)!} x^{2k+p}. \tag{4.37}$$

Since $y(x)$ is a solution of (4.30) for any value of the constant a_0, standard convention is (for reasons that are not apparent at this point) to take

$$a_0 = \frac{1}{2^p p!},$$

after which the solution (4.37) can be rewritten to take the canonical form

$$y(x) = \sum_{k=0}^{\infty} \frac{(-1)^k}{2^{2k+p} k!(k+p)!} x^{2k+p}. \tag{4.38}$$

In case p is not an integer, we can use the gamma function[9] instead of factorials:

$$y(x) = J_p(x) = \sum_{k=0}^{\infty} \frac{(-1)^k}{k!\,\Gamma(k+p+1)} \left(\frac{x}{2}\right)^{2k+p}. \tag{4.39}$$

This solution is called *Bessel's function of the first kind* (of order p) and is denoted by $J_p(x)$. Thankfully, this function (defined in terms of its power series) is common in applied mathematics and therefore is built into most computer packages.[10]

However, (4.38) is but one of the two linearly independent members of the fundamental solution set of the second order equation (4.30). The other is obtained from the $m = -p$ case in (4.32):

$$J_{-p}(x) = \sum_{k=0}^{\infty} \frac{(-1)^k}{2^{2k-p} k!\,\Gamma(k-p+1)} x^{2k-p}.$$

However, we can prove that $J_{-p}(x)$ and $J_p(x)$ are linearly independent only if p is *not* an integer. That is, the general solution to (4.30) when p is not an integer is

$$y(x) = c_1 J_p(x) + c_2 J_{-p}(x), \qquad p \text{ is not an integer.}$$

[9] The *gamma function*, $\Gamma(x)$, interpolates the factorial function to noninteger values. In particular, $\Gamma(n+1) = n!$, for $n = 0, 1, 2, \ldots$ is useful.

[10] In *Mathematica*, $J_p(x)$ is referenced by `BesselJ[p,x]`.

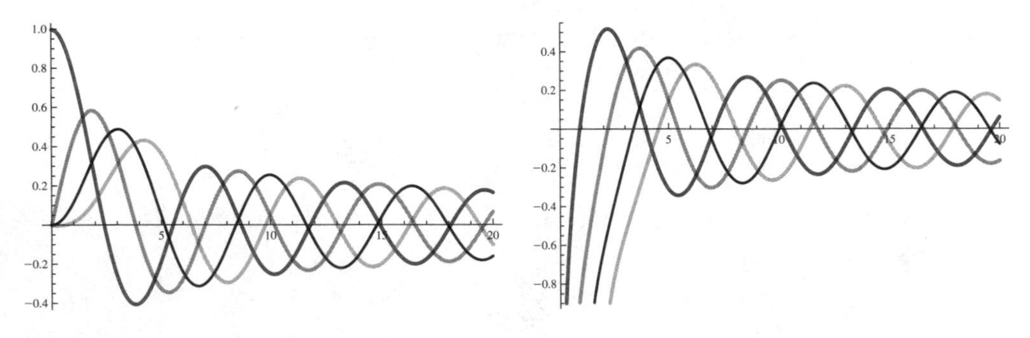

Figure 4.12: (Left) Bessel functions of the first kind, $J_p(x)$, $p = 0, 1, 2, 3$. (Right) Bessel functions of the second kind, $Y_p(x)$, $p = 0, 1, 2, 3$.

If p *is* an integer, we need to construct the second linearly independent solution to (4.30). We could use the Method of Frobenius all over again, but the following approach is easier. Suppose p is not an integer. Then

$$Y_p(x) := \frac{J_p(x)\cos(p\pi) - J_{-p}(x)}{\sin(p\pi)}, \qquad p \text{ not an integer}, \tag{4.40}$$

defines a solution of (4.30) which is linearly independent from $J_p(x)$; see Exercise 2. $Y_p(x)$ is called *Bessel's function of the second kind* (of order p) and is also built into computer packages.[11] We handle the integer p case by defining

$$Y_p(x) := \lim_{q \to p} Y_q(x), \qquad q \text{ is not an integer}. \tag{4.41}$$

This limit in fact exists (this does take some proof), thereby defining $Y_p(x)$ for integer p, so the general solution of (4.30) is given by

$$y(x) = c_1 J_p(x) + c_2 Y_p(x), \qquad p \geq 0.$$

It is important to note that while $J_p(x)$ remains bounded for all x and any given p, the same is not true for $Y_p(x)$ since $\lim_{x \to 0^+} Y_p(x) = -\infty$. See Figure 4.12.

A final property that is important for viewing $J_p(x)$ as a solution of a singular Sturm-Liouville problem is the fact that

$$J_p'(x) = \frac{1}{2}(J_{p-1}(x) - J_{p+1}(x)). \tag{4.42}$$

See Exercise 10. Because of this, the modified boundary condition at $x = 0$,

$$y, y' \text{ bounded as } x \to 0^+,$$

will be satisfied for $y(x) = J_p(x)$.

[11] In *Mathematica*, $Y_p(x)$ is referenced by `BesselY[p,x]`.

Exercises

1. (a) Write out the Sturm-Liouville form of (4.30) on $0 < x < \ell$.
   (b) State the weighted orthogonality relation for this problem.

2. Explain why $Y_p(x)$ in (4.40) is a second solution of (4.30) which is linearly independent from $J_p(x)$.

3. For the equations below, find the general solution in the form $y(x) = c_1 y_1(x) + c_2 y_2(x)$. Write out the first three nonzero terms for $y_1(x)$ and $y_2(x)$.
   (a) $x^2 y'' + xy' + (x^2 - \frac{1}{4})y = 0$
   (b) $x^2 y'' + xy' + (x^2 - \pi)y = 0$

4. Recreate the plots in Figure 4.12.

5. (a) List the numerical values for the first 5 positive zeros of $J_p(x)$ for $p = 0, 1, 2$.
   (b) List the numerical values for the first 5 positive zeros of $Y_p(x)$ for $p = 0, 1, 2$.

6. Use (4.39) to show that $J_0(0) = 1$, but $J_p(0) = 0$ for $p > 0$.

7. (a) Use the identities

$$\Gamma\left(k + \frac{1}{2}\right) = \frac{(2k)!}{2^{2k} k!}\sqrt{\pi}, \quad k = 0, 1, 2, \ldots$$

$$\Gamma\left(k + \frac{1}{2} + 1\right) = \frac{(2k+1)!}{2^{2k+1} k!}\sqrt{\pi}, \quad k = 0, 1, 2, \ldots$$

   to show that

$$J_{1/2}(x) = \sqrt{\frac{2}{\pi x}} \sin x \quad \text{and} \quad J_{-1/2}(x) = \sqrt{\frac{2}{\pi x}} \cos x.$$

   (b) Plot $J_{1/2}(x)$, $J_{-1/2}(x)$, and their "envelope" $\sqrt{\frac{2}{\pi x}}$ on the same coordinate plane.

8. To demonstrate that (4.40) and (4.41) are in fact a practical way to obtain $Y_p(x)$ where p is an integer, we demonstrate this for $p = 1$. Plot $Y_{1.6}(x)$, $Y_{1.3}(x)$, and $Y_{1.1}(x)$ on the same coordinate plane along with $Y_1(x)$.

9. Let p be a nonnegative integer. Use the Ratio Test from calculus to show that the interval of convergence for $J_p(x)$ in (4.38) is $(-\infty, \infty)$.

10. Let $p \geq 0$. Use term-by-term differentiation of the power series representation of $J_p(x)$ to prove (4.42).

4.5 The Gram-Schmidt Procedure

In the previous sections, we saw one way to generate a family of orthogonal functions: by computing the eigenfunctions of regular, periodic, and singular Sturm-Liouville problems. But this is certainly not the only way. In fact, given any set of linearly independent vectors in an inner product space, we can "orthogonalize" them and create a corresponding orthogonal family by the *Gram-Schmidt procedure.*

Definition 4.3. Let u and v be vectors in an inner product. The *projection of v onto u*, denoted $\text{proj}_u(v)$, is the vector

$$\text{proj}_u(v) := \frac{\langle v, u \rangle}{\langle u, u \rangle} u.$$

This projection operator enables us to transform a given set of linearly independent vectors into an orthogonal set of vectors with the following algorithm.

Theorem 4.7 (The Gram-Schmidt Procedure).
Suppose $\{v_1, v_2, \dots, v_n\}$ is a given set of linearly independent vectors in a vector space V equipped with an inner product. Define new vectors $u_1, u_2, \dots, v_n$ via

$$\begin{aligned}
u_1 &:= v_1, \\
u_2 &:= v_2 - \text{proj}_{u_1}(v_2), \\
u_3 &:= v_3 - \text{proj}_{u_1}(v_3) - \text{proj}_{u_2}(v_3), \\
&\vdots \\
u_n &:= v_n - \sum_{k=1}^{n-1} \text{proj}_{u_k}(v_n).
\end{aligned}$$

Then:

(i) $\{u_1, u_2, \dots, u_n\}$ *is an orthogonal set of vectors.*

(ii) $\left\{\frac{u_1}{\|u_1\|}, \frac{u_2}{\|u_2\|}, \dots, \frac{u_n}{\|u_n\|}\right\}$ *is an orthonormal set of vectors.*

(iii) *If* $\dim(V) = n$*, then all three sets form a basis for* V.

(iv) *If* $\{v_i\}$ *is an infinite sequence, then so are* $\{u_i\}$ *and* $\left\{\frac{u_i}{\|u_i\|}\right\}$.

Proof. (i) We prove this by induction. The statement is obviously true for $n = 1$. Suppose $\{u_1, \ldots, u_n\}$ defined above is an orthogonal set. We want to show that $\{u_1, \ldots, u_n, u_{n+1}\}$ is an orthogonal set. Using the inner product properties (IP2) and (IP3) from Section 3.5, for $m \leq n$, we compute

$$\begin{aligned}\langle u_{n+1}, u_m \rangle &= \langle v_{n+1} - \sum_{k=1}^{n} \operatorname{proj}_{u_k}(v_n), u_m \rangle \\ &= \langle v_{n+1}, u_m \rangle - \left\langle \sum_{k=1}^{n} \operatorname{proj}_{u_k}(v_n), u_m \right\rangle \\ &= \langle v_{n+1}, u_m \rangle - \sum_{k=1}^{n} \frac{\langle v_{n+1}, u_k \rangle}{\langle u_k, u_k \rangle} \langle u_k, u_m \rangle,\end{aligned}$$

but in the last term, $\langle u_k, u_m \rangle = 0$ except when $k = m$ since $\{u_1, \ldots, u_n\}$ is an orthogonal set. Therefore, the last line of the equation above becomes

$$\langle u_{n+1}, u_m \rangle = \langle v_{n+1}, u_m \rangle - \frac{\langle v_{n+1}, u_m \rangle}{\langle u_m, u_m \rangle} \langle u_m, u_m \rangle = 0,$$

so $\{u_1, \ldots, u_{n+1}\}$ is in fact an orthogonal set.

(ii), (iv) These follow immediately.

(iii) Any set of n linearly independent vectors will form a basis for an n dimensional space, and all of these are sets of linearly independent vectors. □

It should be noted that the Gram-Schmidt procedure is but one possible way to construct an orthogonal or orthonormal set of vectors. We now illustrate the theorem with a few examples.

Example 4.5.1. Consider the vectors in $\mathbb{R}^3$ given by

$$\mathbf{v}_1 = \begin{bmatrix} 1 \\ 0 \\ 1 \end{bmatrix}, \quad \mathbf{v}_2 = \begin{bmatrix} 1 \\ 2 \\ 3 \end{bmatrix}, \quad \mathbf{v}_3 = \begin{bmatrix} 1 \\ 1 \\ 1 \end{bmatrix}.$$

Since these are linearly independent, we can orthogonalize them via the Gram-Schmidt procedure, where the inner product is the usual dot product. Taking $\mathbf{u}_1 = \mathbf{v}_1$, then

$$\mathbf{u}_2 := \mathbf{v}_2 - \operatorname{proj}_{\mathbf{u}_1}(\mathbf{v}_2) = \mathbf{v}_2 - \frac{\langle \mathbf{v}_2, \mathbf{u}_1 \rangle}{\langle \mathbf{u}_1, \mathbf{u}_1 \rangle} \mathbf{u}_1 = \begin{bmatrix} 1 \\ 2 \\ 3 \end{bmatrix} - 2 \begin{bmatrix} 1 \\ 0 \\ 1 \end{bmatrix} = \begin{bmatrix} -1 \\ 2 \\ 1 \end{bmatrix},$$

and finally

$$\begin{aligned}\mathbf{u}_3 &:= \mathbf{v}_3 - \operatorname{proj}_{\mathbf{u}_1}(\mathbf{v}_3) - \operatorname{proj}_{\mathbf{u}_2}(\mathbf{v}_3) \\ &= \mathbf{v}_3 - \frac{\langle \mathbf{v}_3, \mathbf{u}_1 \rangle}{\langle \mathbf{u}_1, \mathbf{u}_1 \rangle} \mathbf{u}_1 - \frac{\langle \mathbf{v}_3, \mathbf{u}_2 \rangle}{\langle \mathbf{u}_2, \mathbf{u}_2 \rangle} \mathbf{u}_2 \\ &= \begin{bmatrix} 1 \\ 1 \\ 1 \end{bmatrix} - 1 \cdot \begin{bmatrix} 1 \\ 0 \\ 1 \end{bmatrix} - \frac{1}{3} \cdot \begin{bmatrix} -1 \\ 2 \\ 1 \end{bmatrix} = \begin{bmatrix} 1/3 \\ 1/3 \\ -1/3 \end{bmatrix},\end{aligned}$$

so the set $\{\mathbf{u}_1, \mathbf{u}_2, \mathbf{u}_3\}$ forms an orthogonal basis for $\mathbb{R}^3$. Normalizing each vector yields

$$\left\{ \frac{1}{\sqrt{2}}\begin{bmatrix}1\\0\\1\end{bmatrix}, \frac{1}{\sqrt{6}}\begin{bmatrix}-1\\2\\1\end{bmatrix}, \sqrt{3}\begin{bmatrix}1/3\\1/3\\-1/3\end{bmatrix} \right\},$$

which is an orthonormal basis for $\mathbb{R}^3$. ◇

Example 4.5.2. Consider the family of functions $v_n(x) := x^n$, $n = 0, 1, 2, \ldots$ in $L^2[0,1]$. Since they are linearly independent on $[0,1]$, the Gram-Schmidt procedure yields

$$\begin{aligned}
u_0(x) &:= v_0(x) \equiv 1, \\
u_1(x) &:= v_1(x) - \mathrm{proj}_{u_0}(v_1) \\
&= v_1(x) - \frac{\langle v_1, u_0\rangle}{\langle u_0, u_0\rangle} u_0 \\
&= x - \frac{\int_0^1 x \cdot 1\, dx}{\int_0^1 1^2\, dx} \cdot 1 \\
&= x - \frac{1}{2}, \\
u_2(x) &:= v_2(x) - \mathrm{proj}_{u_0}(v_2) - \mathrm{proj}_{u_1}(v_2) \\
&= v_2(x) - \frac{\langle v_2, u_0\rangle}{\langle u_0, u_0\rangle} u_0 - \frac{\langle v_2, u_1\rangle}{\langle u_1, u_1\rangle} u_1 \\
&= x^2 - \frac{\int_0^1 x^2 \cdot 1\, dx}{\int_0^1 1^2\, dx} \cdot 1 - \frac{\int_0^1 x^2 (x - 1/2)\, dx}{\int_0^1 (x-1/2)^2\, dx} \cdot (x - 1/2) \\
&= x^2 - x + \frac{1}{6},
\end{aligned}$$

and finally,

$$\begin{aligned}
u_3(x) &:= v_3(x) - \mathrm{proj}_{u_0}(v_3) - \mathrm{proj}_{u_1}(v_3) - \mathrm{proj}_{u_2}(v_3) \\
&= v_3(x) - \frac{\langle v_3, u_0\rangle}{\langle u_0, u_0\rangle} u_0 - \frac{\langle v_3, u_1\rangle}{\langle u_1, u_1\rangle} u_1 - \frac{\langle v_3, u_2\rangle}{\langle u_2, u_2\rangle} u_2 \\
&= x^3 - \frac{\int_0^1 x^3 \cdot 1\, dx}{\int_0^1 1^2\, dx} \cdot 1 - \frac{\int_0^1 x^3 (x - 1/2)\, dx}{\int_0^1 (x-1/2)^2\, dx} \cdot (x - 1/2) \\
&\qquad - \frac{\int_0^1 x^3 (x^2 - x + 1/6)\, dx}{\int_0^1 (x^2 - x + 1/6)^2\, dx} (x^2 - x + 1/6) \\
&= x^3 - \frac{3}{2}x^2 + \frac{3}{5}x - \frac{1}{20},
\end{aligned}$$

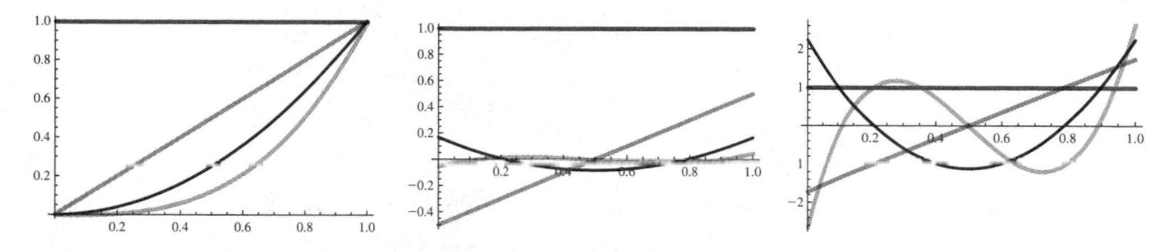

Figure 4.13: (Left) The linearly independent family $\{1, x, x^2, x^3\}$ on $0 \le x \le 1$. (Center) The first four members of the orthogonal family (4.43) resulting from the Gram-Schmidt procedure. (Right) The first four members of the orthonormal family (4.44).

and so on. Therefore,

$$\left\{1, x - \frac{1}{2}, x^2 - x + \frac{1}{6}, x^3 - \frac{3}{2}x^2 + \frac{3}{5}x - \frac{1}{20}, \ldots\right\} \tag{4.43}$$

forms an orthogonal family in $L^2[0,1]$. Normalizing, we see

$$\left\{1, \sqrt{3}\,(2x-1), \sqrt{5}\left(6x^2 - 6x + 1\right), \sqrt{7}\left(20x^3 - 30x^2 + 12x - 1\right), \ldots\right\} \tag{4.44}$$

forms an orthonormal family in $L^2[0,1]$. See Figure 4.13.

Unlike the finite dimensional case, the Gram-Schmidt procedure makes no claim about either of these being a basis for $L^2[0,1]$. ◊

A primary application for the Gram-Schmidt procedure is in converting a basis for an inner product space to an orthogonal (or orthonormal) basis, because once we have an orthogonal basis, all of the L^2 theory is at our disposal.

Exercises

1. Consider $V = \mathbb{R}^4$ with the inner product $\langle \mathbf{u}, \mathbf{v} \rangle := \mathbf{u} \cdot \mathbf{v}$ (the usual dot product).

   (a) Let

   $$\mathbf{v}_1 = \begin{bmatrix} 2 \\ 3 \\ 5 \\ 7 \end{bmatrix}, \mathbf{v}_2 = \begin{bmatrix} 11 \\ 13 \\ 17 \\ 19 \end{bmatrix}, \mathbf{v}_3 = \begin{bmatrix} 23 \\ 29 \\ 31 \\ 37 \end{bmatrix}, \mathbf{v}_4 = \begin{bmatrix} 41 \\ 43 \\ 47 \\ 53 \end{bmatrix}.$$

   Show that $\{\mathbf{v}_1, \ldots, \mathbf{v}_4\}$ forms a basis for $\mathbb{R}^4$, but not an orthogonal basis.

   (b) Use the Gram-Schmidt procedure to generate an orthogonal basis for $\mathbb{R}^4$.

   (c) Compute $\langle \mathbf{v}_1, \mathbf{v}_4 \rangle$, $\text{proj}_{\mathbf{v}_3}(\mathbf{v}_2)$, and $\|\mathbf{v}_1\|$. Interpret each geometrically.

2. Consider the vector space $V := \{2 \times 2 \text{ matrices with real entries}\}$ and define

$$\langle A, B \rangle := \text{tr}\left(A^T B\right),$$

where tr denotes the trace of a matrix (the sum of its diagonal elements) and B^T denotes the transpose of B.

(a) Show that this is an inner product space.

❀ (b) Let $A := \begin{bmatrix} 1 & 2 \\ 3 & 4 \end{bmatrix}$ and $B = \begin{bmatrix} 5 & 6 \\ 7 & 8 \end{bmatrix}$. Compute $\langle A, B \rangle$, $\text{proj}_A(B)$, and $\|A\|$. Interpret each geometrically.

❀ 3. Consider the functions $\{\sin(nx)\}$, $n = 1, 2, \ldots$ as members of $L^2[0, \pi/2]$.

(a) Show that this is not an orthogonal family.

(b) Use the Gram-Schmidt procedure to construct the first three members of an orthogonal family and then normalize to obtain an orthonormal family.

(c) Plot the first three members of the orthogonal family on the same coordinate plane.

❀ 4. Repeat Exercise 3 using $\{\cos^n(x)\}$, $n = 0, 1, 2, \ldots$ as members of $L^2[0, \pi]$.

❀ 5. Repeat Exercise 3 using $\{\cos(nx)\}$, $n = 0, 1, 2, \ldots$ as members of $L^2_w[0, \pi]$ with weight function $w(x) = x$.

❀ 6. Consider the functions $\{\sqrt{nx}\}$, $n = 1, 2, \ldots$ as members of $L^2[0, 1]$.

(a) Show that this is not an orthogonal family.

(b) Attempt the Gram-Schmidt procedure and show that it fails. Explain why.

❀ 7. In this exercise, we derive the Legendre polynomials from the Gram-Schmidt procedure.

(a) Consider the inner product space $L^2[-1, 1]$ with the standard inner product. Apply the Gram-Schmidt procedure to $\{1, x, x^2, x^3, \ldots\}$ to obtain the orthogonal family

$$\left\{1, x, x^2 - \frac{1}{3}, x^3 - \frac{3}{5}x, \ldots\right\}.$$

(b) Write out the first four Legendre polynomials, $P_n(x)$. Although these are not the same as (a), they are proportional. Use the fact that

$$\|P_n\|_w = \sqrt{\frac{2}{2n+1}}, \quad n = 0, 1, 2, \ldots$$

to determine the scaling factor (which will depend on n) that turns (a) into the Legendre polynomials.

(c) From this derivation, can we conclude that the Legendre polynomials form a complete orthogonal family on $-1 \leq x \leq 1$? Explain.

8. In this exercise, we derive the Chebyshev polynomials from the Gram-Schmidt procedure.

   (a) Consider the weighted inner product space $L_w^2[-1,1]$ with weight function $w(x) = (1-x^2)^{-1/2}$. Apply the Gram-Schmidt procedure to $\{1, x, x^2, x^3, \ldots\}$ to obtain the orthogonal family

$$\left\{1, x, x^2 - \frac{1}{2}, x^3 - \frac{3}{4}x, \ldots\right\}.$$

   (b) Write out the first four Chebyshev polynomials, $T_n(x)$. Although these are not the same as (a), they are proportional. Use the fact that

$$\|T_n\|_w = \begin{cases} \sqrt{\pi}, & n = 0, \\ \sqrt{\pi/2}, & n > 0, \end{cases}$$

   to determine the scaling factor (which will depend on n) that turns (a) into the Chebyshev polynomials.

   (c) From this derivation, can we conclude that the Chebyshev polynomials form a complete orthogonal family on $-1 \leq x \leq 1$? Explain.

9. In this exercise, we derive the Hermite polynomials from the Gram-Schmidt procedure.

   (a) Consider the weighted inner product space $L_w^2(-\infty, \infty)$ with weight function $w(x) = e^{-x^2}$. Apply the Gram-Schmidt procedure to

$$\{1, x, x^2, x^3, \ldots\}$$

   to obtain the orthogonal family

$$\left\{1, x, x^2 - \frac{1}{2}, x^3 - \frac{3}{2}x, \ldots\right\}.$$

   (b) Write out the first four Hermite polynomials, $H_n(x)$. Although these are not the same as (a), they are proportional. Use the fact that

$$\|H_n\|_w = \sqrt{2^n n! \sqrt{\pi}}, \quad n = 0, 1, 2, \ldots$$

   to determine the scaling factor (which will depend on n) that turns (a) into the Hermite polynomials.

   (c) From this derivation, can we conclude that the Hermite polynomials form a complete orthogonal family on $(-\infty, \infty)$? Explain.

10. In this exercise, we derive the Laguerre polynomials from the Gram-Schmidt procedure.

   (a) Consider the weighted inner product space $L^2_w(0, \infty)$ with weight function $w(x) = e^{-x}$. Apply the Gram-Schmidt procedure to

   $$\{1, x, x^2, x^3, \dots\}$$

   to obtain the orthogonal family

   $$\left\{1, x - 1, x^2 - 4x + 2, \frac{1}{6}x^3 + \frac{3}{2}x^2 - 3x + 1, \dots\right\}.$$

   (b) Write out the first four Laguerre polynomials, $L_n(x)$. Although these are not the same as (a), they are proportional. Use the fact that

   $$\|L_n\|_w = 1, \quad n = 0, 1, 2, \dots$$

   to determine the scaling factor (which will depend on n) that turns (a) into the Laguerre polynomials.

   (c) From this derivation, can we conclude that the Laguerre polynomials form a complete orthogonal family on $(0, \infty)$? Explain.

11. Why don't we just use the Gram-Schmidt procedure to generate the various orthogonal polynomial families in Exercises 7–10, rather than the Sturm-Liouville approach?

Chapter 5

PDEs in Higher Dimensions

5.1 Nuggets from Vector Calculus

Background: Two Dimensions

A *vector field* $\mathbf{F}$ in $\mathbb{R}^2$ takes each point (x, y) and assigns it to a vector $\mathbf{F}(x, y)$ in $\mathbb{R}^2$:

$$\mathbf{F} := F_1(x, y)\mathbf{i} + F_2(x, y)\mathbf{j} := \langle F_1(x, y), F_2(x, y)\rangle,$$

where $\mathbf{i}$ and $\mathbf{j}$ are the standard unit basis vectors in $\mathbb{R}^2$, and $F_1(x, y)$, $F_2(x, y)$ are the scalar component functions[1] of the vector field $\mathbf{F}$. Think of fluid flowing across a two dimensional plate. At each point (x, y) in the plate, we can associate a velocity vector $\mathbf{F}(x, y)$. Fluid could very well be flowing in different directions and at different speeds throughout the plate, as shown in Figure 5.1.

The *gradient* or *del* operator acting on a scalar function $f := f(x, y)$ is defined by

$$\operatorname{grad} f := \nabla f := \langle f_x, f_y\rangle.$$

This immediately tells us one way to generate a vector field from a scalar function f: start with f, calculate ∇f, and the result is $\langle f_x, f_y\rangle$. This is called the *gradient vector field of f*.

Applications of these concepts are found throughout physics and engineering. For example, if a vector field $\mathbf{F}$ is the gradient of some scalar function f, then $\mathbf{F}$ is said to be a *conservative vector field.* In symbols, this says if $\mathbf{F} = \nabla f$ for some scalar function f, then $\mathbf{F}$ is conservative. Physicists call f the *potential function* of $\mathbf{F}$.

The ∇ operator appears in another context from vector calculus. Recall the *divergence* of a vector field $\mathbf{F} = \langle F_1, F_2\rangle$ is defined by

$$\operatorname{div} \mathbf{F} := \nabla \cdot \mathbf{F} := \frac{\partial F_1}{\partial x} + \frac{\partial F_2}{\partial y}.$$

[1]There is an unfortunate, but ingrained, overlap in notation here: angle brackets are used to denote both the inner product of two vectors or functions as well as the components of a vector field. The context should make it clear which of these we mean.

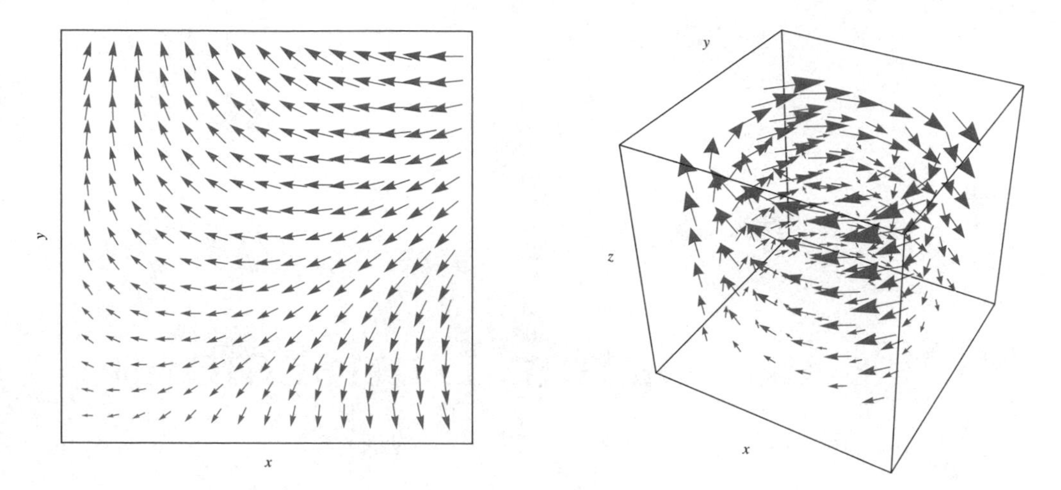

Figure 5.1: (Left) A 2D vector field from fluid dynamics showing the velocity of a fluid in a plate. (Right) A 3D vector field representing fluid flow in a box.

If we think of $\mathbf{F}$ as the velocity of a fluid flowing over a plate, as pictured above, then $\operatorname{div}\mathbf{F}$ represents the outflow[2] of the mass of fluid flowing from the point (x, y). In other words, $\operatorname{div}\mathbf{F}$ measures the tendency of the fluid to "diverge" from the point (x, y).

Suppose we want to measure the divergence of a gradient field in $\mathbb{R}^2$ (this is oftentimes the case in applications). That is, we want to calculate

$$\operatorname{div}\operatorname{grad} f = \operatorname{div}\nabla f = \nabla\cdot\nabla f = \nabla\cdot\langle f_x, f_y\rangle = f_{xx} + f_{yy}.$$

Note the output is a scalar function, which makes sense physically (why?). This operator is called the *Laplacian* of f and is denoted in two ways:

$$\nabla^2 f := f_{xx} + f_{yy} \qquad \text{or} \qquad \Delta f := f_{xx} + f_{yy}.$$

The symbol ∇^2 is more common in physics, while Δ is more common in mathematics.

To see why the Laplacian operator is important, consider the problem of finding the potential function f for a conservative vector field $\mathbf{F}$. We need more information about $\mathbf{F}$ in order to solve for f. In a fluid flow context, $\mathbf{F}$ is *incompressible* if $\operatorname{div}\mathbf{F} = 0$. In the theory of electromagnetism, we call $\mathbf{F}$ *divergence-free* when $\operatorname{div}\mathbf{F} = 0$. In either setting, we are now able to find the potential function f by solving

$$\operatorname{div}\mathbf{F} = 0 = \operatorname{div}\nabla f = \nabla\cdot\nabla f = \Delta f = 0.$$

In summary, the potential function must satisfy

$$\boxed{\Delta f = 0 \quad \text{or equivalently} \quad f_{xx} + f_{yy} = 0.}$$

This PDE is called *Laplace's equation* or the *potential equation.* We will discuss other physical interpretations of Laplace's equation in the next section.

[2]Per unit area in $\mathbb{R}^2$; per unit volume in $\mathbb{R}^3$.

Recap in Three Dimensions

Here's a summary of the topics above, except this time in a three dimensional setting. Let $f := f(x, y, z)$ be a scalar function and $\mathbf{F} := \langle F_1, F_2, F_3 \rangle$ be a vector field:

- $\operatorname{grad} f := \nabla f := \langle f_x, f_y, f_z \rangle$
- $\operatorname{div} \mathbf{F} := \nabla \cdot \mathbf{F} := \dfrac{\partial F_1}{\partial x} + \dfrac{\partial F_2}{\partial y} + \dfrac{\partial F_3}{\partial z}$
- $\Delta f := \operatorname{div} \operatorname{grad} f := \nabla \cdot \nabla f = f_{xx} + f_{yy} + f_{zz}$
- Note: $\nabla \cdot \nabla = \nabla^2 = \Delta$

In $\mathbb{R}^3$, we also have the *curl* of a vector field:

$$\operatorname{curl} \mathbf{F} := \nabla \times \mathbf{F} := \begin{vmatrix} \mathbf{i} & \mathbf{j} & \mathbf{k} \\ \frac{\partial}{\partial x} & \frac{\partial}{\partial y} & \frac{\partial}{\partial z} \\ F_1 & F_2 & F_3 \end{vmatrix} = \left\langle \frac{\partial F_3}{\partial y} - \frac{\partial F_2}{\partial z}, \frac{\partial F_1}{\partial z} - \frac{\partial F_3}{\partial x}, \frac{\partial F_2}{\partial x} - \frac{\partial F_1}{\partial y} \right\rangle.$$

At each point (x, y, z), $\operatorname{curl} \mathbf{F}$ is a vector oriented perpendicular to the plane of circulation that points in the direction of maximum circulation[3] and whose magnitude is the maximum circulation. Intuitively, $\operatorname{curl} \mathbf{F}$ provides the direction the axis of maximum rotation points in, and the magnitude of $\operatorname{curl} \mathbf{F}$ is the magnitude of this maximum rotation. In fluid dynamics, if $\operatorname{curl} \mathbf{F} = 0$ at a point P, we call the fluid *irrotational* at P, meaning that if a small paddle wheel as in Figure 5.2 were placed in the fluid at P, it would move with the fluid, but wouldn't rotate about its axis.

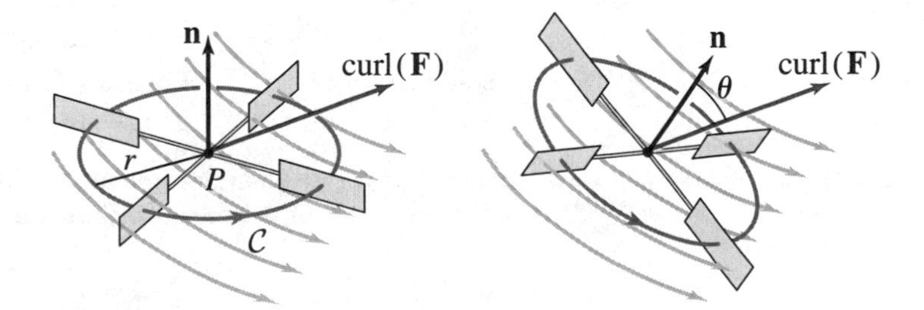

Figure 5.2: A geometric interpretation of $\operatorname{curl} \mathbf{F}$ in terms of the rotation of a fluid.

Note how both the divergence and curl of a vector field can be thought of in terms of a del vector product with $\mathbf{F}$—divergence as a del dotted with the field and curl as del crossed with the field. Finally, keep in mind that $\operatorname{div} \mathbf{F}$ is a scalar quantity, while $\operatorname{curl} \mathbf{F}$ is a vector field.

[3]Meaning it provides the direction in which the axis of maximum rotation points.

Multivariable Integration

- *Area integrals*: In two space dimensions, we denote integrals over a domain D by

$$\iint_D f(x,y)\,dx\,dy \quad \text{or} \quad \iint_D f(x,y)\,dA \quad \text{or just} \quad \int_D f(x,y)\,dA.$$

  The last one is admittedly an abuse of notation, but it is common in the literature. The area element dA should alert the reader that we are in fact computing a double integral.

- *Volume integrals*: In three space dimensions, we have

$$\iiint_D f(x,y,z)\,dx\,dy\,dz \quad \text{or} \quad \iiint_D f(x,y,z)\,dV \quad \text{or just} \quad \int_D f(x,y,z)\,dV.$$

  Again, the volume element dV should alert the reader that we are computing a triple integral.

However, we can also integrate along paths in $\mathbb{R}^2$ or surfaces in $\mathbb{R}^3$ which form the boundary, ∂D, of a domain D (see Figure 5.3). We can integrate both scalar functions and vector fields in each of these manners.

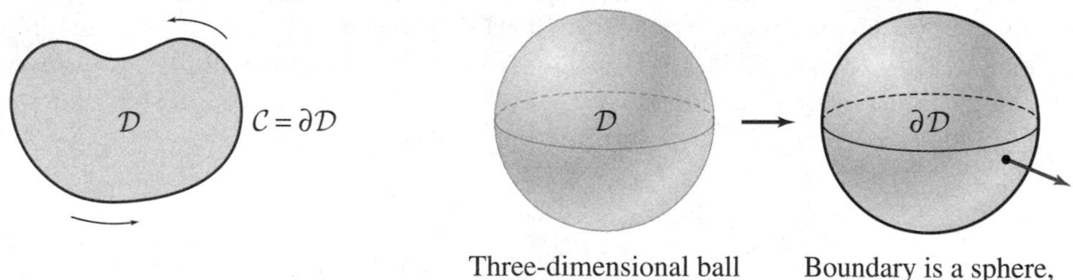

Figure 5.3: (Left) The boundary of a two dimensional region is an oriented curve. (Right) The boundary of a three dimensional solid is the surface or "skin" of the solid, which is oriented by the surface outward normal vector.

- *Line integrals*: If D is two dimensional, then ∂D is one dimensional. For scalar functions f, we have

$$\int_{\partial D} f(x,y)\,ds,$$

  which means a line integral (also called a *path integral*) of the scalar function f along the curve which forms the boundary of D. Here, ds denotes an element of arc length.

On the other hand, the line integral of a vector field $\mathbf{F}$ is given by

$$\int_{\partial D} \mathbf{F} \cdot d\mathbf{s} := \int_{\partial D} \mathbf{F} \cdot \mathbf{t}\, ds,$$

where $\mathbf{t}$ is the unit vector tangent to ∂D. This provides a geometric interpretation for the line integral of a vector field along a curve: it is the integral of the tangential component of the vector field along the curve. See Figure 5.4.

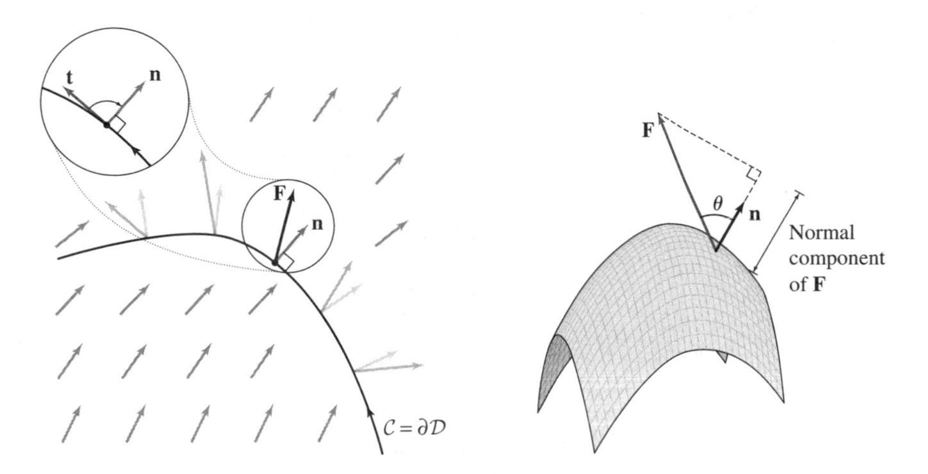

Figure 5.4: (Left) The line integral over a vector field $\mathbf{F}$ integrates the normal component of $\mathbf{F}$ over the curve $\mathcal{C}$. (Right) The surface integral over a vector field $\mathbf{F}$ integrates the outward normal component of $\mathbf{F}$ over the bounding surface ∂D.

- *Surface integrals*: If D is three dimensional, then ∂D is two dimensional. For scalar functions $f(x, y, z)$, we have

  $$\iint_{\partial D} f(x, y, z)\, dS \quad \text{or just} \quad \int_{\partial D} f(x, y, z)\, dS,$$

  which means a surface integral of the scalar function f over the surface which forms the boundary of D. Here, dS denotes an element of surface area.

  On the other hand, the surface integral of a vector field $\mathbf{F}$ is given by

  $$\iint_{\partial D} \mathbf{F} \cdot d\mathbf{S} := \iint_{\partial D} \mathbf{F} \cdot \mathbf{n}\, dS \quad \text{or just} \quad \int_{\partial D} \mathbf{F} \cdot d\mathbf{S} := \int_{\partial D} \mathbf{F} \cdot \mathbf{n}\, dS.$$

  Again, we have a nice geometric representation for the surface integral of a vector field along a bounding surface: it is the integration of the normal component of $\mathbf{F}$ along ∂D (see Figure 5.4). Because of this, these are often called *flux integrals*. If we use the single integral sign notation, the context should make it clear whether the computation is a line or surface integral.

Important Vector Calculus Theorems

The following theorems (and their higher dimensional analogues) will be used to derive PDEs in higher dimensions.

Theorem 5.1 (The Divergence Theorem).
Let $D \subset \mathbb{R}^2$ be a smooth domain with positively oriented boundary ∂D and $\mathbf{F}$ be a smooth vector field. Then

$$\iint_D \operatorname{div} \mathbf{F}\, dA = \int_{\partial D} \mathbf{F} \cdot \mathbf{n}\, ds,$$

or, in more modern notation,

$$\iint_D \nabla \cdot \mathbf{F}\, dA = \int_{\partial D} \mathbf{F} \cdot \mathbf{n}\, ds.$$

The Divergence Theorem is a multidimensional analogue of the Fundamental Conservation Law from equation (1.3): the left-hand side of the equation represents the total divergence of $\mathbf{F}$ in D, and the right-hand side represents the total (outward) flow across the boundary of D. Computationally, the Divergence Theorem is a device for trading an area integral over the domain for a line integral over the boundary in 2D, or in 3D, trading a volume integral over the domain for a surface integral over the boundary.

Theorem 5.2 (Stokes' Theorem).
Let $\mathbf{F}$ be a smooth vector field and $\mathcal{S}$ be a smooth surface in $\mathbb{R}^3$ with bounding curve C. Then

$$\iint_{\mathcal{S}} (\operatorname{curl} \mathbf{F}) \cdot \mathbf{n}\, dS = \int_C \mathbf{F} \cdot \mathbf{t}\, ds,$$

or, in more modern notation,

$$\iint_{\mathcal{S}} (\nabla \times \mathbf{F}) \cdot \mathbf{n}\, dS = \int_C \mathbf{F} \cdot \mathbf{t}\, ds.$$

Be very careful here: the left-hand side is a *surface* integral over $\mathcal{S}$, while the right-hand side is a line integral along the bounding curve C (see Figure 5.5). Consistent use of proper notation is important, since the differentials communicate the difference in each case.

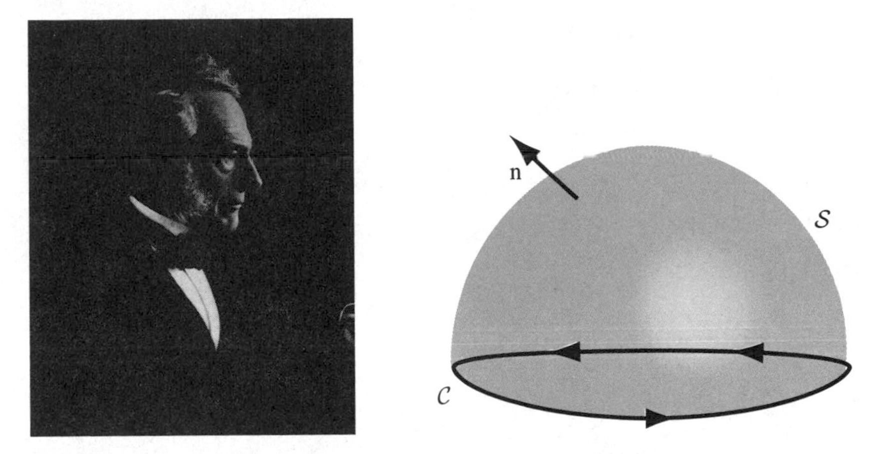

Figure 5.5: (Left) George Stokes (1819–1903) of England made significant contributions to fluid dynamics, optics, and mathematical physics, but was known for being humble, modest, and willing to help younger professors with their difficult theories. (Right) Stokes' Theorem relates the surface integral of curl $\mathbf{F}$ over $\mathcal{S}$ to a line integral of $\mathbf{F}$ along the bounding curve C.

Exercises

1. Justify the notation ∇f for grad f and $\nabla \cdot \mathbf{F}$ for div $\mathbf{F}$ by writing the del operator as a vector of partial derivatives (in $\mathbb{R}^2$ and then again in $\mathbb{R}^3$) and simplifying each expression (formally).

2. (a) Let $f(x, y) = x^2y^3 + \sin(xy)$. Compute ∇f and Δf.

   (b) Let $f(x, y, z) = \sqrt{x^2 + y^2 + z^2}$. Compute grad f and $\nabla^2 f$.

3. Answer the following, assuming the setting is $\mathbb{R}^3$.

   (a) View the gradient as an operator: say, $\mathscr{G} f := \nabla f$ where f is a smooth scalar function. Is $\mathscr{G}$ a linear operator? Prove your assertion.

   (b) View the divergence as an operator: say, $\mathscr{D}\mathbf{F} := \nabla \cdot \mathbf{F}$ where $\mathbf{F}$ is a smooth vector field. Is $\mathscr{D}$ a linear operator? Prove your assertion.

   (c) View the curl as an operator: say, $\mathscr{C}\mathbf{F} := \nabla \times \mathbf{F}$ where $\mathbf{F}$ is a smooth vector field. Is $\mathscr{C}$ a linear operator? Prove your assertion.

4. Consider the vector field $\mathbf{F} = \langle x, y \rangle$.

   (a) Make a basic sketch of $\mathbf{F}$ by hand.

   (b) Compute $\nabla \cdot \mathbf{F}$.

   (c) Explain your answer to (b) in light of (a).

   (d) By trial and error, find a potential function $f(x, y)$ for this $\mathbf{F}$.

   ❁ (e) Plot the vector field.

5. Consider the vector field $\mathbf{F} = \langle y, -x \rangle$.
   (a) Make a basic sketch of $\mathbf{F}$.
   (b) Compute $\nabla \cdot \mathbf{F}$.
   (c) Explain your answer to (b) in light of (a).
   (d) Can you find a potential function $f(x, y)$ for this $\mathbf{F}$?
   ❀ (e) Plot the vector field.
6. Consider the vector field $\mathbf{F} = \left\langle \frac{y}{x^2+y^2}, \frac{-x}{x^2+y^2}, 0 \right\rangle$.
   (a) Show that $\mathbf{F}$ is irrotational everywhere it is defined.
   ❀ (b) Plot the vector field.
   (c) Explain your answer to (a) in light of (b).
7. (a) Show that $\mathbf{a} \times \mathbf{b} = -\mathbf{b} \times \mathbf{a}$. This *anticommutativity* demonstrates that care must be taken with the order in which vectors are crossed.
   (b) Is the vector cross product associative, i.e., is $(\mathbf{a} \times \mathbf{b}) \times \mathbf{c} = \mathbf{a} \times (\mathbf{b} \times \mathbf{c})$?
8. If $f(x, y, z)$ has continuous second order partial derivatives, then mixed partials of f (up to order two) are equal, e.g., $f_{xy} = f_{yx}$, $f_{zy} = f_{yz}$, etc.
   (a) Suppose $f(x, y, z)$ has continuous second order partial derivatives. Show that $\nabla \times (\nabla f) = \mathbf{0}$. That is, the curl of a gradient vector field is always the zero vector.
   (b) If the components of $\mathbf{F}$ have continuous second order partials, show that $\nabla \cdot (\nabla \times \mathbf{F}) = 0$. That is, the divergence of the curl of $\mathbf{F}$ is always zero.
9. Which of the following expressions are meaningful? Of those that are, which are necessarily zero?
   (a) $\nabla \cdot (\nabla f)$
   (b) $\nabla \times (\nabla f)$
   (c) $\nabla(\nabla \times f)$
   (d) $\nabla \cdot (\nabla \times \mathbf{F})$
   (e) $\nabla \times (\nabla \cdot \mathbf{F})$
   (f) $\nabla(\nabla \cdot \mathbf{F})$
10. Carefully state the Divergence Theorem for $\mathbb{R}^3$.
11. Show that if $\mathbf{F}$ is irrotational in a domain $\mathcal{S} \subset \mathbb{R}^3$, then $\int_C \mathbf{F} \cdot \mathbf{t}\, ds = 0$ for any simple closed bounding curve C of $\mathcal{S}$.

12. Suppose $\mathbf{F}$ is a vector field whose components have continuous second order partial derivatives (see Exercise 8). Prove the following.

    (a) If $\mathbf{F} = \langle F_1(x,y), F_2(x,y)\rangle$ is a conservative vector field, then the "cross partials must be equal," by which we mean $\dfrac{\partial F_1}{\partial y} = \dfrac{\partial F_2}{\partial x}$.

    (b) If $\mathbf{F} = \langle F_1(x,y,z), F_2(x,y,z), F_3(x,y,z)\rangle$ is a conservative vector field, then the "cross partials must be equal," by which we mean

$$\frac{\partial F_1}{\partial y} = \frac{\partial F_2}{\partial x}, \quad \frac{\partial F_2}{\partial z} = \frac{\partial F_3}{\partial y}, \quad \frac{\partial F_3}{\partial x} = \frac{\partial F_1}{\partial z}.$$

13. It turns out that if the cross partials of $\mathbf{F}$ are equal, then $\mathbf{F}$ is conservative. Furthermore, we can use the equality of the cross partials to integrate back to obtain a potential function f. We outline this procedure with a specific example.

    (a) Let $\mathbf{F} = \langle 2xy + y^3, x^2 + 3xy^2 + 2y\rangle$. Show that the cross partials of $\mathbf{F}$ are equal.

    (b) Use the fact that $\frac{\partial f}{\partial x} = F_1$ and integration to determine $f(x,y)$ up to an arbitrary function of y.

    (c) Use the fact that $\frac{\partial f}{\partial y} = F_2$ to determine $f(x,y)$ up to an arbitrary constant.

14. Explain why the method outlined in Exercise 13 fails for $\mathbf{F} = \langle y, 0\rangle$. Argue that $\mathbf{F}$ cannot be a conservative vector field.

15. Adapt the process outlined in Exercise 13 to find a potential function for $\mathbf{F} = \langle y, x, z^3\rangle$.

16. Suppose $g(x,y,z)$ is a scalar function and $\mathbf{F}(x,y,z)$ a smooth vector field. Prove the following analogue of the product rule for the del operator in $\mathbb{R}^3$:

$$\nabla \cdot (g\mathbf{F}) = \nabla g \cdot \mathbf{F} + g \nabla \cdot \mathbf{F}.$$

17. Suppose $\mathbf{B}(x,y,z)$ is a magnetic field with an associated potential function $u(x,y,z)$, i.e., $\mathbf{B} = \nabla u$, and u is a solution of Laplace's equation in a domain $\Omega \subset \mathbb{R}^3$. Show that $\iint_{\partial\Omega} \mathbf{B} \cdot \mathbf{n}\, dS = 0$. This is called *Gauss' Law for Magnetism.*

18. Let $\mathbf{F} = \langle x^2yz, -xy^2z, z + 5x\rangle$. Explain why $\mathbf{F}$ cannot be the curl of another vector field $\mathbf{G}$.

19. (a) Suppose $\mathbf{\Phi}(x,y,t)$ is a smooth heat flux vector field at the point (x,y) in a smooth domain D at time t. Argue that the net heat flux normal to the boundary of D equals the integral of the divergence of $\mathbf{\Phi}(x,y,t)$ over D.

    (b) Suppose $\mathcal{S}$ is a smooth surface in $\mathbb{R}^3$ with bounding curve C. Let $\mathbf{k} = \langle 0,0,1\rangle$ be the standard basis vector. Show that

$$\int_C (\mathbf{n} \times \mathbf{k}) \cdot \mathbf{t}\, ds = \iint_S (\nabla \times (\mathbf{n} \times \mathbf{k})) \cdot \mathbf{n}\, dS.$$

5.2 Deriving PDEs in Higher Dimensions

2D Heat Equation

Suppose we want to model the flow of heat energy (or diffusion of a chemical concentration) in a two dimensional domain Ω. To this end, define the following quantities:

$$\begin{aligned} u(x,y,t) &:= \text{ the temperature at the point } (x,y) \text{ in the plate at time } t, \\ \mathbf{\Phi}(x,y,t) &:= \text{ flux vector at the point } (x,y) \text{ and time } t, \\ F(x,y,t) &:= \text{ rate of internal heat generation (source term) at the point } (x,y) \text{ and time } t, \\ c &:= \text{ the constant specific heat of the material of the plate,} \\ \rho &:= \text{ the constant density of the plate.} \end{aligned}$$

Consider a subset D of the domain Ω. The total heat energy[4] in D at time t is given by

$$\iint_D c\rho u(x,y,t)\, dA.$$

Conservation of energy requires

$$\begin{array}{ccccc} \text{time rate of change} & & \text{net flow of energy} & & \text{rate of generation} \\ \text{of total heat energy in } D & = & \text{across } \partial D & + & \text{inside } D. \end{array}$$

In symbols,

$$\frac{d}{dt}\iint_D c\rho u(x,y,t)\, dA = -\int_{\partial D} \mathbf{\Phi}(x,y,t)\cdot \mathbf{n}\, ds + \iint_D F(x,y,t)\, dA, \qquad (5.1)$$

or equivalently,

$$\iint_D c\rho u_t(x,y,t)\, dA + \int_{\partial D} \mathbf{\Phi}(x,y,t)\cdot \mathbf{n}\, ds = \iint_D F(x,y,t)\, dA.$$

In order to write this as a single integral over D, we apply the Divergence Theorem to the flux term, thereby replacing a line integral over ∂D with an area integral over D:

$$\iint_D c\rho u_t(x,y,t)\, dA + \iint_D \operatorname{div} \mathbf{\Phi}(x,y,t)\, dA = \iint_D F(x,y,t)\, dA.$$

Then

$$\iint_D \left[c\rho u_t(x,y,t) + \nabla\cdot \mathbf{\Phi}(x,y,t)\right]\, dA = \iint_D F(x,y,t)\, dA.$$

Since $D \subset \Omega$ was arbitrary, the integrands must be equal, producing a two dimensional analogue of the Fundamental Conservation Law in equation (1.3):

$$c\rho u_t(x,y,t) + \nabla\cdot \mathbf{\Phi}(x,y,t) = F(x,y,t).$$

[4]By definition, heat energy is specific heat times density times temperature.

Just as in the one dimensional derivation, at this point we must rely on constitutive equations for the flux term to obtain a specific model. A standard assumption in heat flow is Fourier's Law, while in chemical diffusion, it is Fick's Law, but mathematically they are identical: $\mathbf{\Phi} = -K\nabla u$, i.e., the flux is negatively proportional to the concentration gradient. Doing so produces

$$\begin{aligned} c\rho u_t + \nabla \cdot (-K\nabla u) &= F \\ c\rho u_t - K\Delta u &= F. \end{aligned}$$

Setting $k := \frac{K}{c\rho}$ (constant) and $f(x,y,t) := \frac{F(x,y,t)}{c\rho}$, we obtain $u_t = k\Delta u + f$. Typically, we have $f \equiv 0$ (no internal generation) which yields

$$\boxed{u_t = k\Delta u \quad \text{or equivalently} \quad u_t = k(u_{xx} + u_{yy}).} \tag{5.2}$$

This is called the *two dimensional heat equation* or the *two dimensional diffusion equation.* Pause for a moment to compare (5.2) with (1.5).

2D Wave Equation

Suppose we want to model the vibrations of membrane stretched across a two dimensional domain Ω such as the plate shown in Figure 5.6. Define the following quantities:

$$\begin{aligned} u(x,y,t) &:= \text{ vertical displacement of the point } (x,y) \text{ in the membrane at time } t, \\ \rho &:= \text{ constant density of the material of the membrane,} \\ \mathbf{F}(x,y,t) &:= \text{ force due to tension at the point } (x,y) \text{ at time } t. \end{aligned}$$

We will make the following (physically realistic) assumptions in our model:

- The deflections of the membrane are "small."
- The only significant motion is vertical; all else is negligible.
- The only force present is due to tension (we ignore friction, external forces, etc.), which is directed in the outward normal direction at (x,y,u) and time t.

Consider a subset $\mathcal{S}$ of the flexed membrane Ω. By the last assumption, the force vector due to tension at a point on the bounding curve C of the surface $\mathcal{S}$ takes the form $\mathbf{F} = \mathbf{t} \times (T\mathbf{n})$, where T is the magnitude of the tension, $\mathbf{t}$ is the unit vector tangent to C, and $\mathbf{n}$ is the surface unit outward normal vector. See Figure 5.6. Therefore, the vertical component of the tension vector is $\mathbf{F} \cdot \mathbf{k}$, where $\mathbf{k} := \langle 0,0,1 \rangle$ is the standard basis vector.

Newton's Second Law on the vertical forces yields

$$\begin{aligned} \iint_{\mathcal{S}} \rho u_{tt}(x,y,t)\, dS &= \int_C (\mathbf{t} \times (T\mathbf{n})) \cdot \mathbf{k}\, ds \\ &= T \int_C (\mathbf{t} \times \mathbf{n}) \cdot \mathbf{k}\, ds. \end{aligned} \tag{5.3}$$

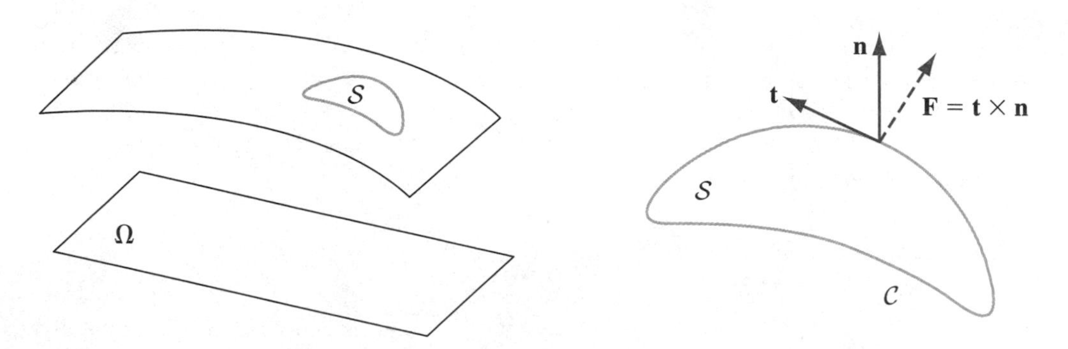

Figure 5.6: Small deflections in a rectangular membrane.

Using the scalar triple product formula from Exercise 3,

$$(\mathbf{a} \times \mathbf{b}) \cdot \mathbf{c} = (\mathbf{b} \times \mathbf{c}) \cdot \mathbf{a},$$

(5.3) can be rewritten in a form better suited for Stokes' Theorem:

$$T \int_{C} (\mathbf{t} \times \mathbf{n}) \cdot \mathbf{k} \, ds = T \int_{C} (\mathbf{n} \times \mathbf{k}) \cdot \mathbf{t} \, ds.$$

An application of Stokes' Theorem allows us to rewrite this line integral along C as a surface integral over $\mathcal{S}$,

$$T \int_{C} (\mathbf{n} \times \mathbf{k}) \cdot \mathbf{t} \, ds = T \iint_{\mathcal{S}} (\nabla \times (\mathbf{n} \times \mathbf{k})) \cdot \mathbf{n} \, dS.$$

With this key step completed, we can revisit (5.3) with both sides expressed as surface integrals

$$\iint_{\mathcal{S}} \rho u_{tt} \, dS = T \iint_{\mathcal{S}} (\nabla \times (\mathbf{n} \times \mathbf{k})) \cdot \mathbf{n} \, dS.$$

Since $\mathcal{S}$ was arbitrary, we have

$$\rho u_{tt} = T(\nabla \times (\mathbf{n} \times \mathbf{k})) \cdot \mathbf{n}. \tag{5.4}$$

At this point, we may be concerned that (5.4) doesn't particularly resemble equation (1.9) from the same juncture in the derivation of the 1D wave equation. Fear not—we only need to simplify the right-hand side of (5.4).

First, the unit surface normal is computed in the standard way:

$$\mathbf{n} = \frac{\langle -u_x, -u_y, 1 \rangle}{\sqrt{u_x^2 + u_y^2 + 1}} \approx \langle -u_x, -u_y, 1 \rangle.$$

The approximation holds since we assumed u, u_x, and u_y were "small" so that $u_x^2 \approx u_y^2 \approx 0$. Then

$$\mathbf{n} \times \mathbf{k} = \langle -u_x, -u_y, 1 \rangle \times \langle 0, 0, 1 \rangle = \begin{vmatrix} \mathbf{i} & \mathbf{j} & \mathbf{k} \\ -u_x & -u_y & 1 \\ 0 & 0 & 1 \end{vmatrix} = \langle -u_y, u_x, 0 \rangle.$$

Since $u := u(x, y, t)$ and thus has no dependence on z,

$$\nabla \times (\mathbf{n} \times \mathbf{k}) = \begin{vmatrix} \mathbf{i} & \mathbf{j} & \mathbf{k} \\ \frac{\partial}{\partial x} & \frac{\partial}{\partial y} & \frac{\partial}{\partial z} \\ -u_y & u_x & 0 \end{vmatrix} = \langle 0, 0, u_{xx} + u_{yy} \rangle.$$

Using these calculations, the vector operations in the right-hand side of (5.4) produce

$$(\nabla \times (\mathbf{n} \times \mathbf{k})) \cdot \mathbf{n} = \langle 0, 0, u_{xx} + u_{yy} \rangle \cdot \langle -u_x, -u_y, 1 \rangle = u_{xx} + u_{yy}.$$

Finally, (5.4) becomes $\rho u_{tt} = T(u_{xx} + u_{yy})$. Setting $c = \sqrt{T/\rho}$, we have

$$\boxed{u_{tt} = c^2(u_{xx} + u_{yy}) \quad \text{or equivalently} \quad u_{tt} = c^2 \Delta u.} \tag{5.5}$$

Here, c is the *wave speed* and Δ is the two dimensional Laplacian operator. Equation (5.5) is called the *two dimensional wave equation.* Pause for a moment to compare (5.5) with the one dimensional case in equation (1.10).

2D Laplace's Equation: A Different Take

In the section on vector calculus, we derived *Laplace's equation* or the *potential equation* for $u := u(x, y)$:

$$\boxed{\Delta u = 0 \quad \text{or equivalently} \quad u_{xx} + u_{yy} = 0.} \tag{5.6}$$

In that context, we saw that the solution u represents the potential function for an incompressible or divergence-free vector field. Solutions to Laplace's equation are called *harmonic functions.*

However, the two dimensional Laplacian operator $\Delta := \frac{\partial^2}{\partial x^2} + \frac{\partial^2}{\partial y^2}$ on the left-hand side of (5.6) played a role in the two dimensional heat equation (5.2) and the two dimensional wave equation (5.5). In fact, if we assume no source term in (5.2), the lineup of two dimensional equations studied thus far is strikingly similar:

$$u_t = k\Delta u \quad \text{or} \quad u_t = k(u_{xx} + u_{yy}), \tag{5.7}$$

$$u_{tt} = c^2 \Delta u \quad \text{or} \quad u_{tt} = c^2(u_{xx} + u_{yy}), \tag{5.8}$$

$$0 = \Delta u \quad \text{or} \quad 0 = u_{xx} + u_{yy}. \tag{5.9}$$

The heat and wave equations are time dependent, while Laplace's equation is time independent. However, the *steady-state*[5] temperature u for a plate (assuming there is one) is obtained from $\lim_{t\to\infty} u(x, y, t)$ in (5.7) and would depend only on the spatial variables:

$$\lim_{t\to\infty} u(x, y, t) = u(x, y).$$

[5]By steady-state, we mean the solution in the limit as $t \to \infty$, after the transient temporal dynamics are gone.

Since this $u(x, y)$ solves (5.7), it also solves (5.9). Therefore, a solution to (5.9) (in the context of heat flow) is the steady-state temperature distribution of a heat flow problem on a plate.

Finally, equations (5.7)–(5.9) show with great clarity why Laplace's equation (and from a more abstract viewpoint, the Laplacian operator) very well may be the most important equation in all of applied mathematics. It is certainly one of the most studied.

Maxwell's Equations

The vector calculus machinery can be exploited in developing PDEs in more than one space dimension, as illustrated in the last few subsections. We will go one step further to see how many fundamental concepts from physics can be described with these same PDEs.

In the theory of electromagnetism, the effects of charged particles in three dimensional space acting on one another result in an electrical vector field $\mathbf{E}(x, y, z, t)$. If the particles are also in motion, a magnetic vector field $\mathbf{B}(x, y, z, t)$ is also generated. The theory of electromagnetism rests on the principle that these vector fields $\mathbf{E}$ and $\mathbf{B}$ obey *Maxwell's equations*,

$$\frac{\partial \mathbf{E}}{\partial t} = c\nabla \times \mathbf{B}, \qquad \nabla \cdot \mathbf{E} = 0, \tag{5.10a,b}$$

$$\frac{\partial \mathbf{B}}{\partial t} = -c\nabla \times \mathbf{E}, \qquad \nabla \cdot \mathbf{B} = 0, \tag{5.10c,d}$$

where c is the speed of light (in a vacuum). This system of four PDEs—two vector equations and two scalar equations in the unknowns $\mathbf{E}$ and $\mathbf{B}$—describes how uninterfered electromagnetic radiation propagates in three dimensional space.

At first, this is quite an intimidating looking set of equations that might seem very difficult to solve. However, the interrelationship between the parts was the brilliance of Maxwell's work (see Figure 5.7) and is very helpful in the solution. Time differentiating each side of (5.10a) yields

$$\begin{aligned} \frac{\partial^2 \mathbf{E}}{\partial t^2} &= \frac{\partial}{\partial t}(c\nabla \times \mathbf{B}) \\ &= c\left(\nabla \times \frac{\partial \mathbf{B}}{\partial t}\right) \\ &= c\,(\nabla \times (-c\nabla \times \mathbf{E})) \\ &= -c^2\left[\nabla \times (\nabla \times \mathbf{E})\right]. \end{aligned} \tag{5.11}$$

This last expression seems intractable, but Lagrange's Formula[6],

$$\nabla \times (\nabla \times \mathbf{F}) = \nabla(\nabla \cdot \mathbf{F}) - (\nabla \cdot \nabla)\mathbf{F},$$

[6]Called the "BAC minus CAB formula" in physics because $\mathbf{a} \times (\mathbf{b} \times \mathbf{c}) = \mathbf{b}(\mathbf{a} \cdot \mathbf{c}) - \mathbf{c}(\mathbf{a} \cdot \mathbf{b})$.

Figure 5.7: James Clerk Maxwell (1831–1879) of Scotland was a mathematician and theoretical physicist who expressed the basic laws of electricity and magnetism in a unified fashion for the first time in his 1861 paper *On Physical Lines of Force* and again (in a more unified manner) in his 1864 paper *A Dynamical Theory of the Electromagnetic Field.* Maxwell is considered by many to be the most influential scientist on 20th century physics. Einstein called Maxwell's research the most profound work that physics had experienced since the time of Newton.

from Exercise 4, is helpful:

$$\begin{aligned}-c^2\left[\nabla\times(\nabla\times\mathbf{E})\right] &= -c^2\left[\nabla(\nabla\cdot\mathbf{E})-(\nabla\cdot\nabla)\mathbf{E}\right]\\ &= -c^2\left[\mathbf{0}-\Delta\mathbf{E}\right]\\ &= c^2\Delta\mathbf{E},\end{aligned}\tag{5.12}$$

where we used (5.10b). Therefore, (5.11) and (5.12) together imply

$$\frac{\partial^2\mathbf{E}}{\partial t^2}=c^2\Delta\mathbf{E}.$$

That is, $\mathbf{E}$ satisfies the 3D wave equation! A similar argument for (5.10c,d) shows $\mathbf{B}$ solves the 3D wave equation as well:

$$\frac{\partial^2\mathbf{B}}{\partial t^2}=c^2\Delta\mathbf{B}.$$

Together, these vector-valued solutions of two wave equations involving the electrical field $\mathbf{E}(x,y,z,t)$ and the magnetic field $\mathbf{B}(x,y,z,t)$ form the solution of Maxwell's equations.

Exercises

1. Explain the presence of the minus sign in (5.1), based on physical reasoning.

2. (a) Carefully state the Divergence Theorem for $\mathbb{R}^3$.

   (b) Using the derivation of the 2D heat equation as a template and assuming no internal generation of heat, derive the 3D heat equation

   $$u_t = k(u_{xx} + u_{yy} + u_{zz}) \quad \text{or equivalently} \quad u_t = k\Delta u,$$

   for the temperature $u := u(x, y, z, t)$ in a cube

   $$\Omega := \{(x, y, z) : 0 < x < a,\ 0 < y < b,\ 0 < z < c\}.$$

   You will need to use (a).

3. The *scalar triple product* formula was used in the derivation of the 2D wave equation. It shows how to compute the vector inner product with a cross product of two other vectors:

   $$\mathbf{a} \cdot (\mathbf{b} \times \mathbf{c}) = \mathbf{b} \cdot (\mathbf{c} \times \mathbf{a}) = \mathbf{c} \cdot (\mathbf{a} \times \mathbf{b}).$$

   Prove this for vectors in $\mathbb{R}^3$.

4. Prove *Lagrange's Formula* in $\mathbb{R}^3$ for rectangular coordinates:

   $$\nabla \times (\nabla \times \mathbf{F}) = \nabla(\nabla \cdot \mathbf{F}) - (\nabla \cdot \nabla)\mathbf{F},$$

   which was used in our work with Maxwell's equation. For the definition of $(\nabla \cdot \nabla)\mathbf{F}$, we mean the vector made up of the Laplacians of the components of $\mathbf{F}$:

   $$(\nabla \cdot \nabla)\mathbf{F} = \Delta\mathbf{F} := \langle \Delta F_1(x, y, z), \Delta F_2(x, y, z), \Delta F_3(x, y, z)\rangle.$$

5. Carefully work out the details of each step in (5.11) and (5.12).

6. Show the magnetic field $\mathbf{B}$ in Maxwell's equations satisfies the 3D wave equation.

7. An expanded version of Maxwell's equations is the following. Let $\mathbf{E}$ denote the electric field, $\mathbf{D}$ the displacement field, $\mathbf{B}$ the magnetic induction field, $\mathbf{H}$ the magnetic field, $\mathbf{J}$ the current density field, and ρ the charge density, which are related via

   $$\begin{aligned} \frac{\partial \mathbf{D}}{\partial t} - \nabla \times \mathbf{H} &= -\mathbf{J}, \qquad & \nabla \cdot \mathbf{D} &= \rho, \\ \frac{\partial \mathbf{B}}{\partial t} + \nabla \times \mathbf{E} &= 0, \qquad & \nabla \cdot \mathbf{B} &= 0. \end{aligned}$$

   Use these to show that the charge density obeys $\dfrac{\partial \rho}{\partial t} + \nabla \cdot \mathbf{J} = 0$.

5.3 Boundary Conditions in Higher Dimensions

Problems in two space dimensions have the spatial variable defined on a domain $\Omega \subset \mathbb{R}^2$. As such, the boundary of the domain consists of a curve lying in the x-y plane.

For specificity, let's consider a rectangular domain, i.e.,

$$\Omega := \{(x, y) : 0 < x < a,\ 0 < y < b\}. \tag{5.13}$$

Then $\partial\Omega$ consists of four line segments: the horizontal edges $y = 0$, $y = b$ and the vertical edges $x = 0$, $x = a$. Let's look at some of the boundary conditions from before.

- *Dirichlet boundary condition*: Specifies the value of the unknown function u on the physical boundary of the domain.

  For $u := u(x, y, t)$, a Dirichlet boundary condition on the top edge of Ω might take the form
  $$u(x, b, t) = T.$$
  For the 2D heat equation (5.2), the physical interpretation is that the temperature on the top edge of the plate is held fixed at T degrees for all time. For the 2D wave equation (5.5), it means the top edge of the membrane is held fixed at a height of T units for all time.

- *Neumann boundary condition*: Specifies the value of the outward normal derivative $\frac{\partial u}{\partial \mathbf{n}} := \nabla u \cdot \mathbf{n}$ on the physical boundary of the domain.

  As in Section 1.3, physically relevant Neumann boundary conditions are typically specified as $-\frac{\partial u}{\partial \mathbf{n}} = R$, consistent with specifying the outward flux on the boundary.[7] For example, on the right edge of Ω, they might take the form
  $$\begin{aligned} -\frac{\partial u}{\partial \mathbf{n}}(a, y, t) = R \quad \text{or equivalently} \quad -\nabla u \cdot \mathbf{n} &= R \quad \text{for } x = a \\ -\langle u_x, u_y\rangle \cdot \langle 1, 0\rangle &= R \quad \text{for } x = a \\ -u_x(a, y, t) &= R, \end{aligned}$$
  while a Neumann condition on the bottom edge would typically have the form
  $$\begin{aligned} -\frac{\partial u}{\partial \mathbf{n}}(x, 0, t) = R \quad \text{or equivalently} \quad -\nabla u \cdot \mathbf{n} &= R \quad \text{for } y = 0 \\ -\langle u_x, u_y\rangle \cdot \langle 0, -1\rangle &= R \quad \text{for } y = 0 \\ u_y(x, 0, t) &= R. \end{aligned}$$

[7] Recall from Section 5.2 that the flux vector is given by $\mathbf{\Phi} = -K\nabla u$ and the minus is crucial for Fourier's Law to hold (for heat to flow "down the gradient" and not up!). Then the outward flux normal to the boundary must take the form $\mathbf{\Phi} \cdot \mathbf{n} > 0$, which is equivalent to saying $-K\nabla u \cdot \mathbf{n} > 0$ or $-K\frac{\partial u}{\partial \mathbf{n}} > 0$. In each of these formulations, we must have $K > 0$ to be physically realistic (for Fourier's Law to hold).

The physical interpretation is consistent with the 1D case: $-\frac{\partial u}{\partial \mathbf{n}} = R > 0$ means a constant normal outward flux of R units on that edge. Of course, if $R < 0$, then a negative outward flux means an inward flux.

- *Robin boundary condition*: Specifies a linear combination of the outward normal derivative $\frac{\partial u}{\partial \mathbf{n}}$ and the unknown function u on the physical boundary of the domain.

  A Robin boundary condition along the left edge of Ω could have the form

$$\frac{\partial u}{\partial \mathbf{n}}(0, y, t) + Ku(0, y, t) = T,$$

  or, equivalently,

$$\begin{aligned} -\nabla u \cdot \mathbf{n} &= K(u - T/K) && \text{for } x = 0 \\ -\langle u_x, u_y \rangle \cdot \langle -1, 0 \rangle &= K(u - T/K) && \text{for } x = 0 \\ u_x(0, y, t) &= K(u(0, y, t) - T/K). \end{aligned}$$

  This is a two dimensional version of Newton's Law of Cooling: the outward flux of heat energy normal to the boundary is proportional to the difference between the temperature on the boundary and the temperature T/K of the surrounding medium. The minus on the normal derivative (or equivalently, $K > 0$) ensures Fourier's Law holds.

Figure 5.8 demonstrates how the various boundary conditions could be imposed on a rectangular domain. Table 5.3 summarizes the mathematical and physical form for each type of boundary condition.

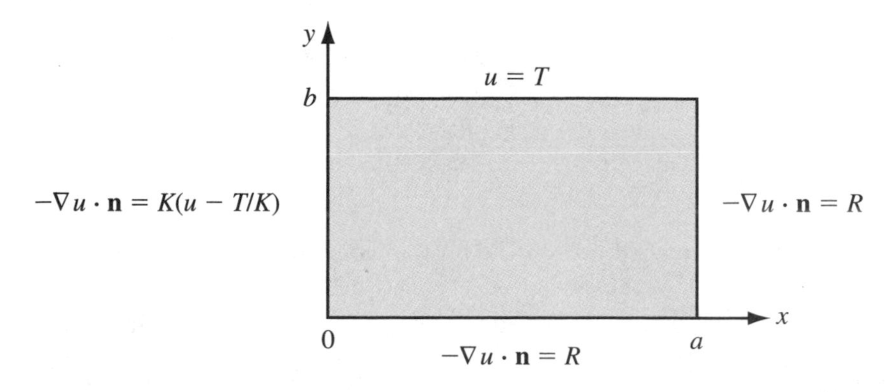

Figure 5.8: The rectangular domain Ω with boundary conditions stated in physical form. The left edge is a Robin boundary condition, indicating an exchange of heat energy with the outside temperature T/K along that edge. The top is a Dirichlet condition, indicating fixed temperature T along that side. The right and bottom edges are Neumann conditions, indicating a fixed flux normal to the boundary.

	Dirichlet	Neumann	Robin
mathematical form	$u = T$	$\frac{\partial u}{\partial \mathbf{n}} = R$ $\nabla u \cdot \mathbf{n} = R$	$\frac{\partial u}{\partial \mathbf{n}} + Ku = T$ $\nabla u \cdot \mathbf{n} + Ku = T$
physical form	$u = T$	$-\frac{\partial u}{\partial \mathbf{n}} = R$ $-\nabla u \cdot \mathbf{n} = R$	$-\frac{\partial u}{\partial \mathbf{n}} = K(u - T/K),\ K > 0$ $-\nabla u \cdot \mathbf{n} = K(u - T/K),\ K > 0$

Table 5.1: Comparison of mathematical versus physical forms for stating boundary conditions. The guiding principle in the physical form of Neumann and Robin boundary conditions is that the left-hand side takes the form of the outward flux normal to the boundary. With Robin boundary conditions, the right-hand side needs to be stated in a way consistent with Fourier's Law.

Exercises

1. Rewrite each boundary condition indicated in Figure 5.8 in its physical form, but without any vector calculus notation (as done at the very end of the examples in the text).

2. Consider Ω from (5.13) in the context of the 2D heat equation.

   (a) Explain whether the Neumann boundary condition $-\nabla u \cdot \mathbf{n} = N$ along the edge $y = b$ is a radiating or absorbing condition based on the sign of N.

   (b) Explain whether the Neumann boundary condition $\frac{\partial u}{\partial \mathbf{n}}(a, y, t) = N$ along the edge $x = a$ is a radiating or absorbing condition based on the sign of N.

   (c) Explain whether the Robin condition $\frac{\partial u}{\partial \mathbf{n}}(x, 0, t) + Ku(x, 0, t) = R$ along the edge $y = 0$ is physically realistic or not based on the sign of K.

3. Consider the rectangular domain Ω from (5.13). Interpret the following boundary conditions in the context of the 2D heat equation.

   (a) $u(0, y, t) = 0,\ u_x(a, y, t) = 0,\ u(x, 0, t) = 0,\ u_y(x, b, t) = 0$

   (b)
   $$\begin{aligned} \nabla u(0, y, t) \cdot \mathbf{n} &= -N < 0, \\ \frac{\partial u}{\partial \mathbf{n}}(a, y, t) + u(a, y, t) &= R_1 > 0, \\ \frac{\partial u}{\partial \mathbf{n}}(x, 0, t) - u(x, 0, t) &= R_2 > 0, \\ u(x, b, t) &= T. \end{aligned}$$

4. Consider Ω from (5.13). Interpret the boundary conditions $u_x(0, y, t) = 0$, $u(a, y, t) = 0$, $u_y(x, 0, t) = 0$, $u(x, b, t) = 0$ in the context of the 2D wave equation.

5.4 Well-Posed Problems: Good Models

We have examined several PDEs arising from mathematical models of physical phenomena such as simple transport, diffusion, waves, and electrostatic potentials. We saw a range of boundary conditions compatible with these physical models. It is time now to put everything together into a mathematical model.

- Does the model have a solution? This is the issue of *existence* of a solution.
- Does the model have at most one solution? This is the issue of *uniqueness* of a solution.
- Does the model give rise to solutions which are empirically useful? By that, we mean that "small" changes in the input data should result in "small" changes to the solution. This is the issue of *stability* of a solution.

If the answer to each of these questions is yes, then we say the mathematical model is *well-posed.* All stable physical phenomena should be modeled by well-posed problems.

There is a similar hierarchy in linear algebra. Consider the system $A\mathbf{x} = \mathbf{b}$, where A is an $m \times n$ matrix, $\mathbf{x}$ is an (unknown) $n \times 1$ column vector, and $\mathbf{b}$ is an $m \times 1$ column vector. If $m > n$, then A has more rows than columns, so the system is *overdetermined*; there is no solution, so existence fails. If $n > m$, then A has more columns than rows, so the system is *underdetermined*; there exist many solutions, so uniqueness fails. On the other hand, A could be $n \times n$, but have an eigenvalue very close to zero. If so, small changes to the input data $\mathbf{b}$ could result in a vastly different solution $\mathbf{x}$; stability fails.

Although a variety of tools exist to establish various components of well-posedness for a particular problem, one is to use vector calculus methods. The following two vector calculus theorems play a key role.

Theorem 5.3 (Green's First Identity).
Suppose $D \subset \mathbb{R}^2$ is a smooth domain with positively oriented boundary ∂D and u, v are twice continuously differentiable functions. Then

$$\int_{\partial D} v \frac{\partial u}{\partial \mathbf{n}}\, ds = \iint_D \nabla v \cdot \nabla u \, dA + \iint_D v \Delta u \, dA \tag{5.14}$$

where

$$\frac{\partial u}{\partial \mathbf{n}} := \nabla u \cdot \mathbf{n}$$

denotes the directional derivative of u in the outward normal direction $\mathbf{n}$. We call $\frac{\partial u}{\partial \mathbf{n}}$ the normal derivative of u.

When $v \equiv 1$, Green's First Identity becomes

$$\int_{\partial D} \frac{\partial u}{\partial \mathbf{n}}\, ds = \iint_D \Delta u\, dA.$$

This is particularly useful since it relates the area integral of Δu to the line integral of the normal derivative of u.

Theorem 5.4 (Green's Second Identity).
Let $D \subset \mathbb{R}^2$ be a smooth domain with positively oriented boundary ∂D and u, v are twice continuously differentiable functions. Then

$$\iint_D (u\Delta v - v\Delta u)\, dA = \int_{\partial D} \left(u\frac{\partial v}{\partial \mathbf{n}} - v\frac{\partial u}{\partial \mathbf{n}} \right) ds.$$

Green's Second Identity hints at a type of inner product symmetry between the Laplacian operator and the normal derivative operator.

The next example shows how we can use Green's Identities to prove uniqueness.

Example 5.4.1. Let $u(x, y)$ be a smooth function and consider the boundary value problem

$$\Delta u = f, \quad \text{in } \Omega, \tag{5.15a}$$
$$u = g, \quad \text{on } \partial\Omega, \tag{5.15b}$$

where f and g are continuous. We can quickly establish that solutions to this problem are unique—without explicitly solving the boundary value problem. To accomplish this, suppose u_1 and u_2 are two solutions to (5.15) and set $w := u_1 - u_2$. Then w solves the boundary value problem

$$\Delta w = 0, \quad \text{in } \Omega, \tag{5.16a}$$
$$w = 0, \quad \text{on } \partial\Omega. \tag{5.16b}$$

Green's First Identity (5.14) with $u = v = w$ becomes

$$\int_{\partial\Omega} w\frac{\partial w}{\partial \mathbf{n}}\, ds = \iint_\Omega \nabla w \cdot \nabla w\, dA + \iint_\Omega w\Delta w\, dA.$$

By (5.16b), the left-hand side above vanishes and by (5.16a), the last term on the right-hand side vanishes also. Thus,

$$0 = \iint_\Omega \nabla w \cdot \nabla w\, dA. \tag{5.17}$$

Now, $\nabla w \cdot \nabla w = |\nabla w|^2 \geq 0$ in Ω, so (5.17) implies $|\nabla w|^2 = 0$ in Ω. Therefore, $w \equiv C$ in Ω for some constant C. But $w = 0$ on $\partial\Omega$ and is smooth, so it must be that $w \equiv 0$ in Ω. That is, $u_1 \equiv u_2$ in Ω, so indeed solutions of (5.15) are unique. ◊

In Chapter 1, we saw that the order of an ODE dictated how many auxiliary conditions—either initial or boundary—we need in order to obtain a unique solution: an nth order linear ODE has an n dimensional basis of fundamental solutions, and therefore the general solution contains n arbitrary constants $c_1, \ldots, c_n$. In order to solve for $c_1, \ldots, c_n$, we need n initial or boundary conditions.

PDEs, however, contain multiple independent variables, so there is an order of the equation with respect to each independent variable. For example, the 1D heat equation

$$u_t = k u_{xx}, \qquad 0 < x < \ell,\ t > 0,$$

is first order in t, but second order in x. A well-posed model for the heat conduction throughout the rod should require one condition in time and two conditions in the spatial variable x.

Intuitively, to solve the 1D heat equation

$$u_t = k u_{xx}, \qquad 0 < x < \ell,\ t > 0,$$

uniquely, we need to be given (or specify)

- boundary conditions at $x = 0$ and $x = \ell$,
- the initial temperature distribution for the entire rod; that is, $u(x, 0) = f(x)$, $0 < x < \ell$.

To solve the 1D wave equation

$$u_{tt} = c^2 u_{xx}, \qquad 0 < x < \ell,\ t > 0,$$

uniquely, we need

- boundary conditions at $x = 0$ and $x = \ell$,
- the initial position of the entire string; that is, $u(x, 0) = f(x)$, $0 < x < \ell$,
- the initial velocity of the entire string; that is, $u_t(x, 0) = g(x)$, $0 < x < \ell$.

Problems in two space dimensions are variations on the same theme. To be specific, we again choose the rectangular domain in $\mathbb{R}^2$ given by

$$\Omega := \{(x, y) : 0 < x < a,\ 0 < y < b\}.$$

This time, to solve the 2D heat equation

$$u_t = k(u_{xx} + u_{yy}), \qquad 0 < x < a,\ 0 < y < b,\ t > 0,$$

uniquely, we need

- boundary conditions at $x = 0$, $x = a$, $y = 0$, and $y = b$,

- the initial temperature distribution for the entire plate; that is, $u(x, y, 0) = f(x, y)$, $0 < x < a$, $0 < y < b$,

whereas the 2D wave equation

$$u_{tt} = c^2(u_{xx} + u_{yy}), \qquad 0 < x < a,\ 0 < y < b,\ t > 0,$$

requires

- boundary conditions at $x = 0$, $x = a$, $y = 0$, and $y = b$,
- the initial position for the entire membrane; that is, $u(x, y, 0) = f(x, y)$, $0 < x < a$, $0 < y < b$,
- the initial velocity for the entire membrane; that is, $u_t(x, y, 0) = g(x, y)$, $0 < x < a$, $0 < y < b$.

Contrast the time-dependent problems above with Laplace's equation in 2D,

$$u_{xx} + u_{yy} = 0, \qquad 0 < x < a,\ 0 < y < b,$$

which requires boundary conditions at $x = 0$, $x = a$, $y = 0$, and $y = b$, but no initial condition at all since there is no time dependence in Laplace's equation.

The lesson here is that we need the same number of initial conditions as the order of the PDE in time, and we need boundary conditions specified along the entire boundary of the domain. The total number of boundary conditions needed is the sum of the orders of the PDE in each spatial variable.

Exercises

1. How many boundary conditions and initial conditions would you expect to need in order to obtain a unique solution to the following PDEs?

   (a) $u_t = Du_{xx} - cu_x + ru$, where $u := u(x, t)$ and D, c, r are constants. This is a diffusion equation with convection and transport.

   (b) $\Delta u = 0$, where Δ is the two-dimensional Laplacian in rectangular coordinates. This is Laplace's equation.

   (c) $u_{tt} = u_{rr} + \frac{1}{r}u_r + \frac{1}{r^2}u_{\theta\theta}$, where $u := u(r, \theta, t)$ is in polar coordinates. This is the wave equation in polar coordinates.

   (d) $\mu\frac{\partial^2 u}{\partial t^2} + \frac{\partial^2}{\partial x^2}\left(EI\frac{\partial^2 u}{\partial x^2}\right) = 0$, where $u := u(x, t)$. This is the beam equation for vibrations in a cantilevered beam.

   (e) $\frac{\partial^4 \varphi}{\partial x^4} + \frac{\partial^4 \varphi}{\partial y^4} + \frac{\partial^4 \varphi}{\partial z^4} + 2\frac{\partial^4 \varphi}{\partial x^2 \partial y^2} + 2\frac{\partial^4 \varphi}{\partial y^2 \partial z^2} + 2\frac{\partial^4 \varphi}{\partial x^2 \partial z^2} = 0$, where $u := u(x, y, z)$. This is the biharmonic equation from continuum mechanics; it plays a role in elasticity theory and Stokes flow.

   (f) $\frac{-h}{8\pi^2 m}\Delta\psi = E\psi$, where $\psi := \psi(x, y, z)$. This is Schrödinger's equation from quantum mechanics.

(g) $\Delta u = 0$, where $\Delta u := \frac{1}{r}\frac{\partial}{\partial r}\left(r\frac{\partial u}{\partial r}\right) + \frac{1}{r^2}\frac{\partial^2 u}{\partial \theta^2} + \frac{\partial^2 u}{\partial z^2}$ and $u := u(r, \theta, z)$ is in cylindrical coordinates. This is Laplace's equation in cylindrical coordinates.

(h) $u_t = \Delta u$, where $\Delta u := \frac{1}{r^2}\frac{\partial}{\partial r}\left(r^2\frac{\partial u}{\partial r}\right) + \frac{1}{r^2 \sin\theta}\frac{\partial}{\partial \theta}\left(\sin\theta\,\frac{\partial u}{\partial \theta}\right) + \frac{1}{r^2 \sin^2\theta}\frac{\partial^2 u}{\partial \varphi^2}$ and $u := u(r, \theta, \varphi, t)$ is in spherical coordinates. This is the heat equation in spherical coordinates.

(i) $\frac{\partial^2 i}{\partial x^2} = LC\frac{\partial^2 i}{\partial t^2} + (RC+GL)\frac{\partial i}{\partial t} + RGi$, where $i := i(x,t)$. This is the telegraph equation and models electrical current across transmission lines.

2. Use Exercise 5.1.16 and the Divergence Theorem to prove Green's First Identity.

3. Explain how Green's First Identity can be viewed as a multivariable analogue to integration by parts. (See equation (2.32) for comparison.)

4. Use Green's First Identity to prove Green's Second Identity.

5. Let $\Omega \subset \mathbb{R}^2$ be a smooth domain, and suppose f and g are continuous. Show that

$$\begin{aligned} \Delta u &= f(x,y), \quad \text{in } \Omega, \\ \frac{\partial u}{\partial \mathbf{n}} &= g(x,y), \quad \text{on } \partial\Omega, \end{aligned}$$

doesn't have a solution unless the source term f and the boundary flux g satisfy the *compatibility condition*

$$\iint_\Omega f\, dA = \int_{\partial\Omega} g\, ds.$$

This shows that the nonhomogeneous Laplace's equation with Neumann boundary conditions is not well-posed in general, but existence is obtained if the compatibility condition holds. (Hint: Use Green's First Identity with $v \equiv 1$.) What is the physical interpretation of the compatibility condition in the context of fluid flow?

6. Assume the compatibility condition for the boundary value problem in Exercise 5 is satisfied for $g \equiv 0$. Are solutions unique? Explain.

7. Suppose Ω is a smooth domain in $\mathbb{R}^2$. Let α and β be constants with $\alpha\beta > 0$. Show that there is at most one solution to the Robin problem

$$\begin{aligned} \Delta u &= f(x,y), \quad \text{in } \Omega, \\ \alpha\frac{\partial u}{\partial \mathbf{n}} + \beta u &= g(x,y), \quad \text{on } \partial\Omega. \end{aligned}$$

Does your conclusion change if we allow $\alpha\beta < 0$? What about if $\alpha = 0$ or $\beta = 0$? (Hint: Show that the difference of any two solutions must be zero. Use Green's First Identity along the way.)

5.5 Laplace's Equation in 2D

Consider the following boundary value problem for Laplace's equation, named in honor of Pierre-Simon Laplace (see Figure 5.9), in 2D for $u(x, y)$:

$$\begin{aligned} u_{xx} + u_{yy} &= 0, && 0 < x < a,\ 0 < y < b, && (5.18a)\\ u(0, y) &= 0, && 0 < y < b, && (5.18b)\\ u(a, y) &= 0, && 0 < y < b, && (5.18c)\\ u(x, 0) &= f(x), && 0 < x < a, && (5.18d)\\ u(x, b) &= g(x), && 0 < x < a. && (5.18e) \end{aligned}$$

Let $u(x, y) = X(x)Y(y)$. The PDE implies $X''(x)Y(y) + X(x)Y''(y) = 0$, or

$$\frac{X''}{X} = -\frac{Y''}{Y} = -\lambda.$$

This gives two spatial ODEs:

$$X'' + \lambda X = 0 \qquad \text{and} \qquad Y'' - \lambda Y = 0.$$

The boundary conditions in the given problem at $x = 0$ and $x = a$ are homogeneous ($X(0) = X(a) = 0$), whereas the boundary conditions at $y = 0$ and $y = b$ are not. Therefore, we will seek out the eigenvalues from the problem

$$X'' + \lambda X = 0, \quad X(0) = X(a) = 0,$$

which we know from our previous work: $\lambda_n = (n\pi/a)^2$, $X_n(x) = \sin(\sqrt{\lambda_n}\, x)$, $n = 1, 2, \ldots$. Then, the general solution of the Y problem is

$$Y_n(y) = c_n \cosh(\sqrt{\lambda_n}\, y) + d_n \sinh(\sqrt{\lambda_n}\, y).$$

By the Superposition Principle,

$$u(x, y) = \sum_{n=1}^{\infty} \left[c_n \cosh(\sqrt{\lambda_n}\, y) + d_n \sinh(\sqrt{\lambda_n}\, y)\right] \sin(\sqrt{\lambda_n}\, x). \tag{5.19}$$

The eigenfunctions are still orthogonal on $0 < x < a$, so our previous methods still lead to formulas for the coefficients once we use the given data at $y = 0$ and $y = b$:

$$\begin{aligned} u(x, 0) &= \sum_{n=1}^{\infty} [c_n \cosh(0) + d_n \sinh(0)] \sin(\sqrt{\lambda_n}\, x) = f(x), \quad 0 < x < a,\\ &= \sum_{n=1}^{\infty} c_n \sin(\sqrt{\lambda_n}\, x) = f(x), \quad 0 < x < a, \end{aligned}$$

with the coefficients given by

$$c_n = \frac{2}{a} \int_0^a f(x) \sin(\sqrt{\lambda_n}\, x)\, dx, \quad n = 1, 2, \ldots. \tag{5.20}$$

Figure 5.9: Pierre-Simon Laplace (1749–1827) of France is most well known for the PDE and integral transform that bear his name, but his contributions went far beyond that. He is considered the father of the French school of mathematics, and one of the greatest mathematicians who ever lived. Because of this, he is referred to as the "French Newton." Laplace was known to be somewhat arrogant, wanting to speak in the Academy on every subject.

Employing the other boundary condition,

$$u(x,b) = \sum_{n=1}^{\infty} \underbrace{\left[c_n \cosh(\sqrt{\lambda_n}\, b) + d_n \sinh(\sqrt{\lambda_n}\, b)\right]}_{\text{Fourier sine coefficient for } g(x)} \sin(\sqrt{\lambda_n}\, x) = g(x), \quad 0 < x < a,$$

which yields

$$\left[c_n \cosh(\sqrt{\lambda_n}\, b) + d_n \sinh(\sqrt{\lambda_n}\, b)\right] = \frac{2}{a} \int_0^a g(x) \sin(\sqrt{\lambda_n}\, x)\, dx, \quad n = 1, 2, \ldots,$$

or equivalently,

$$d_n = \frac{\frac{2}{a}\int_0^a g(x) \sin(\sqrt{\lambda_n}\, x)\, dx - c_n \cosh(\sqrt{\lambda_n}\, b)}{\sinh(\sqrt{\lambda_n}\, b)}, \quad n = 1, 2, \ldots. \tag{5.21}$$

Finally, the solution to (5.18) is assembled from (5.19), (5.20), (5.21). See Figure 5.10.

It is important to note that we had a pair of *homogeneous* boundary conditions *on parallel sides* in the given BVP. Hence, the X problem dictated the eigenvalues, while the Y problem dictated the coefficients. Had the horizontal edges been fixed at $u = 0$, we would have used the Y problem to determine the eigenvalues and the X problem to determine the coefficients.

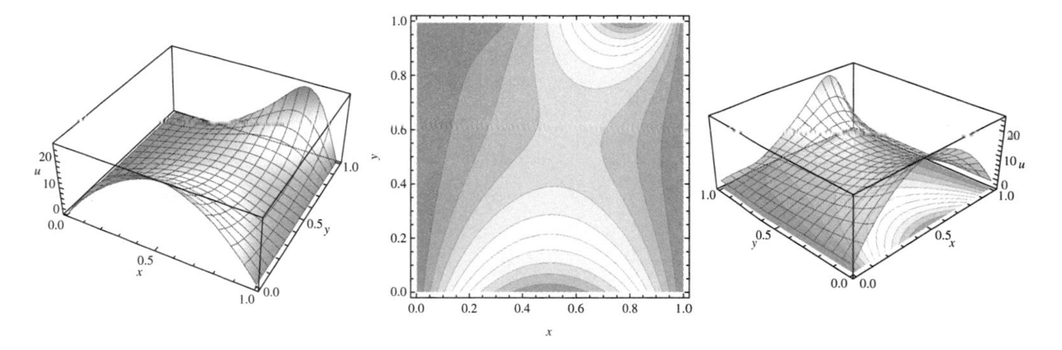

Figure 5.10: (Left) The solution surface for (5.18) using the first 5 terms of the Fourier series solution, where $a = b = 1$, $f(x) = 100x(1-x)$, $g(x) = 300x^4(1-x)$. (Center) A contour plot of the level curves of the solution surface. (Right) The relationship between the solution surface and the contour plot.

Homogenizing the Boundary Conditions

Consider the following BVP for Laplace's equation in 2D for $u(x, y)$:

$$\begin{aligned}
u_{xx} + u_{yy} &= 0, & 0 < x < a,\ 0 < y < b, && (5.22a)\\
u(0, y) &= f(y), & 0 < y < b, && (5.22b)\\
u(a, y) &= g(y), & 0 < y < b, && (5.22c)\\
u(x, 0) &= p(x), & 0 < x < a, && (5.22d)\\
u(x, b) &= q(x), & 0 < x < a. && (5.22e)
\end{aligned}$$

We want to rewrite this as the sum of two problems, each of which is *homogenized* in a useful (and complementary) way. Let $u(x, y) := v(x, y) + w(x, y)$ where

$$\begin{aligned}
v_{xx} + v_{yy} &= 0, & 0 < x < a,\ 0 < y < b, \qquad & w_{xx} + w_{yy} = 0, & 0 < x < a,\ 0 < y < b,\\
v(0, y) &= 0, & 0 < y < b, \qquad & w(0, y) = f(y), & 0 < y < b,\\
v(a, y) &= 0, & 0 < y < b, \qquad & w(a, y) = g(y), & 0 < y < b,\\
v(x, 0) &= p(x), & 0 < x < a, \qquad & w(x, 0) = 0, & 0 < x < a,\\
v(x, b) &= q(x), & 0 < x < a, \qquad & w(x, b) = 0, & 0 < x < a.
\end{aligned}$$

By the Superposition Principle, $u(x, y)$ is the solution to the original problem, (5.22). See Figure 5.11.

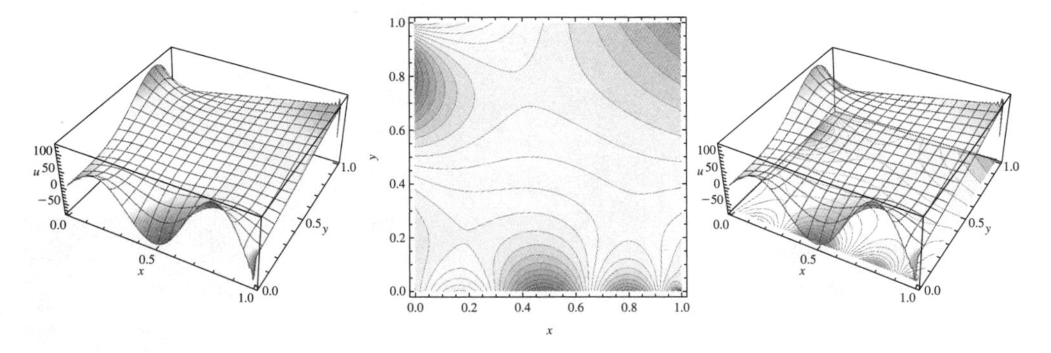

Figure 5.11: (Left) The solution surface for (5.22) using the first 100 terms of the Fourier series, where $a = b = 1$, $f(y) = 1500y^4(1-y)$, $g(y) = 100y$, $p(x) = 50\exp(x)\sin(10x)$, $q(x) = 100x$. (Center) A contour plot of the level curves of the solution surface. (Right) The relationship between the solution surface and the contour plot.

Exercises

1. (a) Solve the boundary value problem

$$\begin{aligned} u_{xx} + u_{yy} &= 0, & 0 < x < 1,\ 0 < y < 1, \\ u(0,y) &= y, & 0 < y < 1, \\ u(1,y) &= -y, & 0 < y < 1, \\ u(x,0) &= 0, & 0 < x < 1, \\ u(x,1) &= 0, & 0 < x < 1. \end{aligned}$$

   (b) What is the physical interpretation of these boundary conditions?

   ❄ (c) Plot the solution surface using the first 5 terms of the Fourier series solution. Plot the level curves as well.

2. (a) Solve the boundary value problem

$$\begin{aligned} u_{xx} + u_{yy} &= 0, & 0 < x < 1,\ 0 < y < 2, \\ u(0,y) &= 0, & 0 < y < 2, \\ u_x(1,y) &= 0, & 0 < y < 2, \\ u_y(x,0) + u(x,0) &= 0, & 0 < x < 1, \\ u(x,2) &= 100, & 0 < x < 1. \end{aligned}$$

   (b) What is the physical interpretation of these boundary conditions?

   ❄ (c) Plot the solution surface using the first 10 terms of the Fourier series solution. Plot the level curves as well.

3. (a) Solve the boundary value problem

$$\begin{aligned} u_{xx} + u_{yy} &= 0, && 0 < x < \pi,\ 0 < y < \pi, \\ u(0, y) &= 0, && 0 < y < \pi, \\ u(\pi, y) &= \cos(y^2), && 0 < y < \pi, \\ u_y(x, 0) &= 0, && 0 < x < \pi, \\ u_y(x, \pi) &= 0, && 0 < x < \pi. \end{aligned}$$

(b) What is the physical interpretation of these boundary conditions?

❀ (c) Plot the solution surface using the first 5 terms of the Fourier series solution. Plot the level curves as well.

4. Set up (but do not solve) the associated v and w problems which homogenize the boundary conditions.

$$\begin{aligned} \Delta u &= 0, && 0 < x < a,\ 0 < y < b, \\ u(0, y) &= f(y), && 0 < y < b, \\ -\nabla u(a, y) \cdot \mathbf{n} &= g(y), && 0 < y < b, \\ -\nabla u(x, 0) \cdot \mathbf{n} &= p(x), && 0 < x < a, \\ u(x, b) &= q(x), && 0 < x < a. \end{aligned}$$

5. Solve the nonhomogeneous boundary value problem

$$\begin{aligned} u_{xx} + u_{yy} &= 0, && 0 < x < a,\ 0 < y < b, \\ u(0, y) &= f(y), && 0 < y < b, \\ u(a, y) &= g(y), && 0 < y < b, \\ u(x, 0) &= p(x), && 0 < x < a, \\ u(x, b) &= q(x), && 0 < x < a. \end{aligned}$$

6. Consider the boundary value problem for $u(x, y)$:

$$\begin{aligned} \Delta u &= 0, && 0 < x < a,\ 0 < y < b, \\ u(0, y) &= f(y), && 0 < y < b, \\ -\nabla u(a, y) \cdot \mathbf{n} &= g(y), && 0 < y < b, \\ -\nabla u(x, 0) \cdot \mathbf{n} &= p(x), && 0 < x < a, \\ u(x, b) &= q(x), && 0 < x < a. \end{aligned}$$

(a) Give a physical interpretation for each line in the problem above.

(b) Solve this boundary value problem.

❀ (c) Let $a = b = 1$ and $f(y) = 8y(1 - y)$, $g(y) \equiv 0$, $p(x) = 3 - x$, and $q(x) = 50x^3(x - 0.5)(1 - x)$. Plot the solution surface using the first 5 terms of the Fourier series solution. Plot the level curves as well.

5.6 The 2D Heat and Wave Equations

Consider the initial-boundary value problem for the 2D heat equation:

$$\begin{aligned}
u_t &= k\Delta u, && 0 < x < a,\ 0 < y < b,\ t > 0, && (5.23a)\\
u(0,y,t) &= u(a,y,t) = 0, && 0 < y < b,\ t > 0, && (5.23b)\\
u(x,0,t) &= u(x,b,t) = 0, && 0 < x < a,\ t > 0, && (5.23c)\\
u(x,y,0) &= f(x,y), && 0 < x < a,\ 0 < y < b. && (5.23d)
\end{aligned}$$

Let $u(x,y,t) = X(x)Y(y)T(t)$. The PDE implies $XYT' = k(X''YT + XY''T)$. Dividing both sides by $kXYT$, we see

$$\frac{T'}{kT} = \frac{X''}{X} + \frac{Y''}{Y} = -\lambda.$$

This last equality is only possible if $X''/X = -\mu$ and $Y''/Y = -\nu$ for constants μ and ν, so

$$T' + \lambda kT = 0, \qquad X'' + \mu X = 0, \qquad Y'' + \nu Y = 0,$$

where $\lambda = \mu + \nu$. Unlike the one dimensional case, the X and Y ODEs above together with the boundary conditions yield *two* eigenvalue problems:

$$\begin{aligned}
X'' + \mu X = 0, &\quad X(0) = 0,\ X(a) = 0,\\
Y'' + \nu Y = 0, &\quad Y(0) = 0,\ Y(b) = 0.
\end{aligned}$$

Fortunately, the eigenvalues and eigenfunctions for each of these are familiar, but they must be solved—and very importantly—*indexed* independently:

$$\begin{aligned}
\mu_n &= (n\pi/a)^2, \qquad X_n(x) = \sin(\sqrt{\mu_n}\,x), \quad n = 1,2,\ldots,\\
\nu_m &= (m\pi/b)^2, \qquad Y_m(y) = \sin(\sqrt{\nu_m}\,y), \quad m = 1,2,\ldots.
\end{aligned}$$

Solving the T problem, we note that λ must be double subscripted since it varies with both n and m:

$$T' + \lambda_{nm}kT = 0, \qquad \lambda_{nm} = \mu_n + \nu_m, \quad n,m = 1,2,\ldots,$$

so that

$$T_{nm} = \exp(-\lambda_{nm}kt), \quad n,m = 1,2,\ldots.$$

By the Superposition Principle, the general solution is

$$\begin{aligned}
u(x,y,t) &= \sum_{n=1}^{\infty}\sum_{m=1}^{\infty} c_{nm}X_n(x)Y_m(y)T_{nm}(t)\\
&= \sum_{n=1}^{\infty}\sum_{m=1}^{\infty} c_{nm}\sin(\sqrt{\mu_n}\,x)\sin(\sqrt{\nu_m}\,y)\exp(-\lambda_{nm}kt).
\end{aligned} \tag{5.24}$$

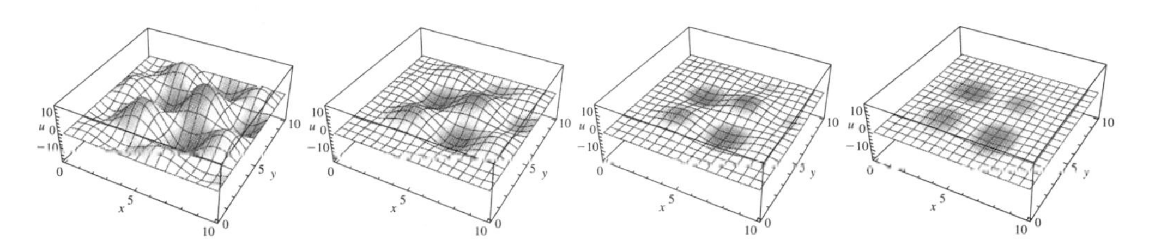

Figure 5.12: The solution surface using the first 5 terms of the double Fourier series, where $a = b = 10$, $k = 0.2$, $f(x, y) = 10 \cos x \cos y$ at times $t = 0, 1.5, 3, 6$.

To obtain the (doubly subscripted) coefficients, we apply the initial condition

$$u(x, y, 0) = \sum_{n=1}^{\infty} \sum_{m=1}^{\infty} c_{nm} \sin(\sqrt{\mu_n}\, x) \sin(\sqrt{\nu_m}\, y) = f(x, y), \qquad 0 < x < a,\ 0 < y < b.$$

From Section 2.4, the eigenfunctions $\{\sin(\sqrt{\mu_n}\, x)\}_{n=1}^{\infty}$ and $\{\sin(\sqrt{\nu_m}\, y)\}_{m=1}^{\infty}$ form orthogonal families on their respective intervals. Therefore, a standard orthogonality argument will determine the coefficient formulas. Let p, q be arbitrary but fixed indices, multiply each side of the equation above by $\sin(\sqrt{\mu_p}\, x)$ and $\sin(\sqrt{\nu_q}\, y)$, and integrate over the respective intervals:

$$\int_0^b \int_0^a \sum_{n=1}^{\infty} \sum_{m=1}^{\infty} c_{nm} \sin(\sqrt{\mu_n}\, x) \sin(\sqrt{\nu_m}\, y) \sin(\sqrt{\mu_p}\, x) \sin(\sqrt{\nu_q}\, y)\, dx\, dy$$
$$= \int_0^b \int_0^a f(x, y) \sin(\sqrt{\mu_p}\, x) \sin(\sqrt{\nu_q}\, y)\, dx\, dy$$

$$\sum_{n=1}^{\infty} \sum_{m=1}^{\infty} c_{nm} \int_0^b \int_0^a \sin(\sqrt{\mu_n}\, x) \sin(\sqrt{\nu_m}\, y) \sin(\sqrt{\mu_p}\, x) \sin(\sqrt{\nu_q}\, y)\, dx\, dy$$
$$= \int_0^b \int_0^a f(x, y) \sin(\sqrt{\mu_p}\, x) \sin(\sqrt{\nu_q}\, y)\, dx\, dy$$

$$c_{pq} \int_0^b \int_0^a \sin^2(\sqrt{\mu_p}\, x) \sin^2(\sqrt{\nu_q}\, y)\, dx\, dy = \int_0^b \int_0^a f(x, y) \sin(\sqrt{\mu_p}\, x) \sin(\sqrt{\nu_q}\, y)\, dx\, dy.$$

Since p, q were chosen arbitrarily, we can restate the coefficient formula as

$$c_{nm} = \frac{\int_0^b \int_0^a f(x, y) \sin(\sqrt{\mu_n}\, x) \sin(\sqrt{\nu_m}\, y)\, dx\, dy}{\int_0^b \int_0^a \sin^2(\sqrt{\mu_n}\, x) \sin^2(\sqrt{\nu_m}\, y)\, dx\, dy}, \qquad n, m = 1, 2, \ldots. \tag{5.25}$$

Combining (5.24), (5.25) we have the solution to (5.23). See Figure 5.12.

Exercises

1. (a) Solve the initial-boundary value problem for the 2D wave equation:

$$\begin{aligned}
u_{tt} &= c^2 \Delta u, && 0 < x < a,\ 0 < y < b,\ t > 0,\\
u(0,y,t) &= u(a,y,t) = 0, && 0 < y < b,\ t > 0,\\
u(x,0,t) &= u(x,b,t) = 0, && 0 < x < a,\ t > 0,\\
u(x,y,0) &= f(x,y), && 0 < x < a,\ 0 < y < b,\\
u_t(x,y,0) &= g(x,y), && 0 < x < a,\ 0 < y < b.
\end{aligned}$$

   (b) What is the physical interpretation of these boundary conditions?

   ✵ (c) Suppose $c = 0.5$, $a = b = 10$, $f(x,y) = 10x(10-x)(10-y)\cos x \cos y$, and $g(x,y) = -1$. Using the first 5 terms of the series solution, animate the dynamics by plotting the solution surface for t values between $t = 0$ and $t = 3.6$ in increments of 0.3.

   (d) Discuss the long term behavior illustrated in (c) in light of (b).

   ✵ (e) Compute the L^2 error between the solution in (c) evaluated at $t = 0$ and the initial function $f(x,y)$.

2. (a) Solve the initial-boundary value problem for the 2D heat equation:

$$\begin{aligned}
u_t &= k \Delta u, && 0 < x < a,\ 0 < y < b,\ t > 0,\\
u_x(0,y,t) &= u(a,y,t) = 0, && 0 < y < b,\ t > 0,\\
u_y(x,0,t) &= u(x,b,t) = 0, && 0 < x < a,\ t > 0,\\
u(x,y,0) &= f(x,y), && 0 < x < a,\ 0 < y < b.
\end{aligned}$$

   (b) What is the physical interpretation of these boundary conditions?

   ✵ (c) Suppose $k = 0.5$, $a = b = 10$, and

$$f(x,y) = \begin{cases} 1, & 4 < x < 6,\ 4 < y < 6,\\ 0, & \text{otherwise.} \end{cases}$$

   Using the first 10 terms of the series solution, animate the dynamics by plotting the solution surface for $t = 0, 0.1, 0.2, 1, 10, 25, 50, 100, 500$.

   (d) Discuss the long term behavior illustrated in (c) in light of (b).

   ✵ (e) Compute the L^2 error between the solution in (c) evaluated at $t = 0$ and the initial function $f(x,y)$.

3. What orthogonality relations were used in the exercises above?

Chapter 6

PDEs in Other Coordinate Systems

6.1 Laplace's Equation in Polar Coordinates

In certain applications, it is more natural to model the physical phenomenon of interest using a coordinate system other than rectangular coordinates. For example, describing vibrations in a circular membrane lends itself to polar coordinates (although it could also be described in rectangular coordinates). On the other hand, heat diffusion in a cylindrical piston or a sphere ball-bearing are naturally suited to cylindrical coordinates[1] and spherical coordinates, respectively.

In this section, our goal is to solve variations of Laplace's equation in polar coordinates. Later in the chapter, we will extend these techniques to Laplace's equation in cylindrical and spherical coordinates.

Laplace's Equation in a Disk

Consider the problem of describing the steady-state heat distribution in a circular plate, where the temperature along the boundary is prescribed. From the earlier chapters, we know that the relevant boundary value problem is

$$u_{xx} + u_{yy} = 0, \qquad x^2 + y^2 < \rho^2, \tag{6.1a}$$

$$u(x, y) = f(x, y), \qquad x^2 + y^2 = \rho^2, \tag{6.1b}$$

where $u := u(x, y)$, ρ is the radius of the disk, and f is the prescribed temperature on the boundary of the disk.

It is more natural to approach this problem using polar coordinates rather than rectangular coordinates. To reformulate (6.1) in polar coordinates, recall that a point

[1] Also called *cylindrical polar coordinates.*

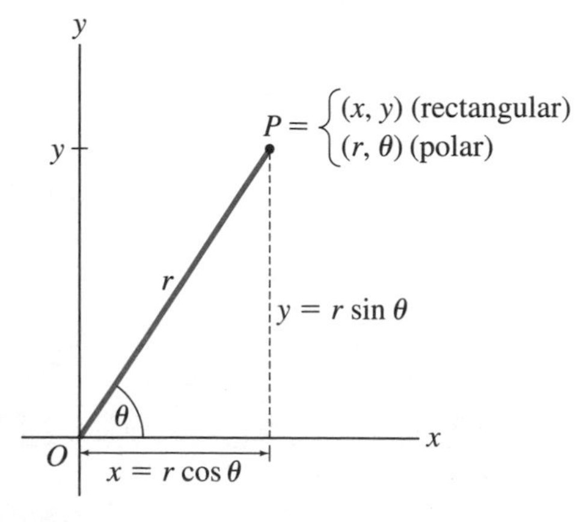

Figure 6.1: Converting between rectangular and polar coordinates.

in the plane can be described in polar coordinates by

$$\begin{aligned} x &= r\cos\theta, \\ y &= r\sin\theta, \end{aligned} \quad \Longleftrightarrow \quad \begin{aligned} r &= \sqrt{x^2+y^2}, \\ \tan\theta &= y/x, \end{aligned} \tag{6.2}$$

as illustrated in Figure 6.1.

To translate (6.1) into its equivalent polar form, we need to convert the expression

$$\Delta u(x,y) := \frac{\partial^2 u}{\partial x^2}(x,y) + \frac{\partial^2 u}{\partial y^2}(x,y),$$

to polar coordinates r and θ using (6.2). In other words, we need to determine the *polar form* of the Laplacian operator. This is an exercise in the chain rule from multivariable calculus (see Exercise 11) and yields

$$\underbrace{u_{xx}(x,y) + u_{yy}(x,y)}_{\text{rectangular form of } \Delta u(x,y)} = \underbrace{u_{rr}(r,\theta) + \frac{1}{r}u_r(r,\theta) + \frac{1}{r^2}u_{\theta\theta}(r,\theta)}_{\text{polar form of } \Delta u(r,\theta)}. \tag{6.3}$$

Now that (6.3) has revealed the polar form of the left-hand side of (6.1), we only need to apply (6.2) to determine the representation of $f(x,y)$ in the new coordinates:

$$f(x,y)\Big|_{x^2+y^2=\rho^2} = f(r\cos\theta, r\sin\theta)\Big|_{r=\rho} = f(\rho\cos\theta, \rho\sin\theta).$$

To avoid confusion as we move between coordinate systems, let's denote the polar form of u by $\tilde{u}$ and the polar form of f by $\tilde{f}$. Then (6.1) translates to polar coordinates as[2]

$$\tilde{u}_{rr}(r,\theta) + \frac{1}{r}\tilde{u}_r(r,\theta) + \frac{1}{r^2}\tilde{u}_{\theta\theta}(r,\theta) = 0, \qquad 0 < r < \rho,\ -\pi < \theta < \theta, \tag{6.4a}$$

$$\tilde{u}(\rho,\theta) = \tilde{f}(\theta), \qquad -\pi < \theta < \pi. \tag{6.4b}$$

[2]It would be fine if the angular range was given as $0 < \theta < 2\pi$ or any other interval of length 2π. Furthermore, any of the endpoints could be included without altering the analysis.

For simplicity of notation, we will use u rather than $\tilde{u}$ in our analysis below. Let $u(r,\theta) = R(r)\Theta(\theta)$. The PDE implies

$$R''\Theta + \frac{1}{r}R'\Theta + \frac{1}{r^2}R\Theta'' = 0.$$

Dividing by $R\Theta$ and multiplying by r^2, we see

$$r^2\frac{R''}{R} + r\frac{R'}{R} + \frac{\Theta''}{\Theta} = 0,$$

and thus

$$r^2\frac{R''}{R} + r\frac{R'}{R} = -\frac{\Theta''}{\Theta} = \lambda,$$

for some constant λ. This leads to the two ODEs

$$\Theta'' + \lambda\Theta = 0, \qquad r^2R'' + rR' = \lambda R,$$

one of which is the eigenvalue problem. The inherent periodicity of this disk problem in polar coordinates requires $\Theta(-\pi) = \Theta(\pi)$ and $\Theta'(-\pi) = \Theta'(\pi)$, i.e., periodic boundary conditions in θ. Therefore, we will use the Θ problem as the eigenvalue problem:

$$\Theta'' + \lambda\Theta = 0, \qquad -\pi < \theta < \pi,$$
$$\Theta(-\pi) = \Theta(\pi), \ \Theta'(-\pi) = \Theta'(\pi).$$

As we have calculated before, the eigenvalues and associated eigenfunctions are

$$\lambda_0 = 0, \qquad \Theta_0(\theta) = 1 \text{ (up to a constant multiple)},$$
$$\lambda_n = n^2, \ n = 1, 2\ldots, \qquad \Theta_n(\theta) = a_n\cos(n\theta) + b_n\sin(n\theta), \ n = 1, 2, \ldots.$$

Knowing the eigenvalues, the R problem then becomes

$$r^2R'' + rR' - n^2R = 0, \ 0 < r < \rho, \quad n = 0, 1, 2\ldots. \tag{6.5}$$

This is a Cauchy-Euler equation (see Section 1.4) with a singular point at $x = 0$ (see Section 4.2), so we impose the modified boundary condition

$$R, R' \text{ bounded as } r \to 0^+. \tag{6.6}$$

Now, the characteristic equation of (6.5) is $r^2 - n^2 = 0$ which has roots $r = \pm n$, $n = 0, 1, 2, \ldots$. Therefore, the solution of (6.5) is $R_n(r) = c_1r^n + c_2r^{-n}$, $n = 1, 2, \ldots$. On the other hand, $n = 0$ is a double root of the characteristic equation, so the solution has the form $R_0(r) = a_0 + b_0\ln r$.

However, since $r^{-n} \to \infty$ as $r \to 0^+$ and $\ln r \to -\infty$ as $r \to 0^+$, the modified boundary condition (6.6) requires us to rule out the r^{-n} and $\ln r$ terms from the solution by taking $c_2 = 0$ and $b_0 = 0$. Thus, the solution of (6.5), (6.6) is (up to a constant multiple)

$$R_0(r) = 1, \qquad R_n(r) = r^n, \ n = 1, 2, \ldots.$$

Superimposing,

$$u(r,\theta) = R_0(r)\Theta_0(\theta) + \sum_{n=1}^{\infty} R_n(r)\Theta_n(\theta) = \frac{1}{2}a_0 + \sum_{n=1}^{\infty} r^n[a_n \cos(n\theta) + b_n \sin(n\theta)]. \quad (6.7)$$

The coefficients are determined by the boundary condition:

$$u(\rho,\theta) = \frac{1}{2}a_0 + \sum_{n=1}^{\infty} \rho^n[a_n \cos(n\theta) + b_n \sin(n\theta)] = f(\theta), \quad -\pi < \theta < \pi.$$

This is a full Fourier series for $f(\theta)$ on $-\pi < \theta < \pi$, where $\rho^n a_n$ plays the role of the Fourier cosine coefficient and $\rho^n b_n$ plays the role of the Fourier sine coefficient. Thus,

$$\begin{aligned} a_0 &= \frac{1}{\pi}\int_{-\pi}^{\pi} f(\theta)\,d\theta, \\ a_n &= \frac{1}{\rho^n\pi}\int_{-\pi}^{\pi} f(\theta)\cos(n\theta)\,d\theta, \\ b_n &= \frac{1}{\rho^n\pi}\int_{-\pi}^{\pi} f(\theta)\sin(n\theta)\,d\theta, \quad n = 1,2,\ldots. \end{aligned} \quad (6.8)$$

The solution is given by (6.7), (6.8). See Figure 6.2.

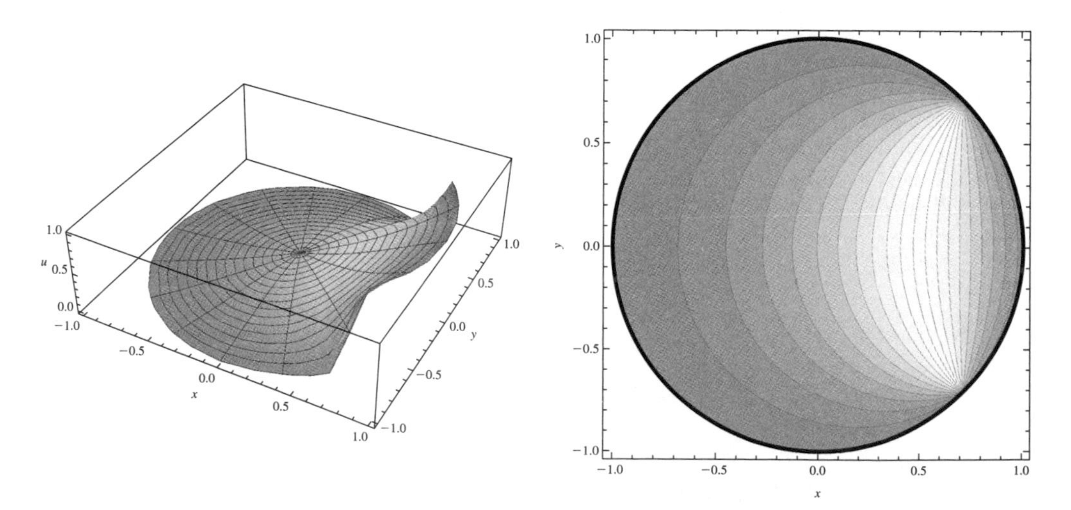

Figure 6.2: (Left) The solution surface of (6.4) using the first 100 terms of the Fourier series solution, where $\rho = 1$ and $f(\theta) = 1$ for $-\pi/4 < \theta < \pi/4$ and 0 elsewhere. (Right) The corresponding contour plot.

Laplace's Equation in an Annulus

We now demonstrate how to solve Laplace's equation on an annular region. Other interesting domains that are naturally described by polar coordinates such as those shown in Figure 6.3 are explored in the exercises.

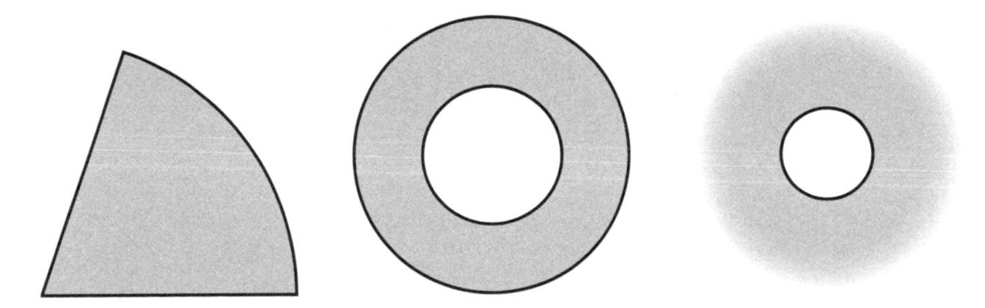

Figure 6.3: Polar coordinates easily describe regions such as a wedge, annulus, or exterior of a disk, or combinations of these. The techniques of this section can be adapted to solve PDEs on these domains.

Consider the boundary value problem for $u := u(r, \theta)$ in the annular region with inner radius ρ_1 and outer radius ρ_2:

$$u_{rr} + \frac{1}{r}u_r + \frac{1}{r^2}u_{\theta\theta} = 0, \qquad 0 < \rho_1 < r < \rho_2,\ -\pi < \theta < \pi, \tag{6.9a}$$

$$u(\rho_1, \theta) = f(\theta), \qquad -\pi < \theta < \pi, \tag{6.9b}$$

$$u(\rho_2, \theta) = g(\theta), \qquad -\pi < \theta < \pi. \tag{6.9c}$$

Letting $u(r, \theta) = R(r)\Theta(\theta)$, separation of variables exactly as before leads to the two problems

$$\Theta'' + \lambda\Theta = 0, \qquad r^2R'' + rR' = \lambda R.$$

Since this is still a full circular geometry, periodic boundary conditions must be imposed on the Θ problem. This yields the same eigenvalue problem from earlier

$$\Theta'' + \lambda\Theta = 0, \qquad -\pi < \theta < \pi,$$
$$\Theta(-\pi) = \Theta(\pi),$$
$$\Theta'(-\pi) = \Theta'(\pi),$$

which has solutions

$$\lambda_0 = 0, \qquad \Theta_0(\theta) = 1 \text{ (up to a constant multiple)},$$
$$\lambda_n = n^2,\ n = 1, 2 \ldots, \qquad \Theta_n(\theta) = a_n\cos(n\theta) + b_n\sin(n\theta),\ n = 1, 2, \ldots.$$

Using our knowledge of the eigenvalues λ_n above, the R problem becomes

$$r^2R'' + rR' - n^2R = 0, \quad n = 0, 1, 2, \ldots.$$

This is the same Cauchy-Euler problem from the disk problem, so it has solutions

$$R_0(r) = c_0 + d_0 \ln r, \qquad R_n(r) = c_n r^n + d_n r^{-n}, \qquad n = 1, 2, \ldots.$$

However, unlike the last problem, $r = 0$ is *not* a singular point of this R problem, because $0 < \rho_1 < r < \rho_2$. Therefore, we don't zero out any of the terms in the R solution.

Superimposing, we get

$$\begin{aligned} u(r,\theta) &= R_0(r)\Theta_0(\theta) + \sum_{n=1}^{\infty} R_n(r)\Theta_n(\theta) \\ &= \frac{1}{2}(c_0 + d_0 \ln r) + \sum_{n=1}^{\infty} [a_n \cos(n\theta) + b_n \sin(n\theta)][c_n r^n + d_n r^{-n}]. \end{aligned} \tag{6.10}$$

Applying the two boundary conditions, we get

$$\begin{aligned} u(\rho_1,\theta) &= \frac{1}{2}(c_0 + d_0 \ln \rho_1) + \sum_{n=1}^{\infty} [a_n \cos(n\theta) + b_n \sin(n\theta)][c_n \rho_1^n + d_n \rho_1^{-n}] \\ &= f(\theta), \quad -\pi < \theta < \pi, \\ u(\rho_2,\theta) &= \frac{1}{2}(c_0 + d_0 \ln \rho_2) + \sum_{n=1}^{\infty} [a_n \cos(n\theta) + b_n \sin(n\theta)][c_n \rho_2^n + d_n \rho_2^{-n}] \\ &= g(\theta), \quad -\pi < \theta < \pi. \end{aligned}$$

If we rewrite these as

$$\begin{aligned} &\frac{1}{2}(c_0 + d_0 \ln \rho_1) + \sum_{n=1}^{\infty} [c_n \rho_1^n + d_n \rho_1^{-n}] a_n \cos(n\theta) + [c_n \rho_1^n + d_n \rho_1^{-n}] b_n \sin(n\theta) \\ &\qquad = f(\theta), \quad -\pi < \theta < \pi, \\ &\frac{1}{2}(c_0 + d_0 \ln \rho_2) + \sum_{n=1}^{\infty} [c_n \rho_2^n + d_n \rho_2^{-n}] a_n \cos(n\theta) + [c_n \rho_2^n + d_n \rho_2^{-n}] b_n \sin(n\theta) \\ &\qquad = g(\theta), \quad -\pi < \theta < \pi, \end{aligned}$$

then we recognize them as full Fourier series expansions of $f(\theta)$ and $g(\theta)$, respectively, on $-\pi < \theta < \pi$. Therefore, the coefficients are determined by the six equations

$$\begin{aligned}
c_0 + d_0 \ln \rho_1 &= \frac{1}{\pi} \int_{-\pi}^{\pi} f(\theta)\, d\theta \\
(c_n \rho_1^n + d_n \rho_1^{-n}) a_n &= \frac{1}{\pi} \int_{-\pi}^{\pi} f(\theta) \cos(n\theta)\, d\theta \\
(c_n \rho_1^n + d_n \rho_1^{-n}) b_n &= \frac{1}{\pi} \int_{-\pi}^{\pi} f(\theta) \sin(n\theta)\, d\theta \\
c_0 + d_0 \ln \rho_2 &= \frac{1}{\pi} \int_{-\pi}^{\pi} g(\theta)\, d\theta \\
(c_n \rho_2^n + d_n \rho_2^{-n}) a_n &= \frac{1}{\pi} \int_{-\pi}^{\pi} g(\theta) \cos(n\theta)\, d\theta \\
(c_n \rho_2^n + d_n \rho_2^{-n}) b_n &= \frac{1}{\pi} \int_{-\pi}^{\pi} g(\theta) \sin(n\theta)\, d\theta,
\end{aligned}$$

in the six unknowns c_0, d_0, $a_n c_n$, $a_n d_n$, $b_n c_n$, and $b_n d_n$. Note that it is not necessary (nor possible) to determine a_n, b_n, c_n, and d_n, but rather just the specific products above. Solving this system produces

$$\begin{aligned}
c_0 &= \frac{\ln(\rho_1) \int_{-\pi}^{\pi} g(\theta)\, d\theta - \ln(\rho_2) \int_{-\pi}^{\pi} f(\theta)\, d\theta}{\pi \ln(\rho_1/\rho_2)} \\
d_0 &= \frac{\int_{-\pi}^{\pi} f(\theta)\, d\theta - \int_{-\pi}^{\pi} g(\theta)\, d\theta}{\pi \ln(\rho_1/\rho_2)} \\
a_n c_n &= \frac{\rho_1^n \int_{-\pi}^{\pi} f(\theta) \cos(n\theta)\, d\theta - \rho_2^n \int_{-\pi}^{\pi} g(\theta) \cos(n\theta)\, d\theta}{\pi(\rho_1^{2n} - \rho_2^{2n})} \\
a_n d_n &= \frac{\rho_1^{2n} \rho_2^n \int_{-\pi}^{\pi} g(\theta) \cos(n\theta)\, d\theta - \rho_1^n \rho_2^{2n} \int_{-\pi}^{\pi} f(\theta) \cos(n\theta)\, d\theta}{\pi(\rho_1^{2n} - \rho_2^{2n})} \\
b_n c_n &= \frac{\rho_1^n \int_{-\pi}^{\pi} f(\theta) \sin(n\theta)\, d\theta - \rho_2^n \int_{-\pi}^{\pi} g(\theta) \sin(n\theta)\, d\theta}{\pi(\rho_1^{2n} - \rho_2^{2n})} \\
b_n d_n &= \frac{\rho_1^{2n} \rho_2^n \int_{-\pi}^{\pi} g(\theta) \sin(n\theta)\, d\theta - \rho_1^n \rho_2^{2n} \int_{-\pi}^{\pi} f(\theta) \sin(n\theta)\, d\theta}{\pi(\rho_1^{2n} - \rho_2^{2n})}.
\end{aligned} \tag{6.11}$$

Finally, rewriting (6.10) as

$$\begin{aligned}
u(r, \theta) = \frac{1}{2}(c_0 + d_0 \ln r) + \sum_{n=1}^{\infty} \big[& a_n c_n r^n \cos(n\theta) + a_n d_n r^{-n} \cos(n\theta) \\
& + b_n c_n r^n \sin(n\theta) + b_n d_n r^{-n} \sin(n\theta) \big],
\end{aligned} \tag{6.12}$$

we conclude that the solution to (6.9) is given by (6.12), (6.11). See Figure 6.4.

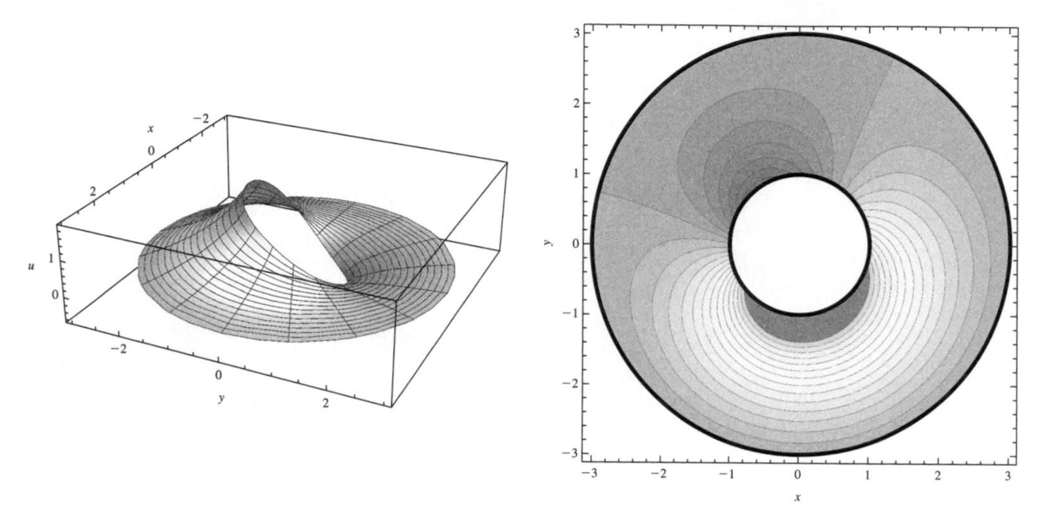

Figure 6.4: (Left) The solution surface using the first 4 terms of the Fourier series solution, where $\rho_1 = 1$, $\rho_2 = 3$, $f(\theta) = \frac{1}{10}(\theta - 1)(\theta^2 - \pi^2)$, $g(\theta) \equiv 0$. (Right) The corresponding contour plot.

Exercises

1. (a) Solve the boundary value problem on the wedge

$$\begin{aligned} u_{rr} + \frac{1}{r}u_r + \frac{1}{r^2}u_{\theta\theta} &= 0, & 0 < r < \rho,\ 0 < \theta < \theta_0, \\ u(r,0) &= 0, & 0 < r < \rho, \\ u(r,\theta_0) &= 0, & 0 < r < \rho, \\ u(\rho,\theta) &= f(\theta), & 0 < \theta < \theta_0. \end{aligned}$$

   (b) State the mathematical and physical boundary conditions for this problem.

   (c) State the orthogonality relation for the eigenfunctions used in (a) and justify your answer.

   ❄ (d) Suppose $\rho = 1$, $\theta_0 = \pi/3$, and $f(\theta) = 66\theta e^{-6\theta}$. Find (numerically) the smallest value of N such that the Nth partial sum of the series solution has L^2 error of less than 10^{-1}.

   ❄ (e) Plot the solution surface and polar contour plot for the N found in (d).

2. (a) Solve the boundary value problem on the exterior of a disk

$$\begin{aligned} u_{rr} + \frac{1}{r}u_r + \frac{1}{r^2}u_{\theta\theta} &= 0, & r > \rho > 0,\ -\pi < \theta < \pi, \\ u(\rho,\theta) &= f(\theta), & -\pi < \theta < \pi. \end{aligned}$$

   (b) State the mathematical and physical boundary conditions for this problem.

(c) Justify the orthogonality of the eigenfunctions used in (a).

✵ (d) Suppose $\rho = 2$ and $f(\theta) = 5|\theta|$. Find (numerically) the smallest value of N such that the Nth partial sum of the series solution has L^2 error of less than 10^{-1}.

✵ (e) Plot the solution surface and polar contour plot for the N found in (d).

3. (a) Solve the boundary value problem on the annulus

$$\begin{aligned} u_{rr} + \frac{1}{r}u_r + \frac{1}{r^2}u_{\theta\theta} &= 0, && 0 < \rho_1 < r < \rho_2,\ -\pi < \theta < \pi, \\ u(\rho_1, \theta) &= f(\theta), && -\pi < \theta < \pi, \\ u_r(\rho_2, \theta) &= 0, && -\pi < \theta < \pi. \end{aligned}$$

(b) State the mathematical and physical boundary conditions for this problem.

(c) Justify the orthogonality of the eigenfunctions used in (a).

✵ (d) Suppose $\rho_1 = 1$, $\rho_2 = 3$, and $f(\theta) = 0.1(\theta - 1)(\theta^2 - \pi^2)$. Find (numerically) the smallest value of N such that the Nth partial sum of the series solution has L^2 error of less than 10^{-1}.

✵ (e) Plot the solution surface and polar contour plot for the N found in (d). Compare with Figure 6.4.

(f) Give a physical interpretation for each line in the given problem.

4. (a) Solve the boundary value problem on the semicircular annulus

$$\begin{aligned} u_{rr} + \frac{1}{r}u_r + \frac{1}{r^2}u_{\theta\theta} &= 0, && 0 < \rho_1 < r < \rho_2,\ 0 < \theta < \pi, \\ u(\rho_1, \theta) &= f(\theta), && 0 < \theta < \pi, \\ u(\rho_2, \theta) &= g(\theta), && 0 < \theta < \pi, \\ u(r, 0) &= 0, && 0 < \rho_1 < r < \rho_2, \\ u(r, \pi) &= 0, && 0 < \rho_1 < r < \rho_2. \end{aligned}$$

(b) State the mathematical and physical boundary conditions for this problem.

(c) Justify the orthogonality of the eigenfunctions used in (a).

✵ (d) Suppose $\rho_1 = 1$, $\rho_2 = 2$, $f(\theta) = 10\theta(\theta - 2\pi/3)(\theta - \pi)$, $g(\theta) = -15|\theta - \pi/2| + 15\pi/2$. Find (numerically) the smallest value of N such that the Nth partial sum of the series solution has L^2 error of less than 10^{-1}.

✵ (e) Plot the solution surface and polar contour plot for the N found in (d).

5. Using the plots generated in part (e) of Exercises 1–4, discuss the location of the absolute maximum and minimum values of the solution on the given domain.

6. (a) Solve for the electrostatic potential in the exterior of the upper half of the unit circle subject to the Neumann boundary condition

$$\begin{aligned} \frac{\partial u}{\partial r}(1, \theta) &= 11\theta(\theta - 2\pi/3)(\theta - \pi), && 0 < \theta < \pi, \\ u(r, 0) &= 0, && r > 1, \\ u(r, \pi) &= 0, && r > 1. \end{aligned}$$

(b) Plot the solution surface and the polar contour plots using the $N = 12$th partial sum.

7. Discuss the general form of regions described by polar coordinates for which the method of separation of variables is applicable.

8. Consider the boundary value problem for $u := u(r, \theta)$ given by

$$\begin{aligned} \Delta u &= 0, && 0 < r < \rho,\ -\pi < \theta < \pi, \\ u(\rho, \theta) &= \sin\theta, && -\pi < \theta < \pi. \end{aligned}$$

Without solving the problem, answer the following.

(a) State a physical interpretation for each line in the problem above in the context of heat conduction.

(b) On what portions of the boundary is heat flowing into or out of the disk?

(c) Is your answer in (b) consistent with a steady-state scenario? Explain.

(d) What is the net flux of heat energy across the boundary?

9. Find the steady-state temperature distribution in the plate shown below.

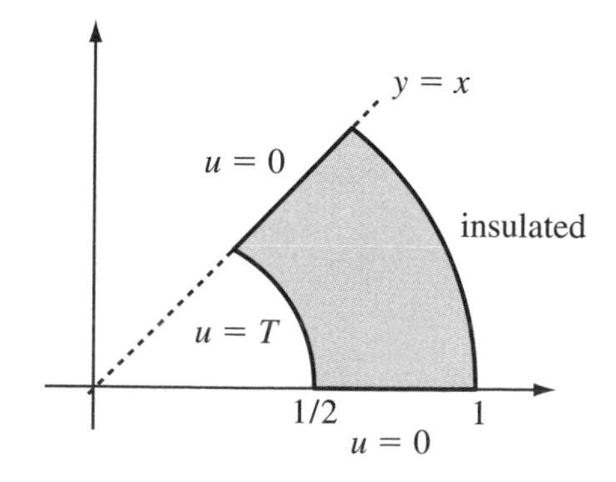

10. (a) Solve the boundary value problem:

$$\begin{aligned} u_{rr} + \frac{1}{r}u_r + \frac{1}{r^2}u_{\theta\theta} &= 0, && 0 < r < 2,\ -\pi/4 < \theta < \pi/4, \\ u(2, \theta) &= f(\theta), && -\pi/4 < \theta < \pi/4, \\ u_\theta(r, -\pi/4) &= 0, && 0 < r < 2, \\ u_\theta(r, \pi/4) &= 0, && 0 < r < 2. \end{aligned}$$

(b) State the mathematical and physical boundary conditions for this problem.

(c) Justify the orthogonality of the eigenfunctions used in (a).

❀ (d) Suppose $f(\theta) = |\theta|$. Find (numerically) the smallest value of N such that the Nth partial sum of the series solution has L^2 error of less than 0.1.

❀ (e) Plot the solution surface and polar contour plot for the N found in (d).

11. **(The Laplacian in Polar Coordinates)** In this exercise, we work through the steps to justify the conversion of the 2D Laplacian in rectangular coordinates to polar coordinates:

$$\begin{aligned}\Delta u(x,y) &= u_{xx}(x,y) + u_{yy}(x,y)\\ &= u_{rr}(r,\theta) + \frac{1}{r}u_r(r,\theta) + \frac{1}{r^2}u_{\theta\theta}(r,\theta)\\ &= \Delta u(r,\theta),\end{aligned}$$

which is mainly an exercise in the multivariable chain rule.[3]

(a) Using (6.2), show that

$$\begin{aligned}\frac{\partial}{\partial x}u(r,\theta) &= \frac{\partial}{\partial x}u(r(x,y),\theta(x,y)) = u_r r_x + u_\theta \theta_x = u_r\cos\theta - u_\theta r^{-1}\sin\theta,\\ \frac{\partial}{\partial y}u(r,\theta) &= \frac{\partial}{\partial y}u(r(x,y),\theta(x,y)) = u_r r_y + u_\theta \theta_y = u_r\sin\theta + u_\theta r^{-1}\cos\theta.\end{aligned}$$

(b) Differentiating again, show that

$$\begin{aligned}\frac{\partial^2}{\partial x^2}u(r,\theta) &= \frac{\partial}{\partial x}[u_r\cos\theta - u_\theta r^{-1}\sin\theta]\\ &= (u_{rr}\cos\theta - u_{r\theta}r^{-1}\sin\theta + u_\theta r^{-2}\sin\theta)\cos\theta\\ &\quad + (u_{r\theta}\cos\theta - u_r\sin\theta - u_{\theta\theta}r^{-1}\sin\theta - u_\theta r^{-1}\cos\theta)(-r^{-1}\sin\theta),\\ \frac{\partial^2}{\partial y^2}u(r,\theta) &= (u_{rr}\sin\theta + u_{r\theta}r^{-1}\cos\theta - u_\theta r^{-2}\cos\theta)\sin\theta\\ &\quad + (u_{r\theta}\sin\theta + u_r\cos\theta + u_{\theta\theta}r^{-1}\cos\theta - u_\theta r^{-1}\sin\theta)(r^{-1}\cos\theta).\end{aligned}$$

(c) Adding the two expressions in (b), show that

$$u_{xx}(x,y) + u_{yy}(x,y) = u_{rr}(r,\theta) + \frac{1}{r}u_r(r,\theta) + \frac{1}{r^2}u_{\theta\theta}(r,\theta).$$

[3]Recall that the multivariable chain rule says $\frac{\partial}{\partial x}f(u(x,y),v(x,y)) = f_u u_x + f_v v_x$.

6.2 Poisson's Formula and Its Consequences

Consider the boundary value problem in a disk:

$$\Delta u = 0, \qquad 0 < r < \rho, \ -\pi < \theta < \pi, \tag{6.13a}$$

$$u(\rho, \theta) = f(\theta), \qquad -\pi < \theta < \pi. \tag{6.13b}$$

We already solved this problem and found

$$u(r, \theta) = \frac{1}{2}a_0 + \sum_{n=1}^{\infty} r^n [a_n \cos(n\theta) + b_n \sin(n\theta)], \tag{6.14}$$

where

$$\begin{aligned} a_0 &= \frac{1}{\pi} \int_{-\pi}^{\pi} f(\theta)\, d\theta, \\ a_n &= \frac{1}{\rho^n \pi} \int_{-\pi}^{\pi} f(\theta) \cos(n\theta)\, d\theta, \\ b_n &= \frac{1}{\rho^n \pi} \int_{-\pi}^{\pi} f(\theta) \sin(n\theta)\, d\theta, \quad n = 1, 2, \ldots. \end{aligned} \tag{6.15}$$

Amazingly, the series (6.14) with coefficients (6.15) can be summed explicitly (see Exercise 1) to obtain the following theorem due to Siméon Poisson.

Theorem 6.1 (Poisson's Integral Formula).
The unique solution of (6.13a), (6.13b) *can be expressed in the integral form*

$$u(r, \theta) = \frac{\rho^2 - r^2}{2\pi} \int_0^{2\pi} \frac{f(\varphi)}{\rho^2 - 2\rho r \cos(\theta - \varphi) + r^2}\, d\varphi.$$

Notice this integral formula for the solution of (6.13a), (6.13b) represents the solution solely in terms of its values on the boundary of the disk. Because of this, Poisson's Integral Formula enables us to prove several powerful theorems about solutions to (6.13) (which are called *harmonic functions*) rather easily.

Theorem 6.2 (Mean Value Property of Harmonic Functions).
Suppose u is a harmonic function in a disk D which is continuous on $D \cup \partial D$. Then the value of u at the center of D is the average of the values of u on ∂D.

Proof. By Poisson's Integral Formula,

$$u(0, \theta) = \frac{\rho^2}{2\pi} \int_0^{2\pi} \frac{f(\varphi)}{\rho^2}\, d\varphi = \frac{1}{2\pi} \int_0^{2\pi} f(\varphi)\, d\varphi,$$

which is the average value of f over the interval $[0, 2\pi]$, as claimed. □

Example 6.2.1. Consider the steady-state temperature distribution in a disk of radius $\rho = 1$, where the boundary condition is the 2π-periodic function

$$f(\theta) = \begin{cases} 1, & |\theta| < \pi/4, \\ 0, & \pi/4 \le |\theta| \le \pi. \end{cases}$$

Without solving the boundary value problem, the Mean Value Property for harmonic functions enables us to compute the temperature at the center of the disk using only the known temperature on the boundary:

$$u(0,0) = \frac{1}{2\pi}\int_{-\pi}^{\pi} f(\theta)\, d\theta = \frac{1}{2\pi}\int_{-\pi/4}^{\pi/4} 1\, d\theta = \frac{1}{4}.$$

See Figure 6.5. ◊

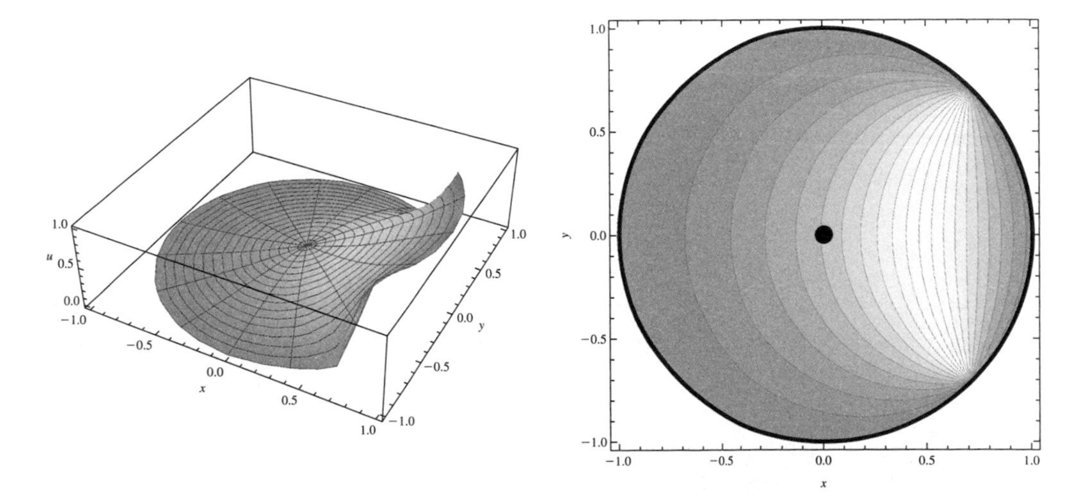

Figure 6.5: (Left) The solution surface using the first 100 terms of the Fourier series solution where $\rho = 1$ and $f(\theta) = 1$ for $-\pi/4 < \theta < \pi/4$ and 0 elsewhere. (Right) The corresponding contour plot. Using the Mean Value Property, the temperature at the center of the disk can be computed directly as $u(0,0) = 1/4$.

We can now use the Mean Value Property to prove a phenomenon that you may have noticed when plotting solution surfaces of Laplace's equation in Section 6.1: that the absolute extrema of a nonconstant harmonic function must occur on the *boundary* of the domain and *nowhere inside*.

Theorem 6.3 (Maximum Principle for Harmonic Functions).
Suppose u is a nonconstant harmonic function in a disk D which is continuous on $D \cup \partial D$. Then all maximum and minimum values of u are obtained on ∂D and nowhere inside D.

Proof. We can show this is true by appealing to the Mean Value Property above. Suppose u attains a maximum at some point called x_M (M for "maximum") inside D. Let's call this maximum value of u at x_M the number M. That is, $u(x) \leq u(x_M) = M$ for all $x \in D$.

Next, put a ball B_1 around x_M which lies entirely in D. By the Mean Value Property, $M = u(x_M) =$ average of u on $\partial B_1 \leq M$. This must be an inequality on the end because the maximum has to be larger than the average on ∂B_1 unless u is a constant. This implies $u(x) \equiv M$ for all $x \in \partial B_1$. This is true for any such circle, so now consider the ball B_2 centered at a point on ∂B_1. Then $u(x) \equiv M$ for all $x \in B_2$. Repeating this argument, we can fill up the domain D with such balls since D is a disk. In this way, we see $u(x) \equiv M$ thoughout D. This says u is a constant function. If u is nonconstant, then it must be that x_M lives in ∂D and not in D. □

Exercises

1. In this exercise, we will derive Poisson's Formula from (6.14), (6.15).

   (a) Since $f(\theta)$ must be a 2π-periodic function, convince yourself that the intervals of integration in (6.15) can be $0 < \theta < 2\pi$ instead of $-\pi < \theta < \pi$, so

$$\begin{aligned} a_0 &= \frac{1}{\pi}\int_0^{2\pi} f(\varphi)\,d\varphi, \\ a_n &= \frac{1}{\rho^n\pi}\int_0^{2\pi} f(\varphi)\cos(n\varphi)\,d\varphi, \\ b_n &= \frac{1}{\rho^n\pi}\int_0^{2\pi} f(\varphi)\sin(n\varphi)\,d\varphi, \end{aligned} \qquad (*)$$

   for $n = 1, 2, \ldots$. (In fact, *any* interval of length 2π would do.) We have replaced the θ with φ as the dummy variable of integration (it could be anything) so as not to confuse the independent variable θ in our solution with this dummy variable of integration.

   (b) Substitute $(*)$ into (6.14) to obtain

$$\begin{aligned} u(r,\theta) = &\frac{1}{2\pi}\int_0^{2\pi} f(\varphi)\,d\varphi \\ &+ \sum_{n=1}^{\infty}\frac{r^n}{\rho^n\pi}\int_0^{2\pi} f(\varphi)\left[\cos(n\varphi)\cos(n\theta) + \sin(n\varphi)\sin(n\theta)\right]d\varphi. \end{aligned}$$

(c) Use the trigonometric identity $\cos(x-y) = \cos x \cos y + \sin x \sin y$ to help in rewriting this last line as

$$u(r,\theta) = \frac{1}{2\pi}\int_0^{2\pi} f(\varphi)\left[1 + 2\sum_{n=1}^{\infty}\left(\frac{r}{\rho}\right)^n \cos(n(\theta-\varphi))\right] d\varphi.$$

(d) Next, we sum the bracketed series above. To accomplish this, recall that $\cos(x) = \frac{1}{2}(e^{ix}+e^{-ix})$, where $i=\sqrt{-1}$. Use this to show

$$1+2\sum_{n=1}^{\infty}\left(\frac{r}{\rho}\right)^n \cos(n(\theta-\varphi)) = 1+\sum_{n=1}^{\infty}\left(\frac{r}{\rho}\right)^n e^{in(\theta-\varphi)}+\sum_{n=1}^{\infty}\left(\frac{r}{\rho}\right)^n e^{-in(\theta-\varphi)}.$$

(e) Notice that each series above is a geometric series. Sum[4] each one to get

$$\begin{aligned} 1+2\sum_{n=1}^{\infty}\left(\frac{r}{\rho}\right)^n \cos(n(\theta-\varphi)) &= 1 + \frac{re^{i(\theta-\varphi)}}{\rho - re^{i(\theta-\varphi)}} + \frac{re^{-i(\theta-\varphi)}}{\rho - re^{-i(\theta-\varphi)}} \\ &= \frac{\rho^2 - r^2}{\rho^2 + r^2 - 2\rho r\cos(\theta-\varphi)}. \end{aligned}$$

(f) Combine the answers from (c) and (e) to obtain Poisson's Formula.

2. Without using a Fourier series, compute the temperature of the disk shown in Figure 6.5 at the points $r = 3/4$, $\theta = 0$ and $r = 1/2$, $\theta = 4\pi/3$.

3. Suppose a disk of radius $\rho = 1$ has its first quadrant boundary kept at $10°$, its second quadrant boundary kept at $20°$, its third quadrant boundary kept at $30°$, and its fourth quadrant boundary kept at $40°$. Find the exact temperature at the center of the disk in the steady-state.

4. Use the plots from Exercises 6.1.1, 6.1.3, and 6.1.4 to confirm the Maximum Principle holds on these domains as well.

5. Suppose u is a harmonic function in the disk D of radius 2 and satisfies $u = 3\sin(2\theta)+1$ on ∂D. *Without solving this boundary value problem*, answer the following.

   (a) Find the maximum value u obtains on $D \cup \partial D$.

   (b) Find the minimum value u obtains on $D \cup \partial D$.

   (c) Find the value of u at the center of D.

6. Use the Maximum Principle to prove that (6.13a), (6.13b) has at most one solution. (Hint: Suppose there are two distinct solutions, say v and w, with $v \neq w$. Let $z = v - w$. What BVP does z solve? What does the Maximum Principle imply about z?)

[4] The sum of this geometric series is $\sum_{n=1}^{\infty} z^n = \frac{z}{1-z}$, $|z|<1$.

6.3 The Wave Equation and Heat Equation in Polar Coordinates

Consider the wave equation in a disk for $u := u(r,\theta,t)$:

$$u_{tt} = c^2\left(u_{rr} + \frac{1}{r}u_r + \frac{1}{r^2}u_{\theta\theta}\right), \qquad 0 < r < \rho,\ -\pi < \theta < \pi,\ t > 0, \tag{6.16a}$$

$$u(\rho,\theta,t) = 0, \qquad -\pi < \theta < \pi,\ t > 0, \tag{6.16b}$$

$$u(r,\theta,0) = f(r,\theta), \qquad 0 < r < \rho,\ -\pi < \theta < \pi, \tag{6.16c}$$

$$u_t(r,\theta,0) = 0, \qquad 0 < r < \rho,\ -\pi < \theta < \pi. \tag{6.16d}$$

Let $u(r,\theta,t) = R(r)\Theta(\theta)T(t)$. Then the PDE yields

$$R\Theta T'' = c^2\left(R''\Theta T + \frac{1}{r}R'\Theta T + \frac{1}{r^2}R\Theta'' T\right).$$

Dividing by $c^2 R\Theta T$ and separating, we see

$$\frac{T''}{c^2 T} = \frac{R''}{R} + \frac{1}{r}\frac{R'}{R} + \frac{1}{r^2}\frac{\Theta''}{\Theta} = -\lambda,$$

which yields the following ODEs,

$$T'' + \lambda c^2 T = 0, \qquad r^2\frac{R''}{R} + r\frac{R'}{R} + \lambda r^2 = -\frac{\Theta''}{\Theta} = \mu,$$

for separation constants λ and μ. The relevant problems then become

$$T'' + \lambda c^2 T = 0, \tag{6.17}$$

$$\Theta'' + \mu\Theta = 0, \qquad \Theta(-\pi) = \Theta(\pi),\ \Theta'(-\pi) = \Theta'(\pi), \tag{6.18}$$

$$r^2 R'' + rR' + (\lambda r^2 - \mu)R = 0, \qquad R(\rho) = 0. \tag{6.19}$$

Equation (6.17) is easily solvable in terms of λ (which we will do later), while the eigenvalue problem (6.18) in μ is also familiar:

$$\mu_0 = 0, \qquad \Theta_0(\theta) = 1 \text{ (up to a constant multiple)}, \tag{6.20}$$

$$\mu_n = n^2, \qquad \Theta_n(\theta) = a_n\cos(n\theta) + b_n\sin(n\theta), \quad n = 1,2,\ldots. \tag{6.21}$$

However, (6.19) does not fall into any category of problem we have encountered to this point. However, we can transform it into a familiar ODE as follows.

Using (6.20) and (6.21) in (6.19), and letting $\lambda = \alpha^2$ we obtain

$$r^2 R'' + rR' + (\alpha^2 r^2 - n^2)R = 0, \quad R(\rho) = 0, \quad n = 0,1,2,\ldots. \tag{6.22}$$

This is Bessel's equation from Chapter 4 with $\lambda = \alpha^2$. Its Sturm-Liouville form is

$$\frac{1}{r}\left[(rR')' - \frac{n^2}{r}R\right] + \lambda R = 0, \quad 0 < r < \rho.$$

Since this equation is singular at $r = 0$, the appropriate modified boundary condition (see Section 4.2) at $r = 0$ is

$$R, R' \text{ bounded as } r \to 0^+. \tag{6.23}$$

To solve (6.22), (6.23), we make the change of variables $r := x/\alpha$ and $y(x) := R(x/\alpha)$ to obtain the standard form for Bessel's equation together with the appropriate modified boundary condition at the singular point $x = 0$:

$$x^2 y''(x) + xy'(x) + (x^2 - n^2)y(x) = 0,$$
$$y, y' \text{ bounded as } x \to 0^+.$$

Bessel's equation was solved in full detail in Section 4.4:

$$y(x) = c_1 J_n(x) + c_2 Y_n(x),$$

where J_n is the Bessel function of the first kind of order n and Y_n is the Bessel function of the second kind of order n. Since $Y_p(x) \to -\infty$ as $x \to 0^+$, the modified boundary condition forces $c_2 = 0$. Thus,

$$y(x) = J_n(x) \text{ (up to a constant multiple)}, \quad n = 0, 1, 2, \ldots.$$

Converting back to the equivalent expression in r, the solution to (6.22), (6.23) is

$$R_n(r) = J_n(\alpha r), \qquad n = 0, 1, 2, \ldots.$$

Next, we need to determine α so that $R(\rho) = J_n(\alpha\rho) = 0$ in (6.22) is satisfied; that is, $\alpha\rho$ must be a zero of $J_n(x)$. Let z_{nm} denote the mth zero of $J_n(x)$. Then $\alpha_{nm}\rho = z_{nm}$ so that $\alpha_{nm} = z_{nm}/\rho$. This yields the radial eigenvalues (in terms of λ) given by

$$\lambda_{nm} = \alpha_{nm}^2 = (z_{nm}/\rho)^2 > 0, \quad n = 0, 1, \ldots, \; m = 1, 2, \ldots,$$

with corresponding eigenfunctions

$$R_{nm}(r) = J_n(\sqrt{\lambda_{nm}}\, r), \quad n = 0, 1, \ldots, \; m = 1, 2, \ldots. \tag{6.24}$$

Furthermore, now that $\lambda = \lambda_{nm}$ are known, (6.17) is easily solved:

$$T_{nm}(t) = c_{nm}\cos(\sqrt{\lambda_{nm}}\, ct) + d_{nm}\sin(\sqrt{\lambda_{nm}}\, ct), \quad n = 0, 1, \ldots, \; m = 1, 2, \ldots. \tag{6.25}$$

Superimposing (6.20), (6.21), (6.24), and (6.25),

$$\begin{aligned} u(r,\theta,t) &= \sum_{m=1}^{\infty} R_{0m}(r)\Theta_0(\theta)T_{0m}(t) + \sum_{n=1}^{\infty}\sum_{m=1}^{\infty} R_{nm}(r)\Theta_n(\theta)T_{nm}(t) \qquad (6.26)\\ &= \sum_{m=1}^{\infty} J_0(\sqrt{\lambda_{0m}}\, r)[c_{0m}\cos(\sqrt{\lambda_{0m}}\, ct) + d_{0m}\sin(\sqrt{\lambda_{0m}}\, ct)] \\ &\quad + \sum_{n=1}^{\infty}\sum_{m=1}^{\infty} J_n(\sqrt{\lambda_{nm}}\, r)[a_{nm}\cos(n\theta) + b_{nm}\sin(n\theta)] \\ &\qquad \times [c_{nm}\cos(\sqrt{\lambda_{nm}}\, ct) + d_{nm}\sin(\sqrt{\lambda_{nm}}\, ct)]. \end{aligned}$$

Applying the initial condition (6.16c),

$$u(r,\theta,0) = \sum_{m=1}^{\infty} J_0(\sqrt{\lambda_{0m}}\, r) c_{0m}$$
$$+ \sum_{n=1}^{\infty}\sum_{m=1}^{\infty} J_n(\sqrt{\lambda_{nm}}\, r)[a_{nm}\cos(n\theta) + b_{nm}\sin(n\theta)]c_{nm} = f(r,\theta).$$

For each fixed r, this is a generalized Fourier series (referred to as a *Fourier-Bessel series*) in the θ variable. The singular Sturm-Liouville theory of Section 4.2 can be used to prove that the eigenfunctions form a complete orthogonal family in $L^2_w[0,\rho]$, where the weight function for Bessel's equation is $w(r) = r$. Therefore, the coefficients are determined by

$$\begin{aligned} c_{0m} &= \frac{\langle J_0(\sqrt{\lambda_{0m}}\, r), f(r,\theta)\rangle_w}{\langle J_0(\sqrt{\lambda_{0m}}\, r), J_0(\sqrt{\lambda_{0m}}\, r)\rangle_w} \\ &= \frac{\int_0^\rho \int_{-\pi}^{\pi} f(r,\theta) J_0(\sqrt{\lambda_{0m}}\, r) r\, d\theta\, dr}{\int_0^\rho \int_{-\pi}^{\pi} J_0^2(\sqrt{\lambda_{0m}}\, r) r\, d\theta dr}, \qquad m = 1,2,\ldots. \end{aligned}$$

Similarly,

$$\begin{aligned} c_{nm}a_{nm} &= \frac{\langle J_n(\sqrt{\lambda_{nm}}\, r)\cos(n\theta), f(r,\theta)\rangle_w}{\langle J_n(\sqrt{\lambda_{nm}}\, r)\cos(n\theta), J_n(\sqrt{\lambda_{nm}}\, r)\cos(n\theta)\rangle_w} \\ &= \frac{\int_0^\rho \int_{-\pi}^{\pi} f(r,\theta) J_n(\sqrt{\lambda_{nm}}\, r)\cos(n\theta) r\, d\theta\, dr}{\int_0^\rho \int_{-\pi}^{\pi} J_n^2(\sqrt{\lambda_{nm}}\, r)\cos^2(n\theta) r\, d\theta\, dr}, \qquad n,m = 1,2,\ldots, \qquad (6.27) \end{aligned}$$

$$\begin{aligned} c_{nm}b_{nm} &= \frac{\langle J_n(\sqrt{\lambda_{nm}}\, r)\sin(n\theta), f(r,\theta)\rangle_w}{\langle J_n(\sqrt{\lambda_{nm}}\, r)\sin(n\theta), J_n(\sqrt{\lambda_{nm}}\, r)\sin(n\theta)\rangle_w} \\ &= \frac{\int_0^\rho \int_{-\pi}^{\pi} f(r,\theta) J_n(\sqrt{\lambda_{nm}}\, r)\sin(n\theta) r\, d\theta\, dr}{\int_0^\rho \int_{-\pi}^{\pi} J_n^2(\sqrt{\lambda_{nm}}\, r)\sin^2(n\theta) r\, d\theta\, dr}, \qquad n,m = 1,2,\ldots. \qquad (6.28) \end{aligned}$$

To apply the initial condition (6.16d), first compute

$$u_t(r,\theta,t) = \sum_{m=1}^{\infty} J_0(\sqrt{\lambda_{0m}}\, r)[-c_{0m}\sqrt{\lambda_{0m}}\, c\sin(\sqrt{\lambda_{0m}}\, ct) + d_{0m}\sqrt{\lambda_{0m}}\, c\cos(\sqrt{\lambda_{0m}}\, ct)]$$
$$+ \sum_{n=1}^{\infty}\sum_{m=1}^{\infty} J_n(\sqrt{\lambda_{nm}}\, r)[a_{nm}\cos(n\theta) + b_{nm}\sin(n\theta)]$$
$$\times [-c_{nm}\sqrt{\lambda_{nm}}\, c\sin(\sqrt{\lambda_{nm}}\, ct) + d_{nm}\sqrt{\lambda_{nm}}\, c\cos(\sqrt{\lambda_{nm}}\, ct)],$$

so that

$$u_t(r,\theta,0) = \sum_{m=1}^{\infty} J_0(\sqrt{\lambda_{0m}}\,r) d_{0m}\sqrt{\lambda_{0m}}\,c + \sum_{n=1}^{\infty}\sum_{m=1}^{\infty} J_n(\sqrt{\lambda_{nm}}\,r)[a_{nm}\cos(n\theta) + b_{nm}\sin(n\theta)] d_{nm}\sqrt{\lambda_{nm}}\,c = 0.$$

Viewing this as an expansion of the zero function, the coefficients must be given by

$$d_{0m}\sqrt{\lambda_{0m}}\,c = 0, \qquad a_{nm}d_{nm}\sqrt{\lambda_{nm}}\,c = 0, \qquad b_{nm}d_{nm}\sqrt{\lambda_{nm}}\,c = 0,$$

which implies $d_{0m} = 0$, $d_{nm} = 0$, $m = 1, 2, \ldots$, $n = 1, 2, \ldots$. (Be sure you understand why.) Since these coefficients are zero, (6.27) and (6.28) can be solved for a_{nm} (in terms of c_{nm} times the appropriate inner product). When a_{nm} is substituted into (6.26), the c_{nm} cancel and we have the general solution to (6.16):

$$u(r,\theta,t) = \sum_{m=1}^{\infty} J_0(\sqrt{\lambda_{0m}}\,r) c_{0m}\cos(\sqrt{\lambda_{0m}}\,ct) + \sum_{n=1}^{\infty}\sum_{m=1}^{\infty} J_n(\sqrt{\lambda_{nm}}\,r)[a_{nm}\cos(n\theta) + b_{nm}\sin(n\theta)]\cos(\sqrt{\lambda_{nm}}\,ct),$$

where the coefficients are given by

$$c_{0m} = \frac{\langle J_0(\sqrt{\lambda_{0m}}\,r), f(r,\theta)\rangle_w}{\langle J_0(\sqrt{\lambda_{0m}}\,r), J_0(\sqrt{\lambda_{0m}}\,r)\rangle_w} = \frac{\int_0^\rho \int_{-\pi}^{\pi} f(r,\theta) J_0(\sqrt{\lambda_{0m}}\,r) r\, d\theta\, dr}{\int_0^\rho \int_{-\pi}^{\pi} J_0^2(\sqrt{\lambda_{0m}}\,r) r\, d\theta dr}, \qquad m = 1, 2, \ldots,$$

$$a_{nm} = \frac{\langle J_n(\sqrt{\lambda_{nm}}\,r)\cos(n\theta), f(r,\theta)\rangle_w}{\langle J_n(\sqrt{\lambda_{nm}}\,r)\cos(n\theta), J_n(\sqrt{\lambda_{nm}}\,r)\cos(n\theta)\rangle_w} = \frac{\int_0^\rho \int_{-\pi}^{\pi} f(r,\theta) J_n(\sqrt{\lambda_{nm}}\,r)\cos(n\theta) r\, d\theta\, dr}{\int_0^\rho \int_{-\pi}^{\pi} J_n^2(\sqrt{\lambda_{nm}}\,r)\cos^2(n\theta) r\, d\theta\, dr}, \qquad n, m = 1, 2, \ldots,$$

$$b_{nm} = \frac{\langle J_n(\sqrt{\lambda_{nm}}\,r)\sin(n\theta), f(r,\theta)\rangle_w}{\langle J_n(\sqrt{\lambda_{nm}}\,r)\sin(n\theta), J_n(\sqrt{\lambda_{nm}}\,r)\sin(n\theta)\rangle_w} = \frac{\int_0^\rho \int_{-\pi}^{\pi} f(r,\theta) J_n(\sqrt{\lambda_{nm}}\,r)\sin(n\theta) r\, d\theta\, dr}{\int_0^\rho \int_{-\pi}^{\pi} J_n^2(\sqrt{\lambda_{nm}}\,r)\sin^2(n\theta) r\, d\theta\, dr}, \qquad n, m = 1, 2, \ldots.$$

See Figure 6.6.

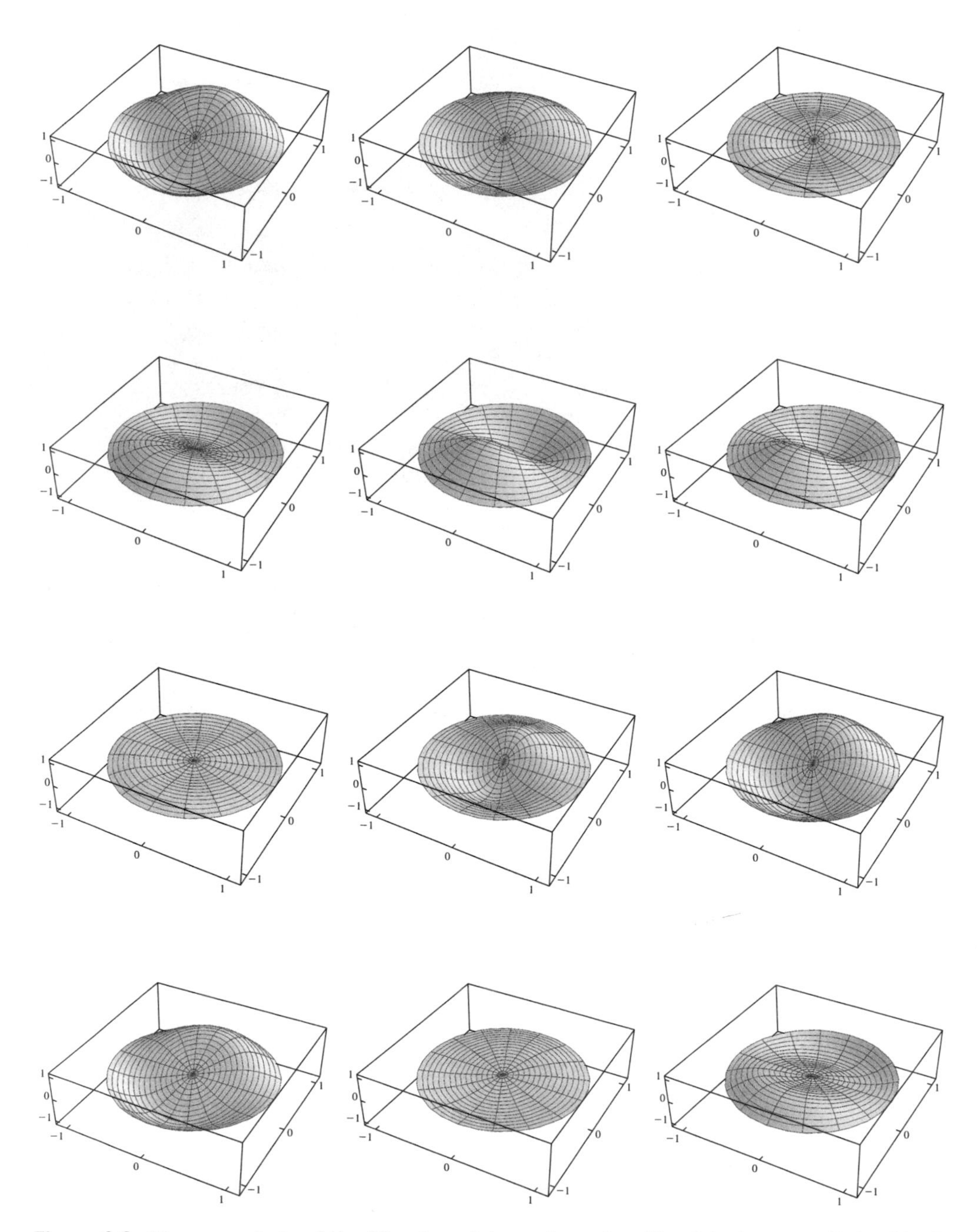

Figure 6.6: Time snapshots of the $N = 3$ partial sum (meaning $N = 3$ is the upper limit of both sums) of the solution to (6.16), where $\rho = c = 1$, $f(r, \theta) = 2.5(1 - r^2) r \sin\theta$.

Exercises

1. (a) Solve this initial-boundary value problem for the vibrations in a wedge:

$$\begin{aligned}
u_{tt} &= c^2(u_{rr} + \frac{1}{r}u_r + \frac{1}{r^2}u_{\theta\theta}), && 0 < r < \rho,\ 0 < \theta < \pi/2,\ t > 0,\\
u(\rho,\theta,t) &= 0, && 0 < \theta < \pi/2,\ t > 0,\\
u(r,0,t) &= 0, && 0 < r < \rho,\ t > 0,\\
u(r,\pi/2,t) &= 0, && 0 < r < \rho,\ t > 0,\\
u(r,\theta,0) &= f(r,\theta), && 0 < r < \rho,\ 0 < \theta < \pi/2,\\
u_t(r,\theta,0) &= 0, && 0 < r < \rho,\ 0 < \theta < \pi/2.
\end{aligned}$$

❃ (b) Take $c = \rho = 1$ and $f(r,\theta) = \theta(\theta - \pi/2)(r-1)$. Using the $N = 3$ partial sum (which means to sum over $0 \le n \le 3$ and $1 \le m \le 3$), plot the time snapshots of the solution for $0 \le t \le 3$ in increments of 0.25 to capture the relevant behavior.

(c) Give a physical interpretation of (a). Discuss (b) in light of this physical interpretation.

2. (a) Solve this initial-boundary value problem for diffusion in a disk:

$$\begin{aligned}
u_t &= k(u_{rr} + \frac{1}{r}u_r + \frac{1}{r^2}u_{\theta\theta}), && 0 < r < \rho,\ -\pi < \theta < \pi,\ t > 0,\\
u(\rho,\theta,t) &= 0, && -\pi < \theta < \pi,\ t > 0,\\
u(r,\theta,0) &= f(r,\theta), && 0 < r < \rho,\ -\pi < \theta < \pi.
\end{aligned}$$

❃ (b) Take $k = \rho = 1$ and $f(r,\theta) = J_0(5r)$. Using the $N = 3$ partial sum (which means to sum over $0 \le n \le 3$ and $1 \le m \le 3$), plot six judiciously spaced time snapshots of the solution which capture the relevant behavior.

(c) Give a physical interpretation of (a). Discuss (b) in light of this physical interpretation.

3. Solve the initial-boundary value problem for diffusion in a semicircle:

$$\begin{aligned}
u_t &= k(u_{rr} + \frac{1}{r}u_r + \frac{1}{r^2}u_{\theta\theta}), && 0 < r < \rho,\ 0 < \theta < \pi,\ t > 0,\\
u(\rho,\theta,t) &= 0, && 0 < \theta < \pi,\ t > 0,\\
u(r,0,t) &= 0, && 0 < r < \rho,\ t > 0,\\
u(r,\pi,t) &= 0, && 0 < r < \rho,\ t > 0,\\
u(r,\theta,0) &= f(r,\theta), && 0 < r < \rho,\ 0 < \theta < \pi.
\end{aligned}$$

6.4 Laplace's Equation in Cylindrical Coordinates

Laplace's equation in polar coordinates can be readily extended to cylindrical coordinates in three space dimensions. Recall that the relationship between three-dimensional rectangular coordinates (x, y, z) and cylindrical coordinates (r, θ, z) is given by the straightforward extension of polar coordinates illustrated in Figure 6.7:

$$\begin{aligned} x &= r\cos\theta, & & & r &= \sqrt{x^2+y^2}, \\ y &= r\sin\theta, & &\iff & \tan\theta &= y/x, \\ z &= z, & & & z &= z. \end{aligned}$$

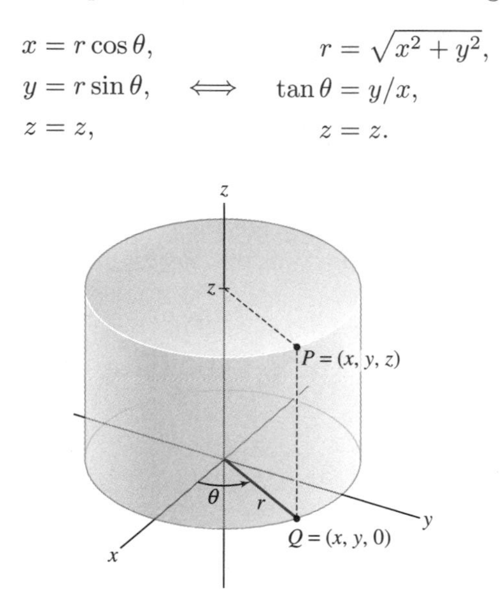

Figure 6.7: Cylindrical coordinates generalize polar coordinates to three space dimensions.

Analogous to Section 6.1, one can convert the 3D rectangular Laplacian to the *cylindrical Laplacian* to obtain

$$\begin{aligned} \Delta u(x,y,z) &= u_{xx}(x,y,z) + u_{yy}(x,y,z) + u_{zz}(x,y,z) \\ &= u_{rr}(r,\theta,z) + \frac{1}{r}u_r(r,\theta,z) + \frac{1}{r^2}u_{\theta\theta}(r,\theta,z) + u_{zz}(r,\theta,z) \\ &= \Delta u(r,\theta,z). \end{aligned}$$

Note that the cylindrical Laplacian has the same form as the polar Laplacian, but with an additional u_{zz} term.

Consider the boundary value problem in a cylinder for $u := u(r, \theta, z)$:

$$u_{rr} + \frac{1}{r}u_r + \frac{1}{r^2}u_{\theta\theta} + u_{zz} = 0, \qquad 0 < r < \rho,\ -\pi < \theta < \pi,\ 0 < z < \ell, \tag{6.29a}$$

$$u(r,\theta,\ell) = 0, \qquad 0 < r < \rho,\ -\pi < \theta < \pi, \tag{6.29b}$$

$$u(\rho,\theta,z) = 0, \qquad -\pi < \theta < \pi,\ 0 < z < \ell, \tag{6.29c}$$

$$u(r,\theta,0) = f(r,\theta), \qquad 0 < r < \rho,\ -\pi < \theta < \pi. \tag{6.29d}$$

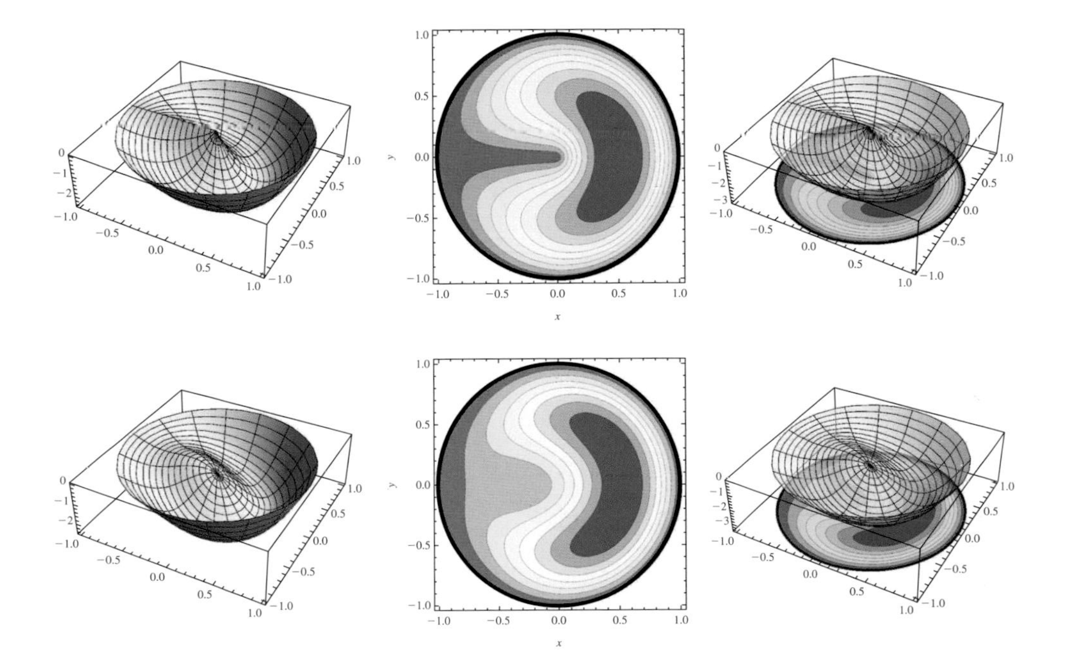

Figure 6.A: Steady-state temperature distribution along the bottom of a cylinder. See Figure 6.8.

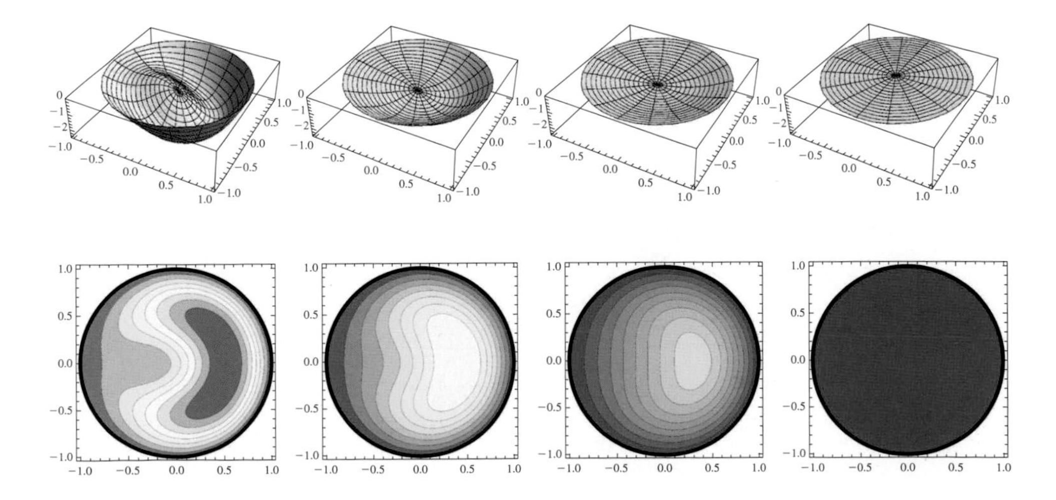

Figure 6.B: Contour plots of the steady-state temperature at various heights within a cylinder. Compare with Figure 6.10.

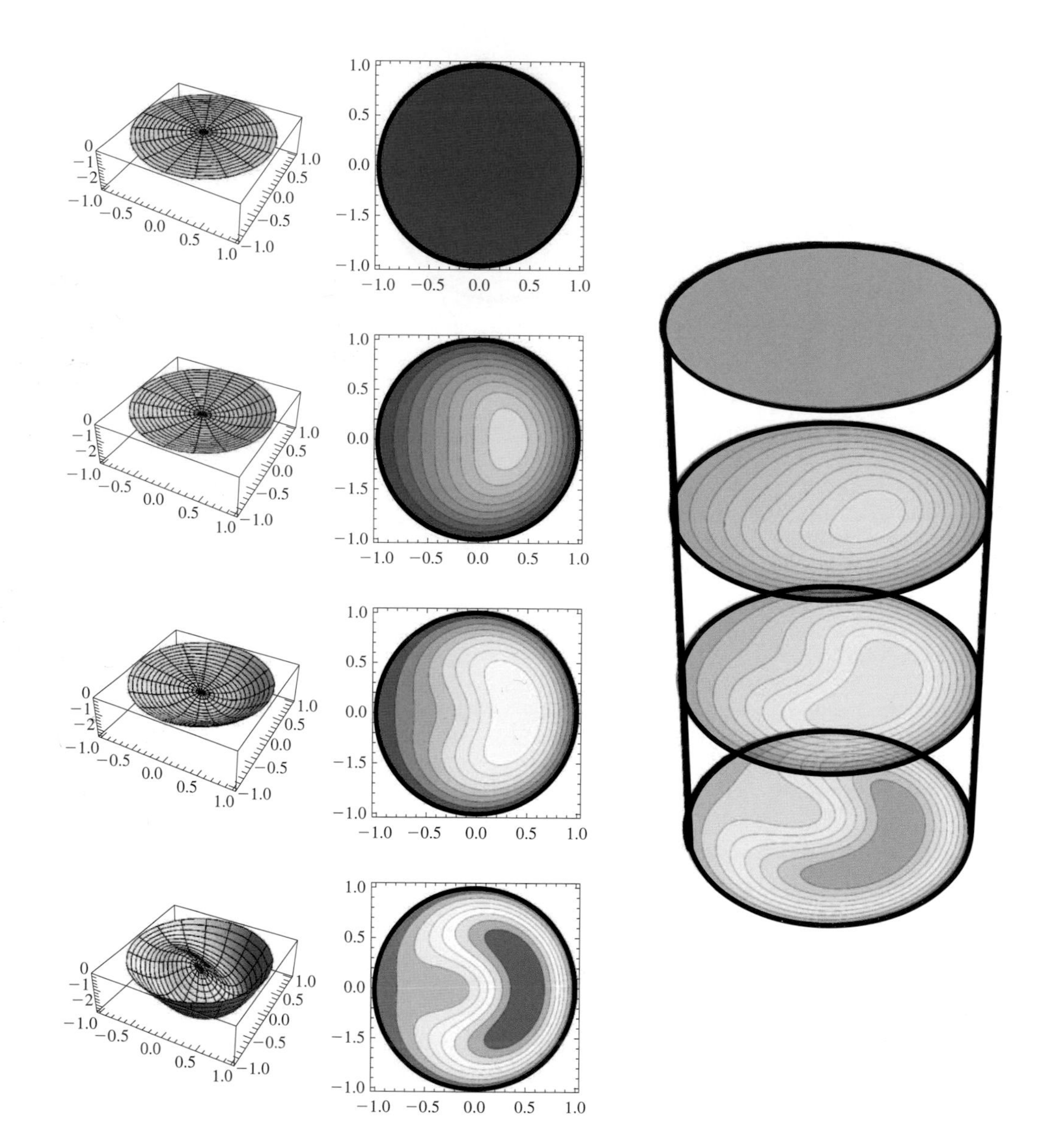

Figure 6.C: Stacking the contour plots from Figures 6.A and 6.B indicating the steady-state temperatures at various heights within the cylinder. Compare with Figure 6.9.

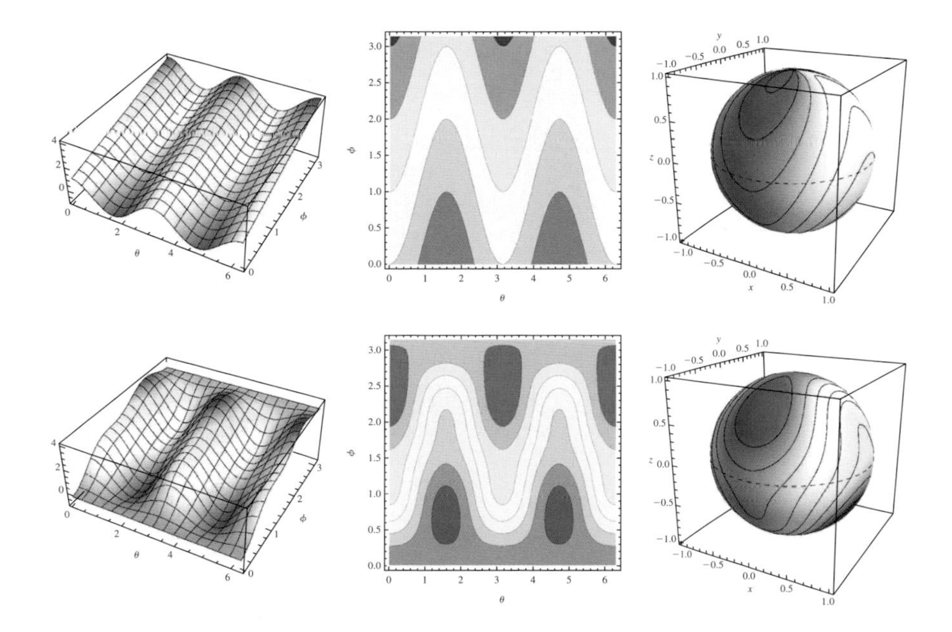

Figure 6.D: Steady-state temperature distribution on a sphere. See Figure 6.12.

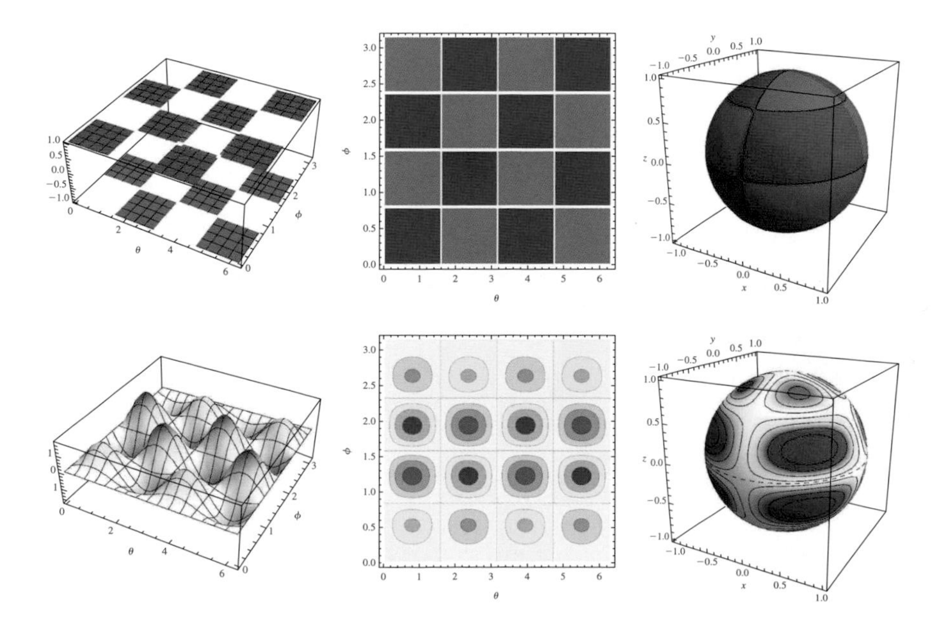

Figure 6.E: Steady-state temperature distribution on a sphere. See Figure 6.15.

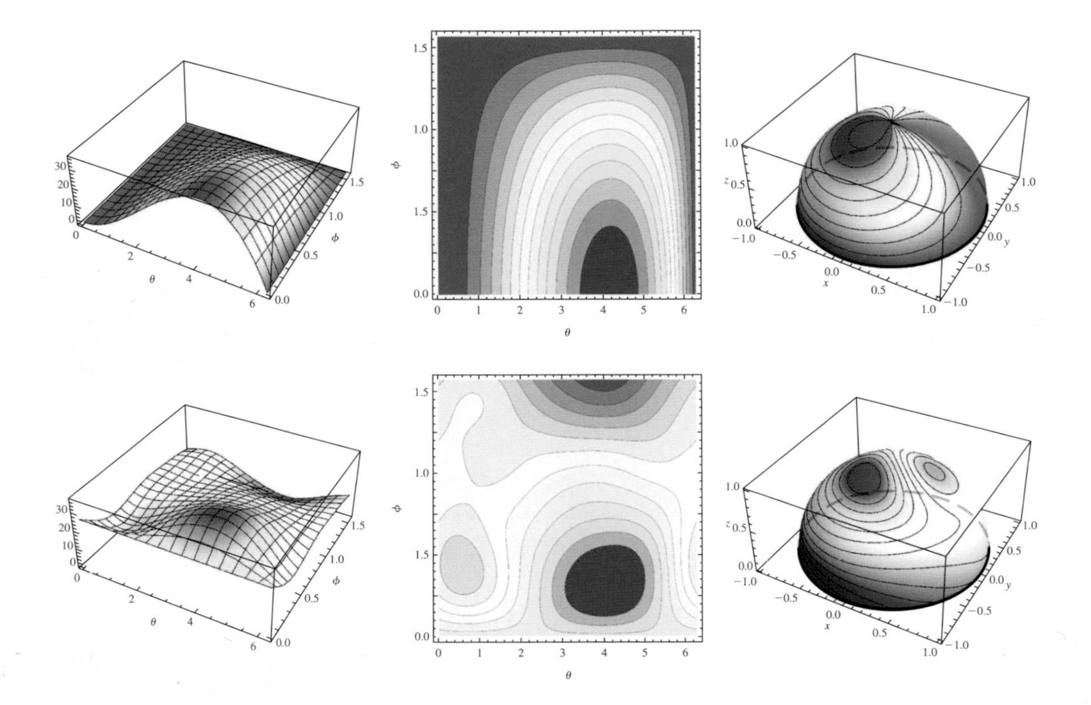

Figure 6.F: Steady-state temperature distribution on a hemisphere. See Figure 6.17.

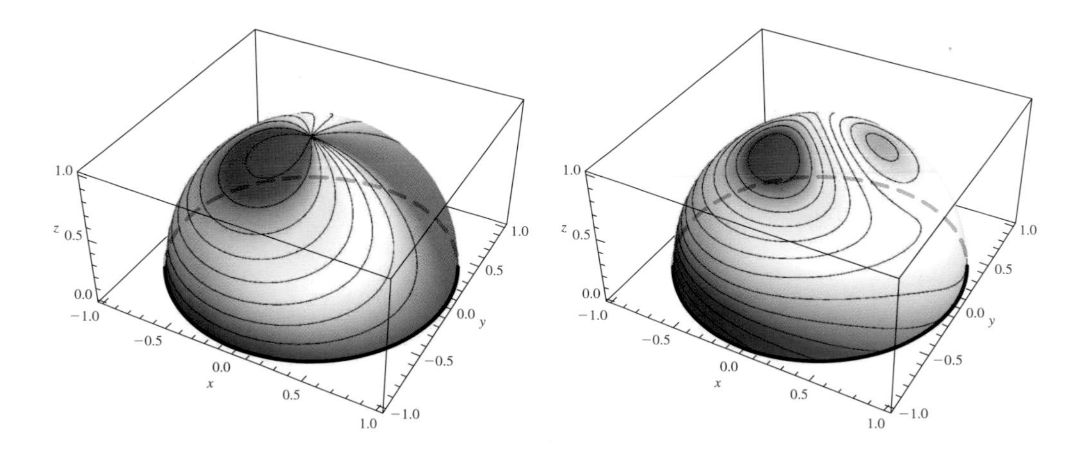

Figure 6.G: A closer look at Figure 6.F.

Let $u(r,\theta,z) = R(r)\Theta(\theta)Z(z)$. Then the PDE yields

$$R''\Theta Z + \frac{1}{r}R'\Theta Z + \frac{1}{r^2}R\Theta''Z + R\Theta Z'' = 0.$$

Dividing by $R\Theta Z$ and separating,

$$\frac{R''}{R} + \frac{1}{r}\frac{R'}{R} + \frac{1}{r^2}\frac{\Theta''}{\Theta} = -\frac{Z''}{Z} = -\lambda,$$

which yields the following ODEs

$$Z'' - \lambda Z = 0, \qquad r^2\frac{R''}{R} + r\frac{R'}{R} + \lambda r^2 = -\frac{\Theta''}{\Theta} = \mu,$$

for separation constants λ and μ. The relevant problems then become

$$Z'' - \lambda Z = 0, \quad Z(\ell) = 0, \tag{6.30}$$

$$\Theta'' + \mu\Theta = 0, \quad \Theta(-\pi) = \Theta(\pi),\ \Theta'(-\pi) = \Theta'(\pi), \tag{6.31}$$

$$r^2R'' + rR' + (\lambda r^2 - \mu)R = 0, \quad R(\rho) = 0. \tag{6.32}$$

The Θ problem (6.31) is a familiar periodic Sturm-Liouville problem:

$$\begin{aligned}
\mu_0 &= 0, & \Theta_0(\theta) &= 1 \text{ (up to a constant multiple)},\\
\mu_n &= n^2, & \Theta_n(\theta) &= A\cos(n\theta) + B\sin(n\theta), \quad n = 1,2,\ldots.
\end{aligned}$$

With these values of μ, the R problem (6.32) is a form of Bessel's equation from Section 4.3. Imposing the appropriate mathematical boundary condition at $x = 0$, we obtain the singular Sturm-Liouville problem

$$\begin{gathered}
r^2R'' + rR' + (\lambda r^2 - n^2)R = 0, \quad 0 < r < \rho,\\
R, R' \text{ bounded as } r \to 0^+, \quad R(\rho) = 0.
\end{gathered} \tag{6.33}$$

As shown in Section 4.3, the eigenvalues and eigenfunctions (up to a constant multiple) for (6.33) are

$$\lambda_{nm} = (z_{nm}/\rho)^2, \quad R_{nm}(r) = J_n(\sqrt{\lambda_{nm}}\,r), \quad n = 0,1,\ldots,\ m = 1,2,\ldots, \tag{6.34}$$

where z_{nm} denotes the mth positive zero of $J_n(x)$.

Turning our attention to the Z problem, we solve (6.30) using our knowledge of λ from (6.34):

$$Z_{nm}(z) = C\cosh(\sqrt{\lambda_{nm}}\,z) + D\sinh(\sqrt{\lambda_{nm}}\,z).$$

However, this can also be written as (see Exercise 4)

$$Z_{nm}(z) = E\sinh(\sqrt{\lambda_{nm}}\,(\ell - z)),$$

and in this form it is easy to see that $Z(\ell) = 0$.

Superimposing and subscripting the coefficients to illustrate the proper dependence on n and m, we obtain

$$\begin{aligned} u(r,\theta,z) &= \sum_{n=0}^{\infty}\sum_{m=1}^{\infty} R_{nm}(r)\Theta_n(\theta)Z_{nm}(z) \\ &= \sum_{n=0}^{\infty}\sum_{m=1}^{\infty} J_n(\sqrt{\lambda_{nm}}\, r)\left[a_{nm}\cos(n\theta)+b_{nm}\sin(n\theta)\right]\sinh(\sqrt{\lambda_{nm}}\,(\ell-z)). \end{aligned} \tag{6.35}$$

To determine the coefficients, we apply the boundary condition (6.29d),

$$\begin{aligned} u(r,\theta,0) &= \sum_{n=0}^{\infty}\sum_{m=1}^{\infty} J_n(\sqrt{\lambda_{nm}}\, r)\left[a_{nm}\cos(n\theta)+b_{nm}\sin(n\theta)\right]\sinh(\sqrt{\lambda_{nm}}\,\ell) \\ &= f(r,\theta), \qquad 0<r<\rho,\ -\pi<\theta<\pi. \end{aligned} \tag{6.36}$$

We can view this as a "two-dimensional" eigenfunction expansion:

1. For each fixed $0<r<\rho$, (6.36) is an eigenfunction expansion of $f(r,\theta)$ in terms of the complete orthogonal family of eigenfunctions, $\{\cos(n\theta),\sin(n\theta)\}_{n=0}^{\infty}$.
2. For each fixed $-\pi<\theta<\pi$, (6.36) is an eigenfunction expansion of $f(r,\theta)$ in the variable r in terms of the complete orthogonal family (with respect to the weight function $w=r$) of eigenfunctions $J_n(\sqrt{\lambda_{nm}}\, r)$.

Because of this, the coefficients are obtained as follows. First, treating the $n=0$ case separately,

$$\begin{aligned} a_{0m} &= \frac{1}{\sinh(\sqrt{\lambda_{0m}}\,\ell)}\cdot\frac{\langle J_0(\sqrt{\lambda_{0m}}\, r), f(r,\theta)\rangle_w}{\langle J_0(\sqrt{\lambda_{0m}}\, r), J_0(\sqrt{\lambda_{0m}}\, r)\rangle_w} \\ &= \frac{1}{\sinh(\sqrt{\lambda_{0m}}\,\ell)}\cdot\frac{\int_{-\pi}^{\pi}\int_0^{\rho} J_0(\sqrt{\lambda_{0m}}\, r) f(r,\theta) r\,dr\,d\theta}{\int_{-\pi}^{\pi}\int_0^{\rho}\left[J_0(\sqrt{\lambda_{0m}}\, r)\right]^2 r\,dr\,d\theta}, \\ b_{0m} &= 0, \quad m=1,2,\ldots, \end{aligned}$$

and then

$$\begin{aligned} a_{nm} &= \frac{1}{\sinh(\sqrt{\lambda_{nm}}\,\ell)}\cdot\frac{\langle J_n(\sqrt{\lambda_{nm}}\, r)\cos(n\theta), f(r,\theta)\rangle_w}{\langle J_n(\sqrt{\lambda_{nm}}\, r)\cos(n\theta), J_n(\sqrt{\lambda_{nm}}\, r)\cos(n\theta)\rangle_w} \\ &= \frac{1}{\sinh(\sqrt{\lambda_{nm}}\,\ell)}\cdot\frac{\int_{-\pi}^{\pi}\int_0^{\rho} J_n(\sqrt{\lambda_{nm}}\, r)\cos(n\theta) f(r,\theta) r\,dr\,d\theta}{\int_{-\pi}^{\pi}\int_0^{\rho}\left[J_n(\sqrt{\lambda_{nm}}\, r)\cos(n\theta)\right]^2 r\,dr\,d\theta}, \\ b_{nm} &= \frac{1}{\sinh(\sqrt{\lambda_{nm}}\,\ell)}\cdot\frac{\langle J_n(\sqrt{\lambda_{nm}}\, r)\sin(n\theta), f(r,\theta)\rangle_w}{\langle J_n(\sqrt{\lambda_{nm}}\, r)\sin(n\theta), J_n(\sqrt{\lambda_{nm}}\, r)\sin(n\theta)\rangle_w} \\ &= \frac{1}{\sinh(\sqrt{\lambda_{nm}}\,\ell)}\cdot\frac{\int_{-\pi}^{\pi}\int_0^{\rho} J_n(\sqrt{\lambda_{nm}}\, r)\sin(n\theta) f(r,\theta) r\,dr\,d\theta}{\int_{-\pi}^{\pi}\int_0^{\rho}\left[J_n(\sqrt{\lambda_{nm}}\, r)\sin(n\theta)\right]^2 r\,dr\,d\theta}, \qquad n,m=1,2,\ldots. \end{aligned}$$

By the Riesz-Fischer Theorem (Theorem 3.9), the series (6.36) converges in the L^2 sense provided, for each fixed $0 < r < \rho$, $f(r,\theta) \in L^2[-\pi,\pi]$ and for each fixed $-\pi < \theta < \pi$, $f(r,\theta) \in L^2_w[0,\rho]$. Said another way, the series converges in the L^2 sense provided

$$\int_{-\pi}^{\pi}\int_{0}^{\rho} f^2(r,\theta) r\,dr\,d\theta < \infty.$$

See Figures 6.8–6.10 for examples of the various ways that we can illustrate the solution to (6.29).

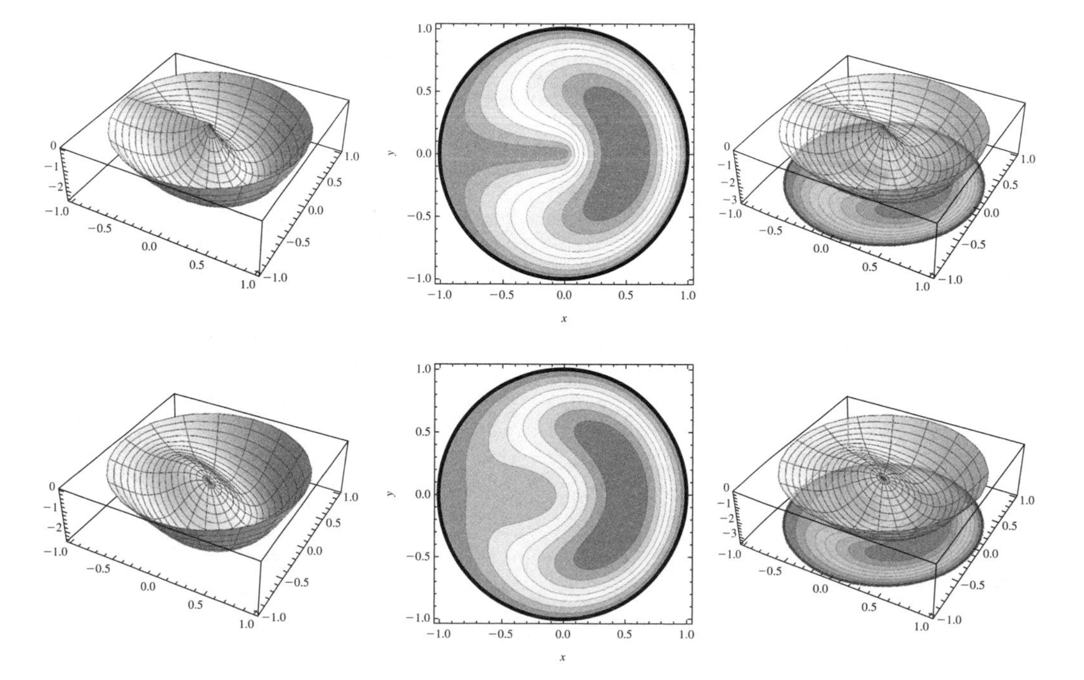

Figure 6.8: (Top) The surface $f(r,\theta) = (1-r)r(\theta^2 - \pi^2)$, $0 < r < 1$, $-\pi < \theta < \pi$ in cylindrical coordinates and its level curves. (Bottom) A cylindrical plot of the $N = 2$ partial sum (meaning $N = 2$ is the upper limit of both sums) of the series solution $u(r,\theta,0)$ in (6.36) and its level curves. Since f is specified when $z = 0$, these plots illustrate the steady-state temperature distribution across the bottom of the cylinder.

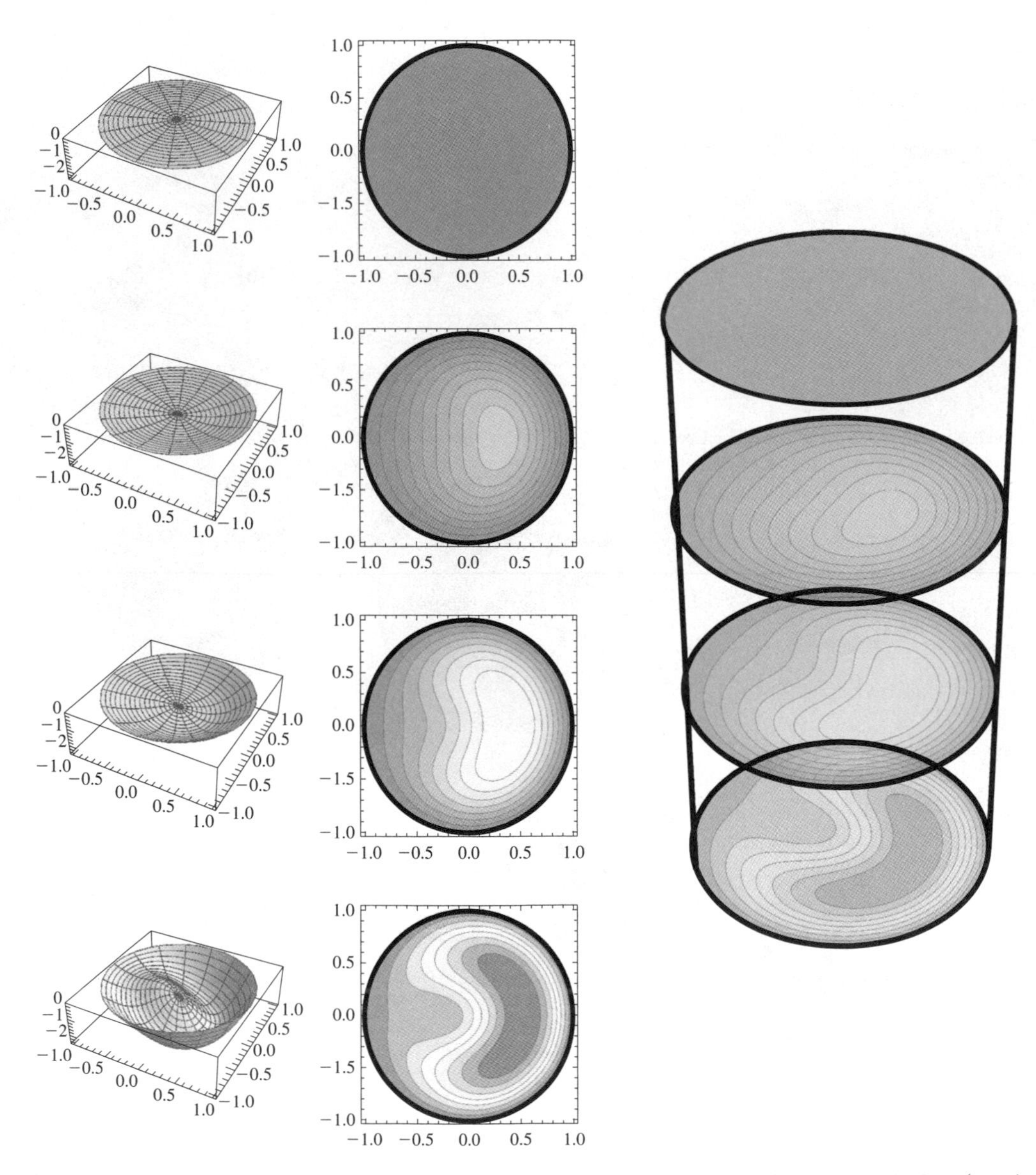

Figure 6.9: (Left) The surface plots of the $N = 2$ partial sum for the series solution (6.35) with $f(r, \theta)$ as in Figure 6.8 for slices of the cylinder at heights $z = 0, 0.25, 0.5, 1.$ (Center) The corresponding polar contour plots. (Right) The polar contour plots positioned at their respective z values inside the cylinder.

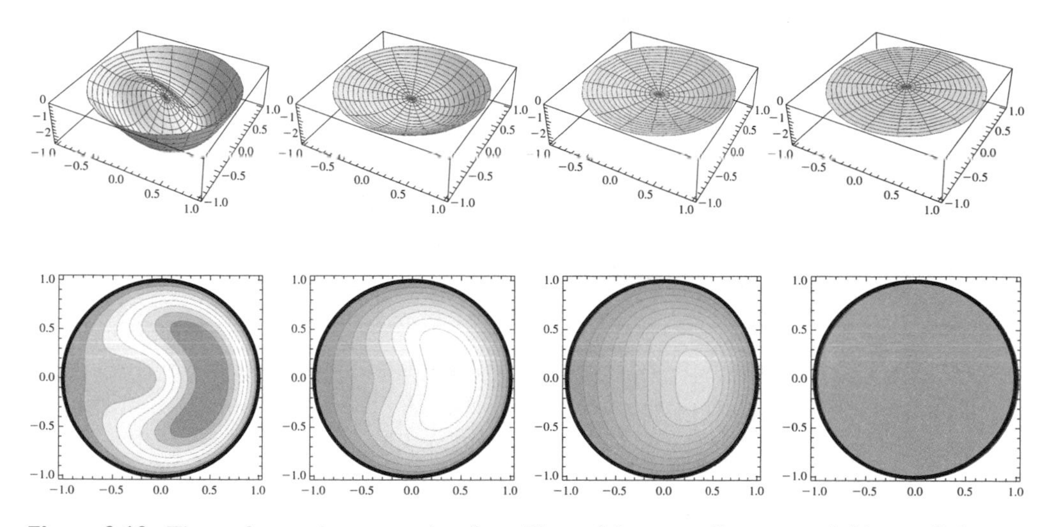

Figure 6.10: The surface and contour plots from Figure 6.9 are usually presented this way (leftmost is the bottom of the cylinder) rather than stacked inside a cylinder.

Exercises

1. (a) Plot the $N = 2$ partial sum (meaning $N = 2$ is the upper limit of both sums) of the solution to (6.29) with $\rho = \ell = 1$, $f(r, \theta) = r\theta^2$ for $z = 0, 0.5, 0.75, 1$.

   (b) Discuss (b) in the context of heat flow.

2. Consider the boundary value problem for $u := u(r, \theta, z)$ given by

$$\begin{aligned} u_{rr} + \frac{1}{r}u_r + \frac{1}{r^2}u_{\theta\theta} + u_{zz} &= 0, && 0 < r < \rho,\ -\pi < \theta < \pi,\ 0 < z < \ell, \\ u(r, \theta, \ell) &= 0, && 0 < r < \rho,\ -\pi < \theta < \pi, \\ u(\rho, \theta, z) &= 0, && -\pi < \theta < \pi,\ 0 < z < \ell, \\ u_z(r, \theta, 0) &= f(r, \theta), && 0 < r < \rho,\ -\pi < \theta < \pi. \end{aligned}$$

   (a) Interpret each part of the problem above in the context of heat flow.

   (b) Solve the boundary value problem by separation of variables.

   (c) Use orthogonality to deduce integral formulas for the coefficients in (b).

   (d) Let $\rho = \ell = 1$ and $f(r, \theta) = 1 - r$. Plot the $N = 3$ partial sum (meaning $N = 3$ is the upper limit of both sums) of the solution and the associated polar contour plots for $z = 0, 0.25, 0.5, 1$.

   (e) Discuss the plots in (d) relative to the answer from (a).

3. Consider the boundary value problem for $u := u(r,\theta,z)$ given by

$$\begin{aligned}
u_{rr} + \frac{1}{r}u_r + \frac{1}{r^2}u_{\theta\theta} + u_{zz} &= 0, && 0<r<\rho,\ 0<\theta<\pi,\ 0<z<\ell,\\
u(\rho,\theta,z) &= 0, && 0<\theta<\pi,\ 0<z<\ell,\\
u(r,0,z) &= 0, && 0<r<\rho,\ 0<z<\ell,\\
u(r,\pi,z) &= 0, && 0<r<\rho,\ 0<z<\ell,\\
u(r,\theta,0) &= 0, && 0<r<\rho,\ 0<\theta<\pi,\\
u(r,\theta,\ell) &= f(r,\theta), && 0<r<\rho,\ 0<\theta<\pi.
\end{aligned}$$

   (a) Interpret each part of the problem above in the context of heat flow.
   (b) Using separation of variables, show that the relevant problems become

$$\begin{aligned}
Z'' - \lambda Z = 0, &\quad Z(0) = 0,\\
\Theta'' + \mu\Theta = 0, &\quad \Theta(0) = \Theta(\pi) = 0,\\
r^2R'' + rR' + (\lambda r^2 - \mu)R = 0, &\quad R(\rho) = 0.
\end{aligned}$$

   (c) Solve the problems above and superimpose to get an infinite series solution.
   (d) Use orthogonality and completeness to deduce integral formulas for the coefficients in (c).
   ❀ (e) Let $\rho = \ell = 1$ and $f(r,\theta) = 4r(1-r)\theta(\pi-\theta)^2$. Plot the $N = 2$ partial sum (meaning $N = 2$ is the upper limit of both sums) of the solution and the associated polar contour plots for $z = 0, 0.5, 0.75, 1$.
   (f) Discuss the plots in (d) relative to the answer from (a).

4. Show that $Z_{nm}(z) = C\cosh(\sqrt{\lambda_{nm}}\,z) + D\sinh(\sqrt{\lambda_{nm}}\,z)$ can be rewritten as $Z_{nm}(z) = E\sinh(\sqrt{\lambda_{nm}}\,(\ell - z))$ when $Z_{nm}(\ell) = 0$ by using the hyperbolic trigonometric identity $\sinh(u-v) = \sinh u\cosh v - \cosh u\sinh v$.

5. Consider the boundary value problem for $u := u(r,\theta,z)$ given by

$$\begin{aligned}
u_{rr} + \frac{1}{r}u_r + \frac{1}{r^2}u_{\theta\theta} + u_{zz} &= 0, && 0<r<\rho,\ -\pi<\theta<\pi,\ 0<z<\ell,\\
u(\rho,\theta,z) &= 0, && -\pi<\theta<\pi,\ 0<z<\ell,\\
u(r,\theta,0) &= f(r,\theta), && 0<r<\rho,\ -\pi<\theta<\pi,\\
u(r,\theta,\ell) &= g(r,\theta), && 0<r<\rho,\ -\pi<\theta<\pi.
\end{aligned}$$

   (a) Interpret each part of the problem above in the context of heat flow.
   (b) Solve the boundary value problem by separation of variables.
   (c) Use orthogonality to deduce integral formulas for the coefficients in (b).
   ❀ (d) Let $\rho = \ell = 1$, $f(r,\theta) = (1-r)|\theta|$, and $g(r,\theta) \equiv 1$. Plot the $N = 2$ partial sum (meaning $N = 2$ is the upper limit of both sums) of the solution and the associated polar contour plots for $z = 0, 0.25, 0.5, 1$.
   (e) Discuss the plots in (d) relative to the answer from (a).

6.5 Laplace's Equation in Spherical Coordinates

Laplace's equation in polar coordinates can also be extended to spherical coordinates in three space dimensions. Recall that the relationship between three dimensional rectangular coordinates (x, y, z) and spherical coordinates (r, θ, φ) is given by

$$\begin{aligned} x &= r\cos\theta\sin\varphi, & & & r &= \sqrt{x^2+y^2+z^2}, \\ y &= r\sin\theta\sin\varphi, & &\Longleftrightarrow & \tan\theta &= y/x, \quad 0 \le \theta \le 2\pi, \\ z &= r\cos\varphi, & & & \cos\varphi &= z/\rho, \quad 0 \le \varphi \le \pi, \end{aligned} \tag{6.37}$$

as illustrated in Figure 6.11. Be careful when dealing with the description of any problem in spherical coordinates because there is no *one* standard way used to express spherical coordinates. Some sources reverse the roles of θ and φ above.

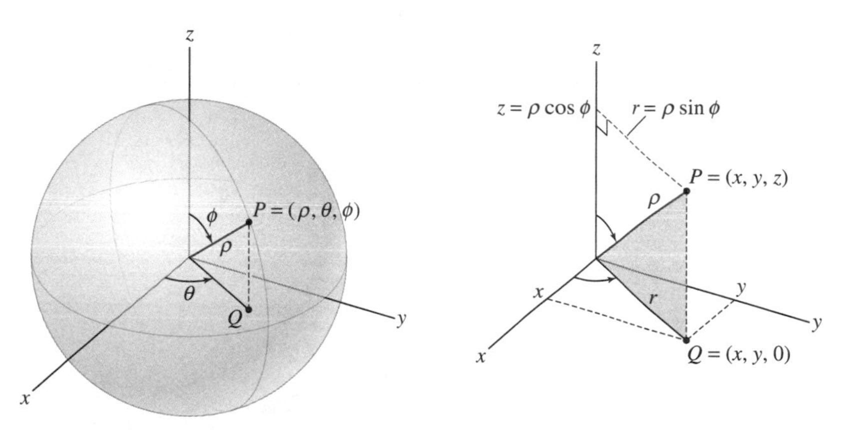

Figure 6.11: The form of spherical coordinates that we adopt takes θ as the *polar angle* of the projection of the point P onto the x-y plane and φ as the *angle of declination* which measures how much the ray through P declines from the vertical. With this convention, θ keeps the same role it played in polar and cylindrical coordinates.

Analogous to Section 6.1—and some tedious work involving the multivariable chain rule—one can convert the 3D rectangular Laplacian to the *spherical Laplacian* to obtain

$$\begin{aligned} \Delta u(x,y,z) &= u_{xx}(x,y,z) + u_{yy}(x,y,z) + u_{zz}(x,y,z) \\ &= u_{rr}(r,\theta,\varphi) + \frac{2}{r}u_r(r,\theta,\varphi) \\ &\qquad + \frac{1}{r^2}\left[u_{\varphi\varphi}(r,\theta,\varphi) + (\cot\varphi)u_\varphi(r,\theta,\varphi) + (\csc^2\varphi)u_{\theta\theta}(r,\theta,\varphi)\right] \\ &= \Delta u(r,\theta,\varphi). \end{aligned}$$

Although there is some similarity between this spherical Laplacian and the cylindrical and polar Laplacians, be careful to note the differences.

The Rotationally Symmetric Case

Before tackling Laplace's equation in the general (r, θ, φ) case, we will consider the simpler situation where u is symmetric with respect to rotations in the θ variable, i.e., u is independent of θ and therefore the Laplacian in this case reduces to

$$\Delta u(r,\varphi) = u_{rr} + \frac{2}{r}u_r + \frac{1}{r^2}\left[u_{\varphi\varphi} + (\cot\varphi)u_\varphi\right].$$

Consider the boundary value problem in a sphere for $u := u(r, \varphi)$:

$$u_{rr} + \frac{2}{r}u_r + \frac{1}{r^2}\left[u_{\varphi\varphi} + (\cot\varphi)u_\varphi\right] = 0, \qquad 0 < r < \rho,\ 0 < \varphi < \pi, \tag{6.38a}$$

$$u(\rho,\varphi) = f(\varphi), \quad 0 < \varphi < \pi. \tag{6.38b}$$

Let $u(r, \varphi) = R(r)\Phi(\varphi)$. Then the PDE yields

$$R''\Phi + \frac{2}{r}R'\Phi + \frac{1}{r^2}\left[R\Phi'' + \cot\varphi\, R\Phi'\right] = 0.$$

Dividing by $R\Phi$ and separating, we obtain

$$r^2\frac{R''}{R} + 2r\frac{R'}{R} = -\left[\frac{\Phi''}{\Phi} + \cot\varphi\,\frac{\Phi'}{\Phi}\right] = \lambda.$$

The relevant problems are then

$$r^2R'' + 2rR' - \lambda R = 0, \tag{6.39}$$

$$\Phi'' + \cot\varphi\,\Phi' + \lambda\Phi = 0. \tag{6.40}$$

We immediately recognize the R problem as a Cauchy-Euler equation, but the Φ problem is not familiar to us. However, the change of variables $x = \cos\varphi$ transforms $\Phi(\varphi)$, $0 < \varphi < \pi$, in (6.49) to the following equation (see Exercise 2 and take $\mu = 0$) in the new dependent variable we will call $y(x)$:

$$(1 - x^2)y''(x) - 2xy'(x) + \lambda y(x) = 0, \quad -1 < x < 1. \tag{6.41}$$

This is Legendre's equation from Section 4.3. The eigenvalues and associated eigenfunctions are

$$\lambda_n = n(n+1), \quad y_n(x) = P_n(x), \quad n = 0, 1, 2, \ldots,$$

where $P_n(x)$ denotes the *Legendre function of the first kind*, also known as the *Legendre polynomial of order* n. Translating this back to the φ variable, we see

$$\lambda_n = n(n+1), \quad \Phi_n(\varphi) = P_n(\cos\varphi), \quad n = 0, 1, 2 \ldots.$$

Now that we know the eigenvalues $\lambda = \lambda_n$, the R problem in (6.47) becomes

$$r^2R'' + 2rR' - n(n+1)R = 0, \quad n = 0, 1, 2, \ldots, \tag{6.42}$$

which is a Cauchy-Euler equation. Since $r = 0$ is a singular point, the required modified boundary condition at $r = 0$ is

$$R, R' \text{ bounded as } r \to 0^+. \tag{6.43}$$

Solving (6.42) subject to (6.43) yields the solutions

$$R_n(r) = c_n r^n, \quad n = 0, 1, 2, \ldots.$$

Superimposing, we get the solution

$$u(r, \varphi) = \sum_{n=0}^{\infty} R_n(r)\Phi_n(\varphi) = \sum_{n=0}^{\infty} c_n r^n P_n(\cos\varphi). \tag{6.44}$$

To determine the coefficients, we must rely on the orthogonality of the underlying eigenfunctions. Now, $\{P_n(x)\}_{n=0}^{\infty}$ are the eigenfunctions of Legendre's equation, which is rooted in a singular Sturm-Liouville problem. From Section 4.3, these eigenfunctions are complete and orthogonal on $-1 < x < 1$ with respect to the weight function $w(x) \equiv 1$ in the Sturm-Liouville form of Legendre's equation. That is,

$$\int_{-1}^{1} P_n(x)P_m(x)\,dx = 0, \quad n \neq m.$$

However, under the original change of variables $x = \cos\varphi$, this integral becomes

$$\int_0^{\pi} P_n(\cos\varphi)P_m(\cos\varphi)(-\sin\varphi)\,d\varphi = 0, \quad n \neq m.$$

Since the right-hand side is zero, we can restate the orthogonality relation above as

$$\int_0^{\pi} P_n(\cos\varphi)P_m(\cos\varphi)\sin\varphi\,d\varphi = 0, \quad n \neq m, \tag{6.45}$$

which says that $\{P_n(\cos\varphi)\}_{n=0}^{\infty}$ are orthogonal on $0 < \varphi < \pi$ with respect to the weight function $w(\varphi) = \sin\varphi$.

We can now find the coefficients in (6.44) by applying the boundary condition (6.38b),

$$u(\rho, \varphi) = \sum_{n=0}^{\infty} c_n \rho^n P_n(\cos\varphi) = f(\varphi), \quad 0 < \varphi < \pi.$$

But this is just a Legendre series from Section 4.3 for $f(\varphi)$ on $0 < \varphi < \pi$. By the orthogonality relation (6.45), the coefficients are given by

$$c_n = \frac{1}{\rho^n}\frac{\langle f, P_n(\cos\varphi)\rangle_w}{\langle P_n(\cos\varphi), P_n(\cos\varphi)\rangle_w} = \frac{1}{\rho^n}\frac{\int_0^{\pi} f(\varphi)P_n(\cos\varphi)\sin\varphi\,d\varphi}{\int_0^{\pi} P_n^2(\cos\varphi)\sin\varphi\,d\varphi}, \quad n = 0, 1, 2, \ldots.$$

By the Riesz-Fischer Theorem (Theorem 3.9), the series (6.44) converges in the L_w^2 sense provided $f(\varphi) \in L_w^2[0, \pi]$. That is, the series converges in the L_w^2 sense provided

$$\int_0^{\pi} f^2(\varphi)\sin\varphi\,d\varphi < \infty.$$

The General Case

Solving the general version of Laplace's equation—without rotational symmetry—is expectedly more complicated since we must account for r, θ, and φ in the separation of variables. However, our familiarity with the process in the rotationally symmetric situation above should make things clearer.

Consider the boundary value problem in a sphere for $u := u(r, \theta, \varphi)$:

$$\Delta u = 0, \qquad 0 < r < \rho,\ 0 < \theta < 2\pi,\ 0 < \varphi < \pi, \tag{6.46a}$$

$$u(\rho, \theta, \varphi) = f(\theta, \varphi), \quad 0 < \theta < 2\pi,\ 0 < \varphi < \pi, \tag{6.46b}$$

where the Laplacian in (6.46a) denotes the full spherical Laplacian,

$$\Delta u = u_{rr} + \frac{2}{r}u_r + \frac{1}{r^2}\left[u_{\varphi\varphi} + (\cot\varphi)u_\varphi + (\csc^2\varphi)u_{\theta\theta}\right].$$

Let $u(r,\theta,\varphi) = R(r)\Theta(\theta)\Phi(\varphi)$. Then the PDE yields

$$R''\Theta\Phi + \frac{2}{r}R'\Theta\Phi + \frac{1}{r^2}\left[R\Theta\Phi'' + \cot\varphi\, R\Theta\Phi' + \csc^2\varphi\, R\Theta''\Phi\right] = 0.$$

Dividing by $R\Theta\Phi$, and separating the radial part from the angular parts, we obtain

$$r^2\frac{R''}{R} + 2r\frac{R'}{R} = -\left[\frac{\Phi''}{\Phi} + \cot\varphi\,\frac{\Phi'}{\Phi} + \csc^2\varphi\,\frac{\Theta''}{\Theta}\right] = \lambda.$$

Multiplying the latter equation by $\sin^2\varphi$, simplifying, and separating Θ and Φ,

$$\sin^2\varphi\,\frac{\Phi''}{\Phi} + \cos\varphi\sin\varphi\frac{\Phi'}{\Phi} + \sin^2\varphi\,\lambda = -\frac{\Theta''}{\Theta} = \mu.$$

The relevant problems are then

$$r^2R'' + 2rR' - \lambda R = 0, \tag{6.47}$$

$$\Theta'' + \mu\Theta = 0, \quad \Theta(0) = \Theta(2\pi),\ \Theta'(0) = \Theta'(2\pi), \tag{6.48}$$

$$\Phi'' + \cot\varphi\,\Phi' + (\lambda - \mu\csc^2\varphi)\Phi = 0. \tag{6.49}$$

We immediately recognize the R problem as a Cauchy-Euler equation. Meanwhile, the Θ problem (6.31) is a familiar periodic Sturm-Liouville problem:

$$\mu_0 = 0, \qquad \Theta_0(\theta) = 1 \text{ (up to a constant multiple)}, \tag{6.50a}$$

$$\mu_m = m^2, \qquad \Theta_m(\theta) = A\cos(m\theta) + B\sin(m\theta), \quad m = 1, 2, \ldots. \tag{6.50b}$$

Even with these values of μ, the Φ problem is not familiar to us. However, the change of variables $x = \cos\varphi$ transforms (6.49) to the following equation (see Exercise 2) in a new dependent variable that we will call $y(x)$:

$$(1-x^2)y''(x) - 2xy'(x) + \left(\lambda - \frac{m^2}{1-x^2}\right)y = 0, \quad -1 < x < 1. \tag{6.51}$$

This is called the *associated Legendre's equation* because it becomes Legendre's equation for $m = 0$, but generalizes Legendre's equation to nonzero μ. The power series method reveals that the eigenvalues in (6.51) are $\lambda_n = n(n+1)$, $n = 0, 1, 2 \ldots$. The two linearly independent power series solutions[5] for these values of λ are

- $P_n^m(x)$, called the *associated Legendre's functions of the first kind*, and
- $Q_n^m(x)$, called the *associated Legendre's functions of the second kind.*

However, there is a more tractable representation of the associated Legendre functions in terms of the standard Legendre functions from Section 4.3, the details of which are discussed in Exercise 3:

$$P_n^m(x) := (-1)^m(1-x^2)^{m/2}\frac{d^m}{dx^m}P_n(x), \quad -1 < x < 1, \tag{6.52}$$

$$Q_n^m(x) := (-1)^m(1-x^2)^{m/2}\frac{d^m}{dx^m}Q_n(x), \quad -1 < x < 1. \tag{6.53}$$

The $(-1)^m$ factor just produces a constant multiple of the solutions found in Exercise 3 and is standard[6] when defining the associated Legendre functions. Several key properties of the associated Legendre functions are outlined in Exercise 4.

Based on the discussion above, we conclude that the general solution of (6.51) is

$$y(x) = c_1 P_n^m(x) + c_2 Q_n^m(x), \quad -1 < x < 1.$$

However, the Sturm-Liouville form of (6.51) (see Exercise 5) indicates that $x = \pm 1$ are singular points of (6.51), so the modified boundary conditions there are

$$y, y' \text{ remain bounded as } x \to -1^+ \text{ and } x \to 1^-. \tag{6.54}$$

From (6.52), (6.53) and Section 4.3, $P_n^m(x)$ is bounded on $-1 \le x \le 1$ and, therefore, satisfies (6.54), but $Q_n^m(x)$ becomes unbounded as $x \to \pm 1$, so we must take $c_2 = 0$. Thus, the eigenvalues and eigenfunctions for the associated Legendre's equation subject to (6.54) are

$$\begin{aligned} \lambda_n &= n(n+1), \quad n = 0, 1, 2, \ldots, \\ y_n(x) &= P_n^m(x), \qquad n = 0, 1, 2 \ldots, \quad m = 0, 1, \ldots, n, \end{aligned}$$

where the range on m is due to the fact that $P_n^m(x) \equiv 0$ for $m > n$. See Exercise 4.

Converting back to the φ variable, we obtain the eigenvalues and eigenfunctions of (6.49) subject to the modified boundary conditions at the singular points $\varphi = 0, \pi$:

$$\lambda_n = n(n+1), \qquad n = 0, 1, 2, \ldots, \tag{6.55a}$$

$$\Phi_{nm}(\varphi) = P_n^m(\cos\varphi), \quad n, = 0, 1, 2 \ldots, \quad m = 0, 1, \ldots, n. \tag{6.55b}$$

[5]In *Mathematica*, the commands are `LegendreP[n,m,x]` and `LegendreQ[n,m,x]`.

[6]The $(-1)^m$ factor is included in *Mathematica*'s definition for `LegendreP` and `LegendreQ`.

Returning to (6.47) with the values established in (6.55a), we have the same Cauchy-Euler problem as in the rotationally symmetric case. Therefore, the solutions (up to a constant multiple) are

$$R_n(r) = r^n, \quad n = 0, 1, 2, \ldots. \tag{6.56}$$

Superimposing (6.56), (6.55b), and (6.50), and subscripting the coefficients to illustrate the proper dependence on n and m, we obtain

$$\begin{aligned} u(r,\theta,\varphi) &= \sum_{n=0}^{\infty}\sum_{m=0}^{\infty} R_n(r)\Theta_{nm}(\theta)\Phi_{nm}(\varphi) \\ &= \sum_{n=0}^{\infty}\sum_{m=0}^{n} r^n \left[a_{nm}\cos(m\theta) + b_{nm}\sin(m\theta)\right] P_n^m(\cos\varphi), \end{aligned} \tag{6.57}$$

Applying the boundary condition (6.46b),

$$u(\rho,\theta,\varphi) = \sum_{n=0}^{\infty}\sum_{m=0}^{n} \rho^n \left[a_{nm}\cos(m\theta) + b_{nm}\sin(m\theta)\right] P_n^m(\cos\varphi) = f(\theta,\varphi),$$

$$0 < \theta < 2\pi, \; 0 < \varphi < \pi.$$

Now, appealing to the fact that the eigenfunctions form complete orthogonal families with respect to their weight functions ($w(\theta) \equiv 1$ for the Θ family and $w(\varphi) = \sin\varphi$ for the Φ family) on the appropriate intervals, we can compute the coefficients as we have before. This yields

$$\begin{aligned} a_{nm} &= \frac{1}{\rho^n}\frac{\langle f(\theta,\varphi), P_n^m(\cos\varphi)\cos(m\theta)\rangle_w}{\langle P_n^m(\cos\varphi)\cos(m\theta), P_n^m(\cos\varphi)\cos(m\theta)\rangle_w} \\ &= \frac{1}{\rho^n}\frac{\int_0^\pi\int_0^{2\pi} f(\theta,\varphi)P_n^m(\cos\varphi)\cos(m\theta)\sin\varphi\,d\theta\,d\varphi}{\int_0^\pi\int_0^{2\pi}\left[P_n^m(\cos\varphi)\cos(m\theta)\right]^2\sin\varphi\,d\theta\,d\varphi}, \qquad n, m = 0, 1, 2, \ldots, \\ b_{n0} &= 0, \qquad n = 0, 1, 2, \ldots, \\ b_{nm} &= \frac{1}{\rho^n}\frac{\langle f(\theta,\varphi), P_n^m(\cos\varphi)\sin(m\theta)\rangle_w}{\langle P_n^m(\cos\varphi)\cos(m\theta), P_n^m(\cos\varphi)\sin(m\theta)\rangle_w} \\ &= \frac{1}{\rho^n}\frac{\int_0^\pi\int_0^{2\pi} f(\theta,\varphi)P_n^m(\cos\varphi)\sin(m\theta)\sin\varphi\,d\theta\,d\varphi}{\int_0^\pi\int_0^{2\pi}\left[P_n^m(\cos\varphi)\sin(m\theta)\right]^2\sin\varphi\,d\theta\,d\varphi}, \qquad n = 0, 1, 2, \ldots, \; m = 1, 2, \ldots. \end{aligned}$$

We treated the $m = 0$ case separately since the $\sin(m\theta)$ term vanishes in that case.

By the Riesz-Fischer Theorem (Theorem 3.9), the series (6.36) converges in the L_w^2 sense provided, for each fixed $0 < \theta < 2\pi$, $f(\theta,\varphi) \in L_w^2[0,\pi]$ and for each fixed $0 < \varphi < \pi$, $f(\theta,\varphi) \in L^2[0,2\pi]$. That is, the series converges in the L_w^2 sense provided

$$\int_0^\pi\int_0^{2\pi} f^2(\theta,\varphi)\sin\varphi\,d\theta\,d\varphi < \infty.$$

See Figures 6.12, 6.13, 6.15, and 6.16 for plots of solutions to these types of problems.

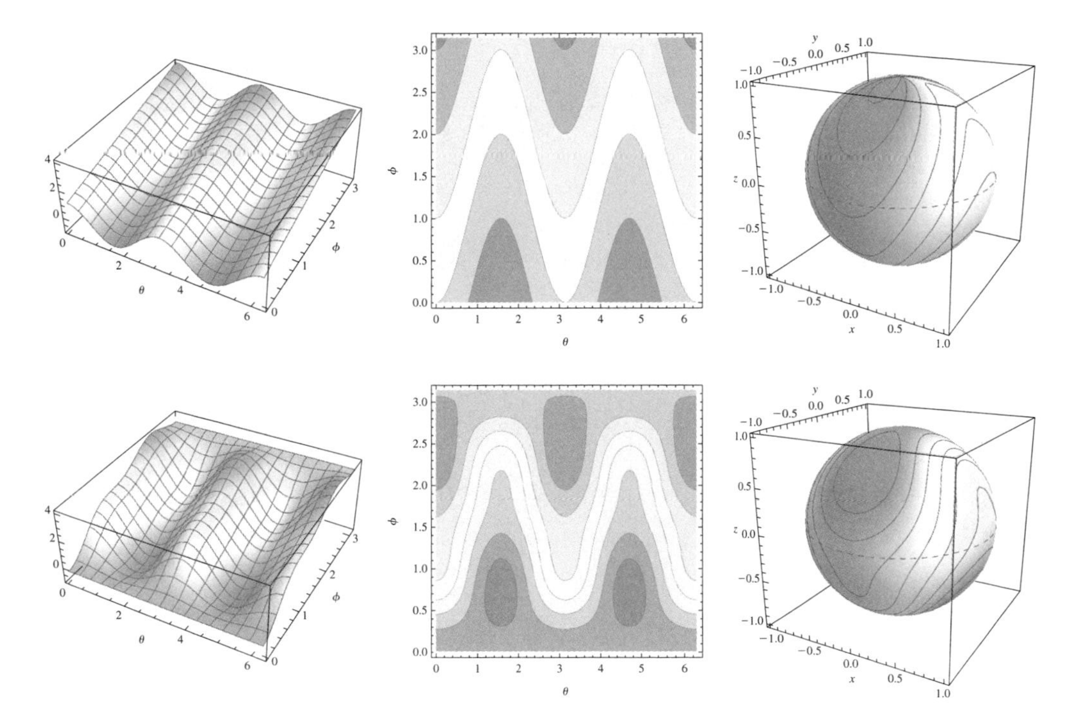

Figure 6.12: (Top) The surface $f(\theta, \varphi) = \cos(2\theta) + \varphi$, the corresponding contour plot, and the contour plot wrapped onto the sphere. (Bottom) The $N = 5$ partial sum of the solution to (6.46) evaluated at $r = 1$, for the given $f(\theta, \varphi)$ and $\rho = 1$.

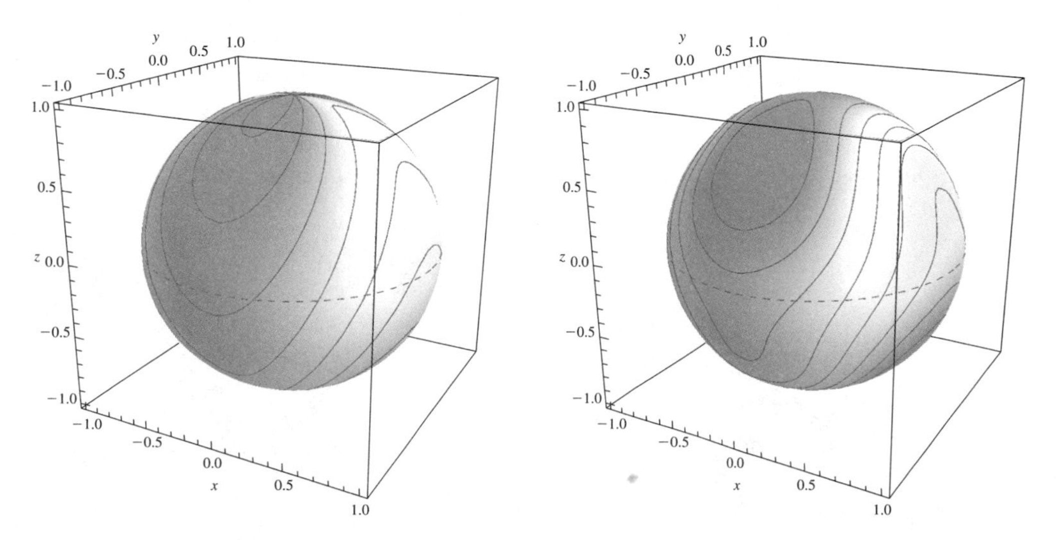

Figure 6.13: A closer look comparing $f(\theta, \varphi)$ and the $N = 5$ partial sum of (6.57) at $r = 1$.

Spherical Harmonics Formulation

Another formulation of the solution (6.57) is to write the θ component in its complex exponential form and then combine it with $P_n^m(\cos\varphi)$ to define the new family[7]

$$e^{im\theta}P_n^m(\cos\varphi), \quad n = 0, 1, 2, \ldots, \quad m = 0, \pm 1, \ldots, \pm n,$$

If we normalize this family with respect to the appropriate two dimensional weighted inner product used above, we obtain what are called the *spherical harmonic functions*, or *spherical harmonics*, given by[8]

$$Y_n^m(\theta, \varphi) := \frac{e^{im\theta}P_n^m(\cos\varphi)}{\|e^{im\theta}P_n^m(\cos\varphi)\|}. \tag{6.58}$$

See Figure 6.14. The norm in (6.58) is in terms of the complex, weighted inner product (see Exercise 3.5.15), i.e.,

$$\begin{aligned}\|e^{im\theta}P_n^m(\cos\varphi)\|^2 &= \langle e^{im\theta}P_n^m(\cos\varphi), e^{im\theta}P_n^m(\cos\varphi)\rangle_w \\ &= \int_0^\pi \int_0^{2\pi} e^{im\theta}P_n^m(\cos\varphi)\overline{e^{im\theta}P_n^m(\cos\varphi)}\sin\varphi\, d\theta\, d\varphi \\ &= \int_0^\pi \int_0^{2\pi} \left(e^{im\theta}P_n^m(\cos\varphi)\right)^2 \sin\varphi\, d\theta\, d\varphi.\end{aligned}$$

We can rewrite the solution (6.57) in terms of the spherical harmonics as

$$u(r, \theta, \varphi) = \sum_{n=0}^{\infty} \sum_{m=-n}^{n} c_{nm} r^n Y_n^m(\theta, \varphi). \tag{6.59}$$

We find the coefficients by applying the boundary condition (6.46b),

$$u(\rho, \theta, \varphi) = \sum_{n=0}^{\infty} \sum_{m=-n}^{n} c_{nm} \rho^n Y_n^m(\theta, \varphi) = f(\theta, \varphi), \quad 0 < \theta < 2\pi,\ 0 < \varphi < \pi.$$

This is an expansion of $f(\theta, \varphi)$ in terms of the complete ortho*normal* family $\{Y_n^m(\theta, \varphi)\}$ with respect to the complex, weighted inner product, i.e., the orthogonality relation is

$$\int_0^\pi \int_0^{2\pi} Y_n^m(\theta, \varphi)\overline{Y_p^q(\theta, \varphi)}\sin\varphi\, d\theta\, d\varphi = \begin{cases} 1, & n = p,\ m = q, \\ 0, & \text{otherwise.} \end{cases}$$

Therefore, the coefficients are[9]

$$\begin{aligned}c_{nm} &= \frac{1}{\rho^n}\langle f(\theta, \varphi), Y_n^m(\theta, \varphi)\rangle_w \\ &= \frac{1}{\rho^n}\int_0^\pi \int_0^{2\pi} f(\theta, \varphi)\overline{Y_n^m(\theta, \varphi)}\sin\varphi\, d\theta\, d\varphi, \quad n = 0, 1, 2, \ldots, \quad m = 0, \pm 1, \ldots, \pm n.\end{aligned}$$

---

[7] We define $P_n^{-m} = (-1)^m \frac{(n-m)!}{(n+m)!} P_n^m$.

[8] In *Mathematica*, the command to match the formulation of spherical coordinates (6.37) used here is `SphericalHarmonicY[n,m,φ,θ]`.

[9] The denominators are one because the family was normalized when defined.

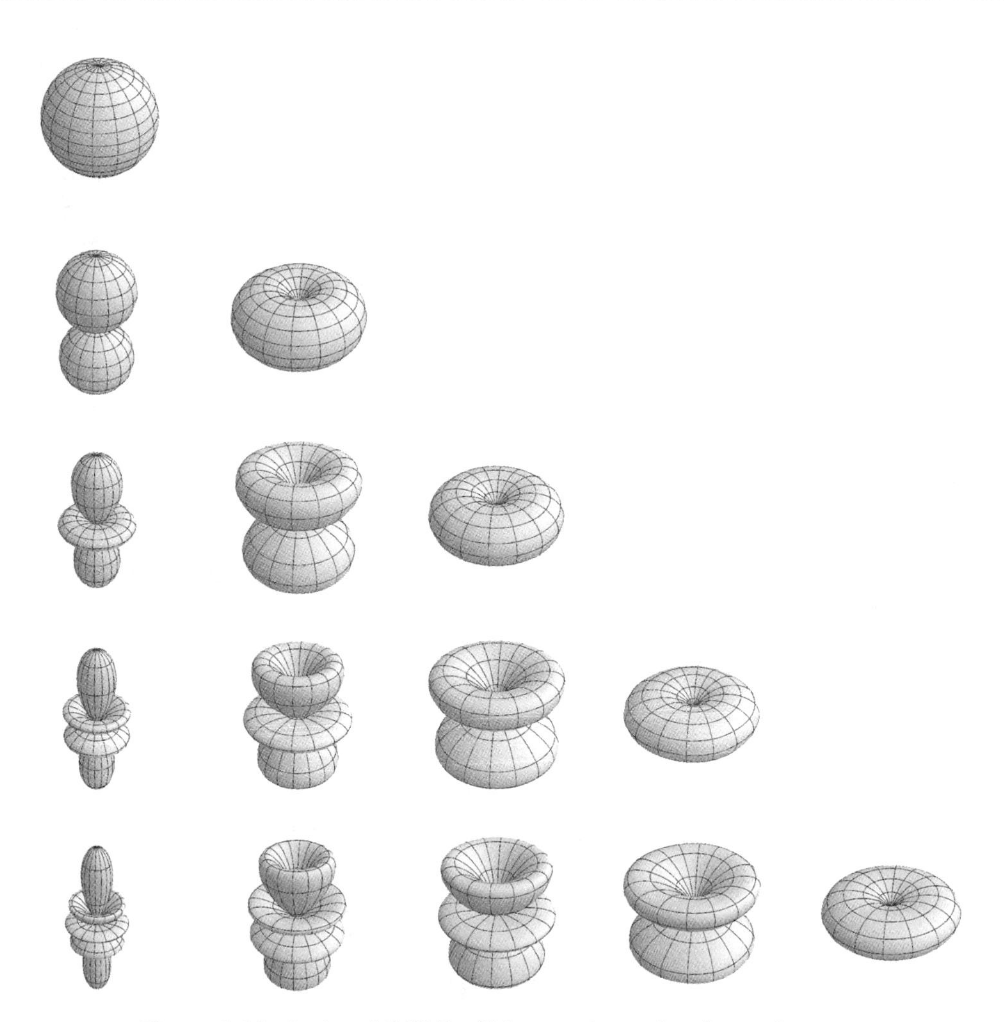

Figure 6.14: A plot of $|Y_n^m(\theta, \varphi)|$ for $n = 0, \ldots, 4$ and $m = 0, \ldots, n$.

By the Riesz-Fischer Theorem (Theorem 3.9), the series (6.59) converges in the L_w^2 sense provided, for each fixed $0 < \theta < 2\pi$, $f(\theta, \varphi) \in L_w^2[0, \pi]$ and for each fixed $0 < \varphi < \pi$, $f(\theta, \varphi) \in L^2[0, 2\pi]$. In other words, the series converges in the L_w^2 sense provided

$$\int_0^{\pi} \int_0^{2\pi} f^2(\theta, \varphi) \sin \varphi \, d\theta \, d\varphi < \infty.$$

See Figures 6.12, 6.13, 6.15, and 6.16.

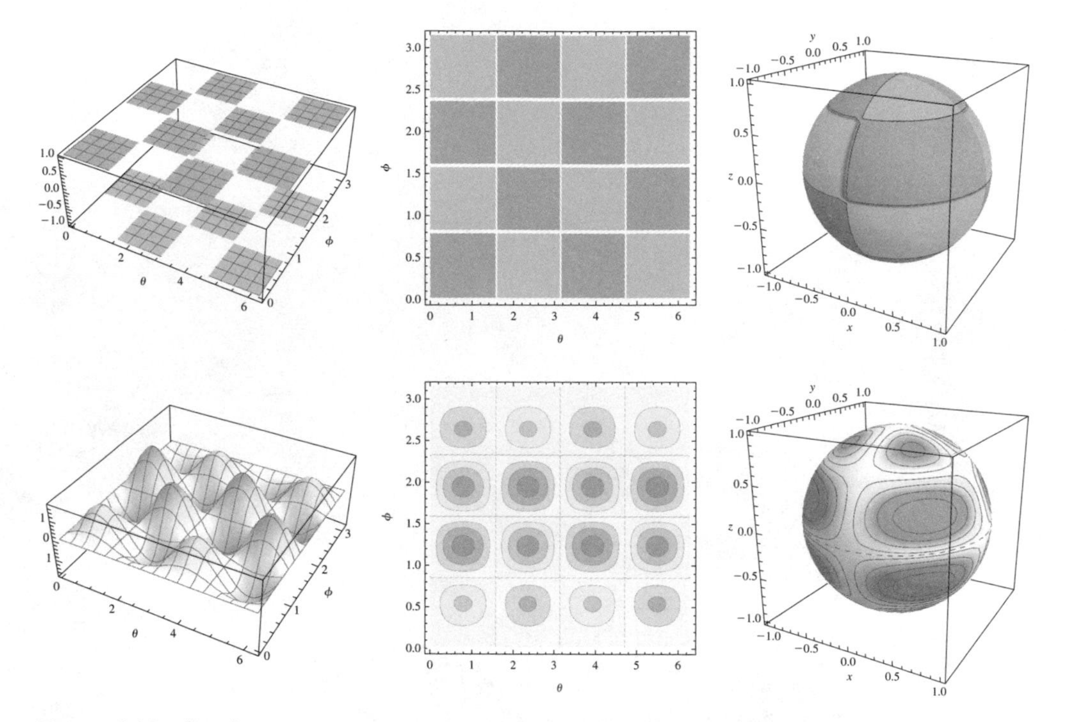

Figure 6.15: (Top) The surface for $f(\theta, \varphi) = \pm 1$ on the various rectangular patches, the corresponding contour plot, and the contour plot wrapped onto the sphere. (Bottom) The $N = 5$ partial sum of the solution to (6.46) evaluated at $r = 1$, for the given $f(\theta, \varphi)$ and $\rho = 1$.

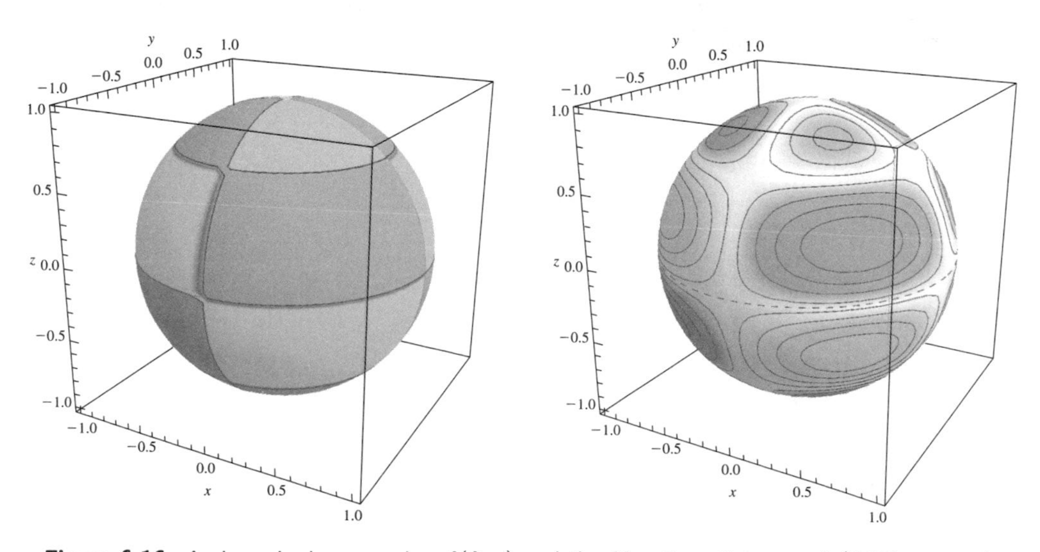

Figure 6.16: A closer look comparing $f(\theta, \varphi)$ and the $N = 5$ partial sum of (6.57) at $r = 1$.

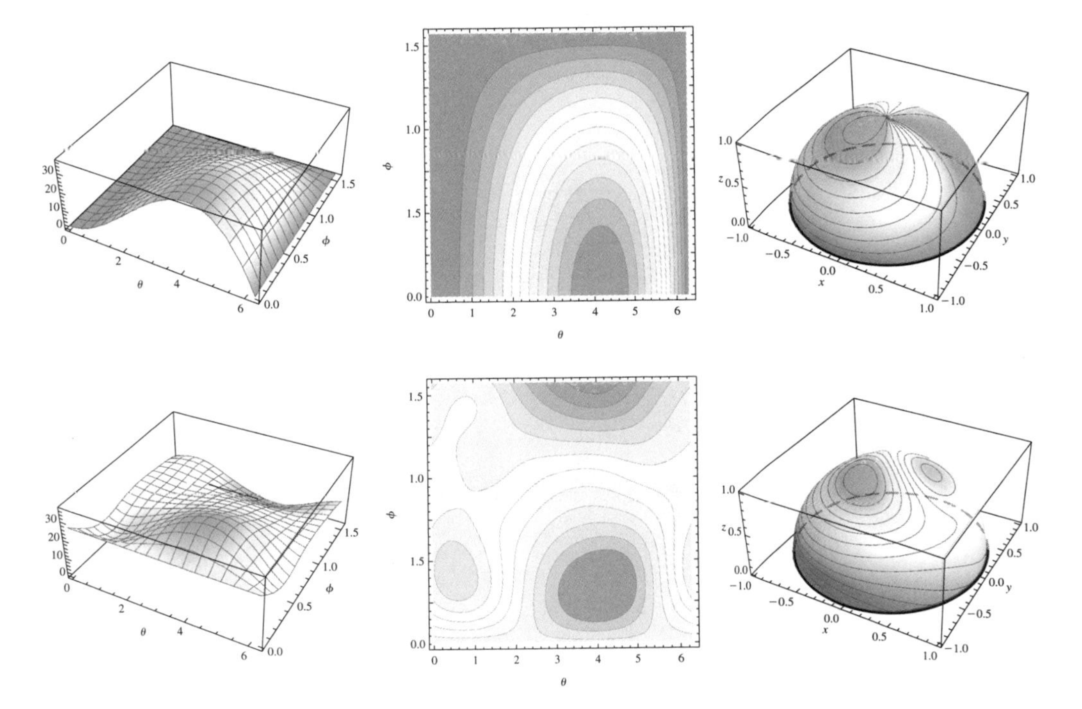

Figure 6.17: (Top) The surface $f(\theta,\varphi) = \theta^2(2\pi - \theta)\cos\varphi$, the corresponding contour plot, and the contour plot wrapped onto the hemisphere. (Bottom) The $N = 3$ partial sum of the solution in Exercise 11 evaluated at $r = 1$, for the given $f(\theta,\varphi)$ and $\rho = 1$.

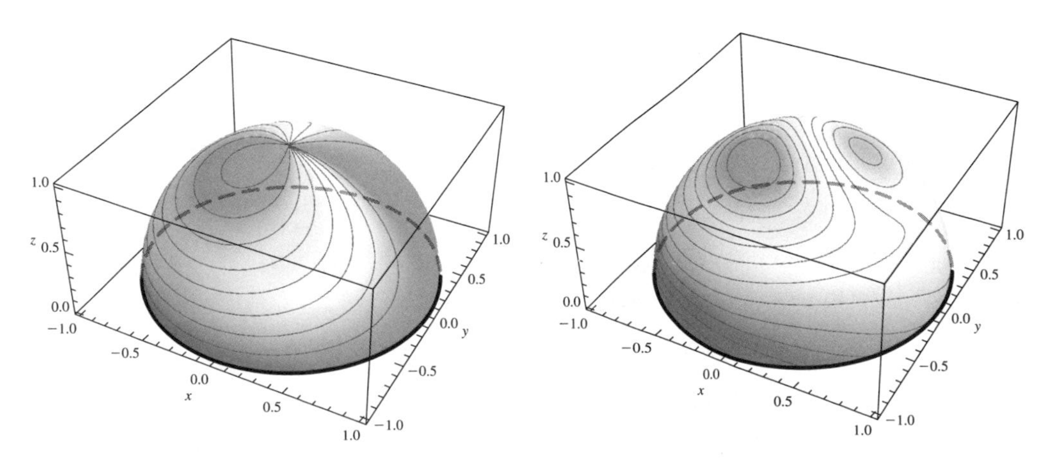

Figure 6.18: A comparison of $f(\theta,\varphi)$ and the $N = 3$ partial sum from Exercise 11 at $r = 1$.

Exercises

1. Write out and plot $P_n^m(x)$, $-1 < x < 1$, for $n = 0, 1, 2, 3$ and $m = 0, 1, \dots, n$.

2. Verify that the equation

$$\Phi'' + \cot\varphi\,\Phi' + (\lambda - \mu\csc^2\varphi)\Phi = 0, \quad 0 < \varphi < \pi,$$

is transformed by the change of variables $x = \cos\varphi$ to the equation

$$(1-x^2)y''(x) - 2xy'(x) + \left(\lambda - \frac{\mu}{1-x^2}\right)y = 0, \quad -1 < x < 1.$$

3. (a) Make the change of variables

$$y = (1-x^2)^{m/2}v(x), \quad -1 < x < 1,$$

so that (6.51) becomes

$$(1-x^2)v''(x) - 2(m+1)xv'(x) + (n-m)(n+m+1)v(x) = 0. \qquad (*)$$

(b) Use (but do not prove) the *Leibniz formula* for computing the nth derivative of a product,

$$\frac{d^n}{dx^n}[f(x)g(x)] = \sum_{k=0}^{n}\binom{n}{k}\frac{d^{n-k}}{dx^{n-k}}f(x)\frac{d^k}{dx^k}g(x), \quad \binom{n}{k} := \frac{n!}{k!(n-k)!},$$

to show that differentiating Legendre's equation (6.41) m times yields $(*)$.

(c) Conclude that $P_n^m(x)$ and $Q_n^m(x)$ given by (6.52), (6.53) are indeed solutions of (6.51).

4. Use (6.52), (6.53) to establish the following:

   (a) $P_n^m(x) \equiv 0$, $m > n$

   (b) $P_n^m(\pm 1) = 0$, $m \neq 0$

   (c) $P_n^m(x)$ is an even function if and only if $n + m$ is even.

   (d) $P_n^m(x)$ is an odd function if and only if $n + m$ is odd.

5. (a) Put the associated Legendre's equation in Sturm-Liouville form.

   (b) Discuss the nature of any singular points and state the appropriate modified boundary conditions at those points.

6. (a) Show that $P_{2n}(0) = \dfrac{(-1)^n(2n)!}{2^{2n}(n!)^2}$, $n = 0, 1, 2, \dots$.

   (b) Show that $P_{2n+1}(0) = 0$, $n = 0, 1, \dots$.

7. Consider the boundary value problem in the rotationally symmetric case for $u(r, \varphi)$ in the exterior of a sphere of radius ρ:

$$u_{rr} + \frac{2}{r}u_r + \frac{1}{r^2}\left[u_{\varphi\varphi} + (\cot\varphi)u_\varphi\right] = 0, \qquad 0 < \rho < r,\ 0 < \varphi < \pi,$$
$$u(\rho, \varphi) = f(\varphi), \quad 0 < \varphi < \pi.$$

   (a) Separate variables to find the relevant ODEs in r and φ.
   (b) State the mathematical and physical boundary conditions for this problem.
   (c) Solve the given boundary value problem, using orthogonality to deduce integral formulas for the coefficients.

8. Recreate the plots in Figure 6.14.

9. (a) Assuming rotational symmetry, find the steady-state temperature $u(r, \varphi)$ in a hemisphere of radius ρ subject to the boundary conditions

$$u(r, \pi/2) = 0, \qquad 0 < r < \rho,$$
$$u(\rho, \varphi) = f(\varphi), \quad 0 < \varphi < \pi/2.$$

   (Hint: Use Exercise 6 for the boundary condition on the base.)
   (b) Plot the $N = 3$ partial sum of the solution in (a) at $r = \rho$, where $f(\varphi) = -10\varphi(\varphi - \pi/2)$ and $\rho = 1$.
   (c) Using (b), generate contour plots for the temperature in the hemisphere at heights $z = 0, 0.2, 0.4, 0.6, 0.8$.
   (d) Compute the L^2_w error in (b).

10. (a) Assuming rotational symmetry, find the steady-state temperature $u(r, \varphi)$ in a hemisphere of radius ρ subject to the boundary conditions

$$u_\varphi(r, \pi/2) = 0, \qquad 0 < r < \rho,$$
$$u(\rho, \varphi) = f(\varphi), \quad 0 < \varphi < \pi/2.$$

   (Hint: Use Exercise 6 for the boundary condition on the base.)
   (b) Plot the $N = 3$ partial sum of the solution in (a) at $r = \rho$, where $f(\varphi) = -10\varphi(\varphi - \pi/2)$ and $\rho = 1$.
   (c) Using (b), generate contour plots for the temperature in the hemisphere at heights $z = 0, 0.2, 0.4, 0.6, 0.8$.
   (d) Compute the L^2_w error in (b).

11. (a) *Without* assuming rotational symmetry, find the steady-state temperature $u(r, \theta, \varphi)$ in a hemisphere of radius ρ subject to the boundary conditions

$$u(r, \theta, \pi/2) = 0, \qquad 0 < r < \rho,\ 0 < \theta < 2\pi,$$
$$u(\rho, \theta, \varphi) = f(\theta, \varphi), \quad 0 < \theta < 2\pi,\ 0 < \varphi < \pi/2.$$

   (Hint: Use Exercise 6 for the boundary condition on the base.)

❄ (b) Plot the $N = 3$ partial sum of the solution in (a) at $r = \rho$, where $f(\theta, \varphi) = \theta^2(2\pi - \theta)\cos\varphi$ and $\rho = 1$. See Figures 6.17 and 6.18.

❄ (c) Using (b), generate contour plots for the temperature in the hemisphere at heights $z = 0, 0.2, 0.4, 0.6, 0.8$.

❄ (d) Compute the L^2_w error in (b).

12. (a) *Without* assuming rotational symmetry, find the steady-state temperature $u(r, \theta, \varphi)$ in a sphere of radius $\rho = 1$ subject to the boundary condition

$$u(\rho, \theta, \varphi) = \sin(\theta + \varphi), \quad 0 < \theta < 2\pi,\ 0 < \varphi < \pi.$$

❄ (b) Plot the $N = 3$ partial sum of the solution in (a).

❄ (c) Using (b), generate contour plots for the temperature in the sphere at heights $z = -0.75, -0.5, -0.25, 0.25, 0.5, 0.75$.

❄ (d) Compute the L^2_w error in (b).

13. (a) *Without* assuming rotational symmetry, find the steady-state temperature $u(r, \theta, \varphi)$ in a sphere of radius $\rho = 1$ subject to the boundary condition

$$u(\rho, \theta, \varphi) = f(\theta, \varphi), \quad 0 < \theta < 2\pi,\ 0 < \varphi < \pi,$$

where

$$f(\theta, \varphi) = \begin{cases} 1, & 0 < \theta < \pi,\ 0 < \varphi < \pi/2, \\ 1, & \pi < \theta < 2\pi,\ \pi/2 < \varphi < \pi, \\ -1, & \pi < \theta < 2\pi,\ 0 < \varphi < \pi/2, \\ -1, & 0 < \theta < \pi,\ \pi/2 < \varphi < \pi. \end{cases}$$

❄ (b) Plot the $N = 3$ partial sum of the solution in (a).

❄ (c) Compute the L^2_w error in (b).

14. Explain why we take $m = 0, \pm 1, \ldots, \pm n$ in (6.59) rather than $m = 0, 1, \ldots, n$.

Chapter 7

PDEs on Unbounded Domains

7.1 The Infinite String: d'Alembert's Solution

Consider the vibrations of an infinitely long string modeled by the 1D wave equation on $-\infty < x < \infty$:

$$u_{tt} = c^2 u_{xx}, \qquad -\infty < x < \infty,\ t > 0, \tag{7.1a}$$

$$u(x,0) = f(x), \qquad -\infty < x < \infty, \tag{7.1b}$$

$$u_t(x,0) = g(x), \qquad -\infty < x < \infty. \tag{7.1c}$$

No boundary conditions are stated since the spatial domain has no physical boundary. Therefore, we will not tackle (7.1) using separation of variables but with a different technique due to d'Alembert (see Figure 7.1).

First, note that (7.1a) has solutions of the form $u(x,t) = F(x-ct)$ and $u(x,t) = G(x+ct)$ for any suitably smooth (single variable) functions F and G. To verify this, use the chain rule:

$$\begin{aligned} u(x,t) = F(x-ct) \implies u_t &= F'(x-ct) \cdot \frac{\partial}{\partial t}(x-ct) = F'(x-ct) \cdot (-c) \\ u_{tt} &= c^2 F''(x-ct) \\ u_x &= F'(x-ct) \cdot \frac{\partial}{\partial x}(x-ct) = F'(x-ct) \cdot 1 \\ c^2 u_{xx} &= c^2 F''(x-ct) \\ u_{tt} - c^2 u_{xx} &= c^2 F''(x-ct) - c^2 F''(x-ct) = 0. \end{aligned}$$

Thus, $u(x,t) = F(x-ct)$ is a solution of (7.1a). Similarly, $u(x,t) = G(x+ct)$ is a solution of (7.1a) (Exercise 1).

Second, we claim that *every* solution of (7.1a) can be written in the form

$$u(x,t) = \underbrace{F(x-ct)}_{\text{right traveling wave}} + \underbrace{G(x+ct)}_{\text{left traveling wave}}, \tag{7.2}$$

Figure 7.1: Jean le Rond d'Alembert (1717–1783) of France published the first article on the solution of the vibrating string problem in 1747. He was a controversial figure who formed rivalries with several of his contemporaries, most notably Clairaut and Euler.

that is, the general solution of (7.1a) is given by (7.2). To confirm this, we use a change of variables to transform the given PDE to one that can be solved simply by integration.

Motivated by (7.2), consider the change of variables

$$\xi := x - ct, \qquad \eta := x + ct. \tag{7.3}$$

Then $u(x,t) = U(\xi(x,t), \eta(x,t))$. Using the multivariable chain rule, we compute the partial derivatives

$$\begin{aligned}
u_t &= U_\xi \xi_t + U_\eta \eta_t = U_\xi \cdot (-c) + U_\eta \cdot c = -c[U_\xi - U_\eta] \\
u_{tt} &= -c[(U_{\xi\xi}\xi_t + U_{\xi\eta}\eta_t) - (U_{\eta\xi}\xi_t + U_{\eta\eta}\eta_t)] \\
&= -c[(U_{\xi\xi}(-c) + U_{\xi\eta} \cdot c) - (U_{\eta\xi}(-c) + U_{\eta\eta} \cdot c)] \\
&= c^2[U_{\xi\xi} - 2U_{\xi\eta} + U_{\eta\eta}] \\
u_x &= U_\xi \xi_x + U_\eta \eta_x = U_\xi + U_\eta \\
u_{xx} &= U_{\xi\xi}\xi_x + U_{\xi\eta}\eta_x + U_{\eta\xi}\xi_x + U_{\eta\eta}\eta_x \\
&= U_{\xi\xi} + 2U_{\xi\eta} + U_{\eta\eta}.
\end{aligned}$$

Therefore, $u_{tt} = c^2 u_{xx}$ becomes (after some rearrangement and simplification)

$$U_{\xi\eta} = 0.$$

This equation can be solved by directly integrating, first with respect to η and then with respect to ξ, to obtain

$$U(\xi, \eta) = F(\xi) + G(\eta),$$

where F and G are any twice-continuously differentiable functions of a single variable. Finally, we translate back from ξ-η coordinates to x-t coordinates using (7.3) to obtain the general solution of (7.1a), as claimed:

$$u(x,t) = F(x-ct) + G(x+ct), \quad -\infty < x < \infty,\ t > 0. \tag{7.4}$$

But how do the initial conditions (7.1b), (7.1c) specify the functions F and G in the general solution? First, (7.1b) yields

$$u(x,0) = F(x) + G(x) = f(x), \qquad -\infty < x < \infty. \tag{7.5}$$

On the other hand,

$$u_t(x,t) = F'(x-ct)\cdot(-c) + G'(x+ct)\cdot c,$$

so (7.1c) yields

$$\begin{aligned} u_t(x,0) = -cF'(x) + c\,G'(x) &= g(x) \\ -F'(x) + G'(x) &= \frac{1}{c}g(x) \\ \int_0^x (-F'(s) + G'(s))\,ds &= \frac{1}{c}\int_0^x g(s)\,ds \\ F(x) - G(x) &= F(0) - G(0) - \frac{1}{c}\int_0^x g(s)\,ds. \end{aligned} \tag{7.6}$$

Together, (7.5) and (7.6) constitute two linear equations in the unknowns $F(x)$ and $G(x)$. Solving, we get

$$F(x) = \frac{1}{2}f(x) + \frac{1}{2}(F(0) - G(0)) - \frac{1}{2c}\int_0^x g(s)\,ds, \tag{7.7}$$

$$G(x) = \frac{1}{2}f(x) - \frac{1}{2}(F(0) - G(0)) + \frac{1}{2c}\int_0^x g(s)\,ds. \tag{7.8}$$

Finally, combining (7.4) with (7.7) and (7.8),

$$\boxed{u(x,t) = \frac{1}{2}[f(x-ct) + f(x+ct)] + \frac{1}{2c}\int_{x-ct}^{x+ct} g(s)\,ds,} \tag{7.9}$$

which is called *d'Alembert's solution to the wave equation.*

Example 7.1.1. Consider (7.1) with initial position $f(x) = \frac{1}{1+x^2}$ and initial velocity $g(x) \equiv 0$ with $c = 1$. d'Alembert's solution simplifies nicely since the integral terms vanishes:

$$u(x,t) = \frac{1}{2}\left[\frac{1}{1+(x-t)^2} + \frac{1}{1+(x+t)^2}\right].$$

Figure 7.2 shows the solution surface as well as some time snapshots of the solution. The initial waveform indeed decomposes into a left traveling wave and right traveling wave, each of half the initial height and each traveling with speed $c = 1$. ◊

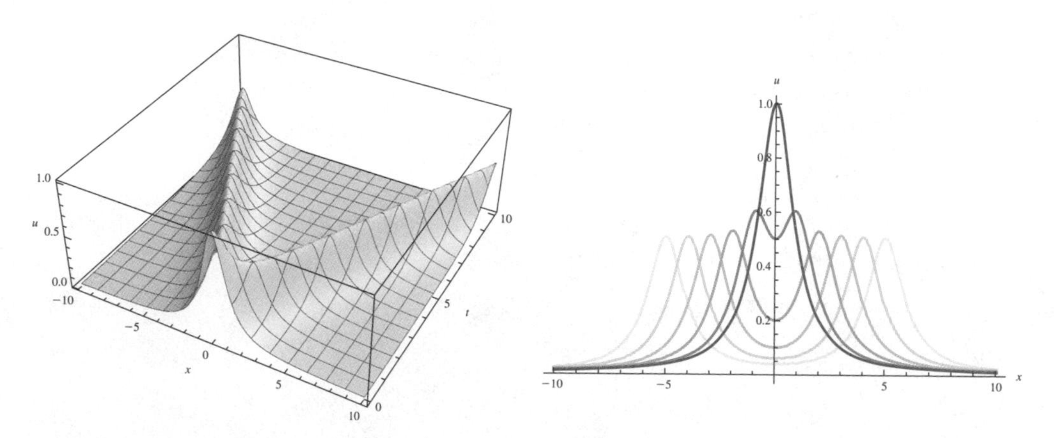

Figure 7.2: (Left) A plot of d'Alembert's solution for initial data $f(x) = \frac{1}{1+x^2}$, $g(x) \equiv 0$ with $c = 1$. (Right) Time slices at $t = 0, 1, 2, 3, 4, 5$ demonstrate how the initial wave decomposes into a left traveling wave and right traveling wave, each with unit speed.

Example 7.1.2. Consider (7.1) with $c = 1$, but this time reversing the roles of the initial position and initial velocity from the last example: $f(x) \equiv 0$ and $g(x) = \frac{1}{1+x^2}$. The physical interpretation here is an infinite string which is initially at rest but subjected to a positive initial velocity centered at the origin. d'Alembert's solution yields

$$u(x,t) = \frac{1}{2}\int_{x-t}^{x+t} \frac{1}{1+s^2}\,ds = \frac{1}{2}[\arctan(x+t) - \arctan(x-t)].$$

Figure 7.3 illustrates how the initial disturbance is propagated along the string. $\diamondsuit$

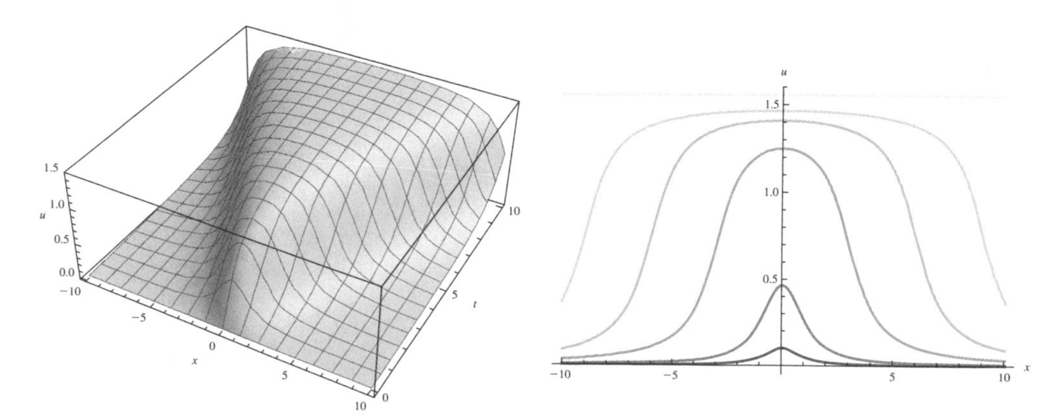

Figure 7.3: (Left) A plot of d'Alembert's solution for initial data $f(x) \equiv 0$, $g(x) = \frac{1}{1+x^2}$ with $c = 1$. (Right) Time slices at $t = 0.1, 0.5, 3, 6, 9, 50$ contrast significantly with those of Figure 7.2.

Example 7.1.3. Combining the ideas from the last two examples, consider (7.1) with $c = 1$ where the initial position of the string is $f(x) = \frac{1}{1+x^2}$ and the initial velocity is $g(x) = \frac{-1}{4(1+x^2)}$. d'Alembert's solution yields

$$u(x,t) = \frac{1}{2}\left[\frac{1}{1+(x-t)^2} + \frac{1}{1+(x+t)^2}\right] - \frac{1}{8}\int_{x-t}^{x+t} \frac{1}{1+s^2}\,ds$$
$$= \frac{1}{2}\left[\frac{1}{1+(x-t)^2} + \frac{1}{1+(x+t)^2}\right] - \frac{1}{8}[\arctan(x+t) - \arctan(x-t)].$$

Figure 7.4 illustrates how the initial disturbance is propagated along the string. $\diamondsuit$

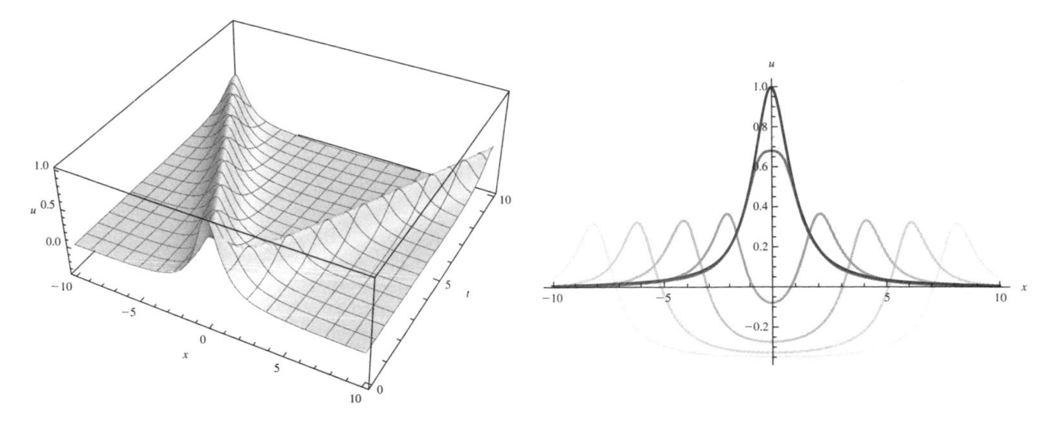

Figure 7.4: (Left) A plot of d'Alembert's solution for initial data $f(x) = \frac{1}{1+x^2}$, $g(x) = \frac{-1}{4(1+x^2)}$, with $c = 1$. (Right) Time slices at $t = 0, 0.5, 2, 4, 6, 8$ contrast significantly with those of Figures 7.2 and 7.3.

Throughout this chapter, we make use of the *unit step function*, also called the *Heaviside function*, which is defined as

$$h(x) := \begin{cases} 1, & x \geq 0, \\ 0, & x < 0. \end{cases} \tag{7.10}$$

Exercises

1. Show that $G(x + ct)$ is a solution of (7.1a).
2. Solve the system of equations (7.5), (7.6) for $F(x)$ and $G(x)$ to obtain (7.7), (7.8).
3. (a) Use Figure 7.2 to justify our saying that the left and right traveling waves move with unit speed.
   (b) How would Figure 7.2 change if $c = 1/2$ or $c = 3$?
4. Consider (7.1) with initial data $f(x) = h(x+1) - h(x-1)$, where $h(x)$ is the unit step function from (7.10) and $g(x) \equiv 0$. Take $c = 1$.

(a) Write out d'Alembert's solution and simplify.

(b) Plot the solution surface for $-10 < x < 10$ and $0 < t < 10$. Plot the time snapshots when $t = 0, 0.25, 0.5, 1, 3, 5$ for $-6 < x < 6$ to demonstrate the dynamics of the solution.

5. Consider (7.1) with $c = 1$ and initial data $f(x) = \begin{cases} \sin x, & |x| < \pi, \\ 0, & |x| \geq \pi, \end{cases}$ and $g(x) \equiv 0$.

   (a) Write out d'Alembert's solution and simplify.

   (b) Plot the solution surface for $-3\pi < x < 3\pi$ and $0 < t < 10$. Plot the time snapshots when $t = 0, 1, 2, 3, 4, 5$ for $-3\pi < x < 3\pi$ to demonstrate the dynamics of the solution.

   (c) Explain the "kinks" in the solution that appear for $t \approx 1$, but then disappear for $t \approx \pi$.

6. Each surface below is a plot of d'Alembert's solution of the same initial value problem, only with different values of c. Which one has $c = 1$? $c = 1/3$? $c = 2$? Justify your answer.

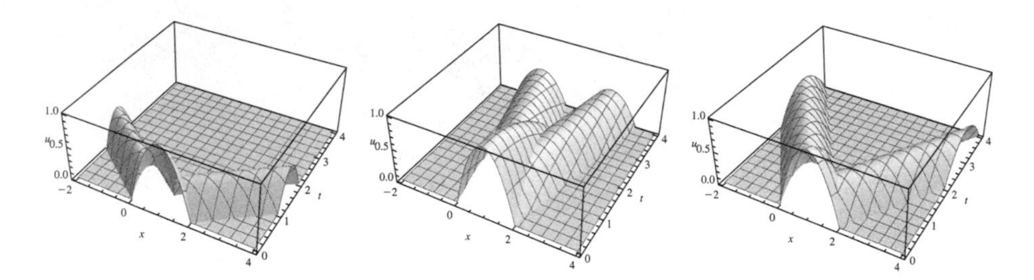

7. Consider (7.1) with initial data $f(x) \equiv 0$, $g(x) = h(x+1) - h(x-1)$, where $h(x)$ is the unit step function from (7.10). Take $c = 1$.

   (a) Write out d'Alembert's solution and simplify.

   (b) Plot the solution surface for $-10 < x < 10$ and $0 < t < 10$. Plot the time snapshots when $t = 0, 1, 2, 3, 4, 5$ for $-10 < x < 10$ to demonstrate the dynamics of the solution.

8. Consider (7.1) with $c = 1$ and initial data $f(x) = \dfrac{x}{1 + x^2}$, $g(x) = 4xe^{-x^2}$.

   (a) Write out d'Alembert's solution and simplify.

   (b) Plot the solution surface for $-8 < x < 8$ and $0 < t < 10$. Plot the time snapshots when $t = 0, 1, 2, 3, 4, 5$ for $-8 < x < 8$ to demonstrate the dynamics of the solution.

9. Shown below are contour plots of d'Alembert's solution of the same initial value problem, only with different values of c. Which one has $c = 1$? $c = 1/2$? $c = 3$? Justify your answer.

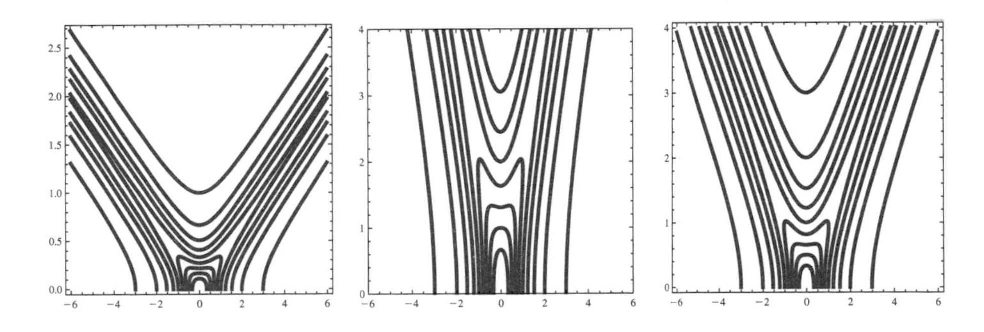

10. (a) Refer back to Exercise 4. Are the discontinuities in f smoothed out or retained in the solution? Justify the statement "the discontinuities in f are propagated with unit speed."

    (b) Refer back to Exercise 7. Are the discontinuities in g smoothed out or retained in the solution?

    (c) Explain (a) versus (b).

11. In Section 1.1, we verified that $u(x,t) = f(x-t)$, where f is an arbitrary differentiable function, is a solution of $u_t + u_x = 0$. In this exercise, we show every solution of $u_t + u_x = 0$ has that form.

    (a) Make the change of variables $\xi = x - t$, $\eta = x + t$. Use the multivariable chain rule to show $u_t + u_x = 0$ becomes $U_\eta = 0$.

    (b) Solve $U_\eta = 0$. Convert back to x-t coordinates to conclude $u(x,t) = f(x-t)$.

12. Consider the initial value problem

$$\begin{aligned} u_t + u_x &= 0, & -\infty < x < \infty,\ t > 0, \\ u(x,0) &= f(x), & -\infty < x < \infty. \end{aligned}$$

    (a) What is the physical interpretation of this problem?

    ❄ (b) Using Exercise 11, plot the solution to this problem for $f(x) = e^{-x^2}$. Plot enough time snapshots to demonstrate the dynamics of the solution.

    ❄ (c) Repeat (b) with $f(x) = h(x+1) - h(x-1)$, where $h(x)$ is the unit step function.

    ❄ (d) Repeat (b) with $f(x) = \dfrac{x}{1+x^2}$.

7.2 Characteristic Lines

A key step in finding d'Alembert's solution of the wave equation was the transformation from the standard coordinates x-t to the special coordinates ξ-η (also called *characteristic coordinates*) given by $\xi = x - ct$, $\eta = x + ct$. This allowed us to solve the wave equation on an infinite domain with the plan outlined in Figure 7.5.

$$\begin{array}{ccc} u_{tt} = c^2 u_{xx} & \xrightarrow{\xi := x-ct,\ \eta := x+ct} & U_{\xi\eta} = 0 \\ \uparrow & & \downarrow \text{integrate twice} \\ u(x,t) = F(x-ct) + G(x+ct) & \xleftarrow{\xi := x-ct,\ \eta := x+ct} & U(\xi,\eta) = F(\xi) + G(\eta) \end{array}$$

Figure 7.5: The strategy of characteristic coordinates is to turn a complicated PDE into one that is simpler to solve. The change of variables that accomplishes this defines the characteristic coordinates.

However, the characteristic coordinates tell us more about the wave equation than just a formula for the general solution,

$$u(x,t) = F(x-ct) + G(x+ct).$$

The structure of the general solution reveals a special geometry based on the quantities $x \pm ct$, which we explore next.

For any point $(x_1, 0)$, consider the lines passing through $(x_1, 0)$ given by $x - ct = x_1$ and $x + ct = x_1$ (as shown in Figure 7.6). Since x_1 is fixed, the value of F along $x - ct = x_1$ must be the constant $F(x_1)$. Similarly, the value of G along $x + ct = x_1$ must be the constant $G(x_1)$. Therefore, the solution u must be constant on these lines as well. Since the characteristic coordinates naturally lead us to the two families of lines in x-t space on which any solution to the wave equation must be constant, we call $x \pm ct = K$ the *characteristic lines* or simply the *characteristics* of the PDE $u_{tt} = c^2 u_{xx}$.

Example 7.2.1. Consider the infinite domain problem

$$\begin{aligned} u_{tt} &= c^2 u_{xx}, && -\infty < x < \infty,\ t > 0, && (7.11a) \\ u(x,0) &= f(x), && -\infty < x < \infty, && (7.11b) \\ u_t(x,0) &= 0, && -\infty < x < \infty, && (7.11c) \end{aligned}$$

which we know has solution

$$u(x,t) = \frac{1}{2}[f(x-ct) + f(x+ct)], \qquad -\infty < x < \infty,\ t > 0.$$

Suppose $c = 1$ and

$$f(x) = \begin{cases} 1, & |x| < 1, \\ 0, & |x| \geq 1, \end{cases}$$

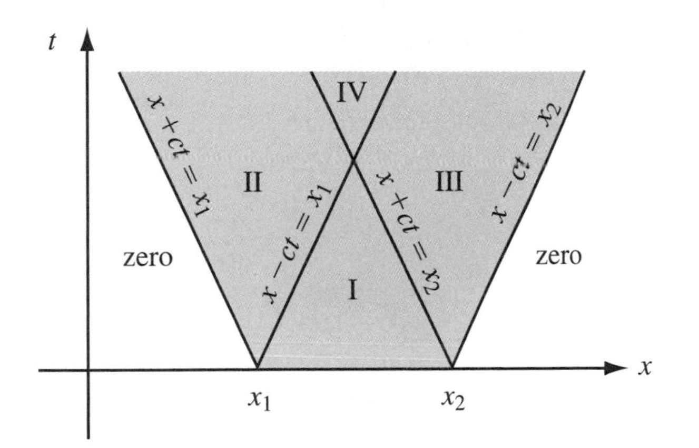

Figure 7.6: The region of influence for the interval $x_1 < x < x_2$ is divided into four subregions based on the characteristic lines emanating from the endpoints of the interval. If u is initially zero outside of $x_1 < x < x_2$, then u must be zero outside the region of influence.

then the family of characteristic lines is given by $x \pm t = K$ for any constant K. Since $f(x)$ is nonzero only on $-1 < x < 1$, we are particularly interested in the characteristic lines passing through $(-1, 0)$ and $(1, 0)$ in the x-t plane, which are given by $x \pm t = \pm 1$. These four lines divide the x-t plane into several regions of interest, as shown in Figure 7.6. We can determine the value of the solution u to (7.11) at any (x, t) geometrically by tracing the characteristic lines passing through (x, t) back to the x axis.

- If (x, t) is in region I, both characteristic lines will intersect the x axis somewhere in $-1 < x < 1$. Since $f(x) = 1$ here, $u = \frac{1}{2}(1 + 1) = 1$ throughout region I.

- If (x, t) is in region II, the positive[1] characteristic line will intersect the x axis in $x < -1$ (where $f(x) = 0$), while the negative characteristic line will intersect the x axis in $-1 < x < 1$ (where $f(x) = 1$). We conclude $u = \frac{1}{2}(0 + 1) = \frac{1}{2}$ throughout region II.

- If (x, t) is in region III, the positive characteristic line will intersect the x axis in $-1 < x < -1$ (where $f(x) = 1$), while the negative characteristic line will intersect the x axis in $x > 1$ (where $f(x) = 0$). We conclude $u = \frac{1}{2}(1 + 0) = \frac{1}{2}$ throughout region III.

- If (x, t) is in region IV, the positive characteristic line will intersect the x axis in $x < -1$ (where $f(x) = 0$), while the negative characteristic line will intersect the x axis in $x > 1$ (where $f(x) = 0$). We conclude $u = \frac{1}{2}(0 + 0) = 0$ throughout region IV.

[1]The *positive* characteristic lines are the ones with positive slopes, i.e., $x - ct = K$. The *negative* characteristic lines are the ones with negative slopes, i.e., $x + ct = K$.

- If (x,t) is anywhere else, both characteristic lines will intersect the x axis outside $-1 < x < 1$, where $f(x) = 0$. We conclude $u = \frac{1}{2}(0+0) = 0$ here.

Figure 7.7 shows the solution surface for various values of c. Indeed, f is the appropriate constant on each of the regions described above. $\diamondsuit$

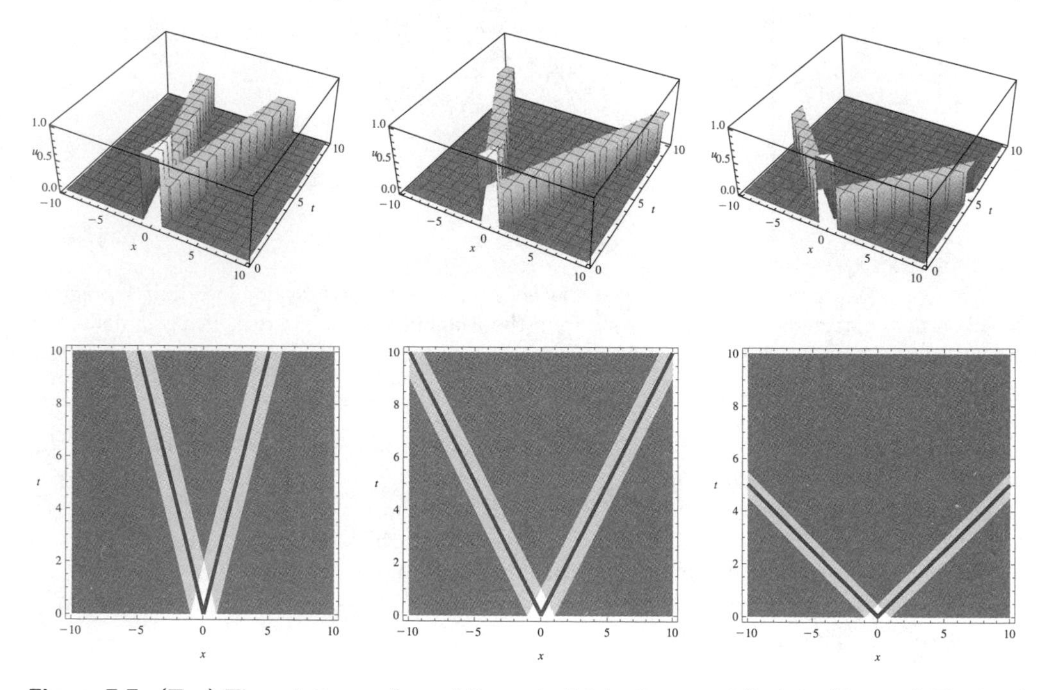

Figure 7.7: (Top) The solution surface of Example 7.2.1 when $c = 1/2, 1, 2$. (Bottom) The level curves indicate the regions in x-t space on which u is constant. The pair of characteristic lines which pass through the origin are shown, and their slopes are determined by c.

More generally, at time $t = 0$, let I denote an interval of x values outside of which f and g are zero. For each $x \in I$, there is a positive and negative characteristic line emanating from $(x, 0)$. The value of the solution u at any (x,t) between these characteristic lines is determined by the values of f and g at $(x, 0)$. The union of all these regions bounded by the characteristics emanating from any $x \in I$ is called the *region of influence of I.* The name makes sense because the solution at any point (x,t) "feels" the effects of f and g if and only if (x,t) is in the region of influence.

This concept can be framed another way. Given a point (x_1, t_1), d'Alembert's solution shows that the value of the solution at (x_1, t_1) depends on the value of f at the points where the characteristics intersect the x axis as well as *all values* of g between these intersection points. For this reason, we call the interval $x_1 - ct_1 < x < x_1 + ct_1$ the *domain of dependence* for the point (x_1, t_1). See Figure 7.8.

For example, Figure 7.7 shows that the range of influence of the interval $-1 < x < 1$ when $c = 1$ is the region bounded by the lines $x + t = -1$ and $x - t = 1$. The domain of dependence of the point $(0, 1/2)$ is $-1/2 < x < 1/2$ when $c = 1$.

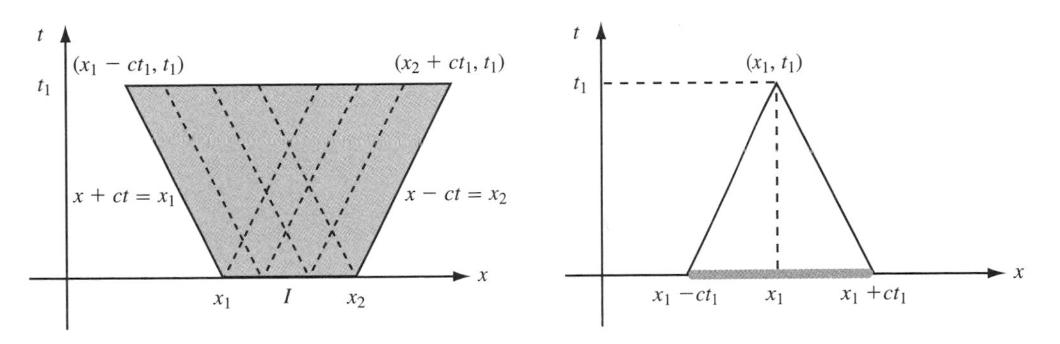

Figure 7.8: (Left) The region of influence for an interval I is the set of all (x, t) between the outermost characteristic lines emanating from the endpoints of I. (Right) The domain of dependence for a point (x_1, t_1) is the interval $x_1 - ct_1 < x < x_1 + ct_1$ which is shaded. Geometrically, the domain of dependence is found by tracing the pair of characteristics passing through (x_1, t_1) back to the x axis.

Exercises

1. (a) Suppose $f(x)$ is not identically zero on $-\ell < x < \ell$, and zero otherwise. Suppose $g(x) \equiv 0$. How long will it take the initial waveform $f(x)$ to fully decompose into a left traveling wave and right traveling wave?

   (b) Suppose instead that $f(x)$ is not identically zero on $a < x < b$, but zero otherwise, while $g(x) \equiv 0$. How long will it take the initial waveform $f(x)$ to fully decompose into a left traveling wave and right traveling wave?

2. Find and sketch the triangular domain of dependence and region of influence of the given PDE at the specified point.

   (a) $u_{tt} = 4u_{xx}$, $(x_1, t_1) = (5, 4)$

   (b) $u_{tt} = 3u_{xx}$, $(x_1, t_1) = (0, 5)$

3. Consider the infinite domain problem for $u := u(x, t)$ given by

$$\begin{aligned} u_{tt} &= u_{xx}, && -\infty < x < \infty,\ t > 0, \\ u(x, 0) &= f(x), && -\infty < x < \infty, \\ u_t(x, 0) &= 0, && -\infty < x < \infty, \end{aligned}$$

where $f(x) = \begin{cases} 2, & |x| \le 3, \\ 0, & |x| > 3. \end{cases}$ Using only the characteristic lines, find $u(0, 2)$, $u(0, 4)$, $u(5, 5)$, $u(10, 6)$, and $u(-5, 3)$.

7.3 The Semi-infinite String: The Method of Reflections

Dirichlet Boundary Condition

We now consider the vibrations of an infinitely long string which is fixed at one end:

$$
\begin{aligned}
u_{tt} &= c^2 u_{xx}, && 0 < x < \infty,\ t > 0, && (7.12a)\\
u(0,t) &= 0, && t > 0, && (7.12b)\\
u(x,0) &= f(x), && 0 < x < \infty, && (7.12c)\\
u_t(x,0) &= g(x), && 0 < x < \infty. && (7.12d)
\end{aligned}
$$

We call this a *semi-infinite* problem since $0 < x < \infty$ rather than $-\infty < x < \infty$, as before. Our goal is to incorporate the Dirichlet boundary condition (7.12b) into d'Alembert's solution from the last section.

To begin, recall that d'Alembert's solution

$$u(x,t) = \frac{1}{2}[f(x-ct) + f(x+ct)] + \frac{1}{2c}\int_{x-ct}^{x+ct} g(s)\,ds, \tag{7.13}$$

is valid in the context of (7.12) only for $0 < x < \infty$ since $f(x)$ and $g(x)$ are defined only for positive inputs. Therefore, (7.13) solves (7.12) only when $x \pm ct > 0$. Since $x, t, c > 0$, we see $x + ct > 0$ always holds. If $x - ct > 0$, then (7.13) still holds. But how do we proceed when $x - ct \leq 0$, since (7.13) is not valid?

Regardless of the sign of $x \pm ct$, our previous work showed that the general solution to (7.12a) is given by the sum of a right traveling wave and a left traveling wave, i.e.,

$$u(x,t) = F(x-ct) + G(x+ct). \tag{7.14}$$

Moreover, we found explicit formulas for F and G in (7.7) and (7.8).

Now suppose $x - ct \leq 0$. We need to extend the F above to a suitable $\tilde{F}$, which is to be determined next. Applying (7.12b),

$$u(0,t) = \tilde{F}(-ct) + G(ct) = 0.$$

Let $w := -ct$ so $\tilde{F}(w) = -G(-w)$ for $w \leq 0$, which reveals how $\tilde{F}$ should be defined. Therefore,

$$
\begin{aligned}
u(x,t) &= \tilde{F}(x-ct) + G(x+ct)\\
&= -G(ct-x) + G(x+ct)\\
&= \frac{1}{2}[f(x+ct) - f(ct-x)] + \frac{1}{2c}\int_{ct-x}^{x+ct} g(s)\,ds, \quad x - ct \leq 0,
\end{aligned} \tag{7.15}
$$

where we used (7.8) for the last equality.

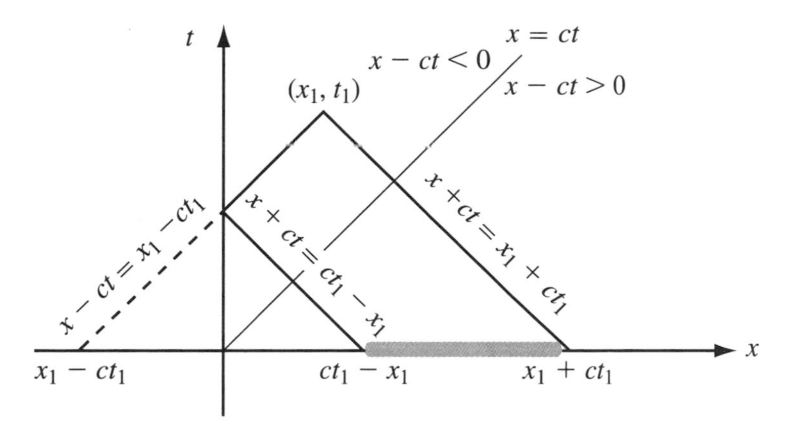

Figure 7.9: For a semi-infinite string subject to a Dirichlet boundary condition, the solution at (x_1, t_1) behind the leading waveform is determined by the values of f at the endpoints of the domain of dependence $I := \{x : ct_1 - x_1 < x < x_1 + ct_1\}$ as well as all values of g on I. Note how the positive characteristic line through (x_1, t_1) reflects off the physical boundary at $x = 0$.

Finally, we combine (7.13) for $x - ct > 0$ with (7.15) for $x - ct \leq 0$ to obtain the solution of (7.12) that we seek:

$$u(x,t) = \begin{cases} \frac{1}{2}[f(x+ct) - f(ct-x)] + \dfrac{1}{2c}\displaystyle\int_{ct-x}^{x+ct} g(s)\,ds, & x - ct \leq 0, \\ \frac{1}{2}[f(x+ct) + f(x-ct)] + \dfrac{1}{2c}\displaystyle\int_{x-ct}^{x+ct} g(s)\,ds, & x - ct > 0. \end{cases} \tag{7.16}$$

There is an interesting geometric connection between the general d'Alembert's formula (7.13) for the infinite string and (7.16) for the semi-infinite string, subject to a Dirichlet boundary condition. If (x_1, t_1) lies ahead of the waveform propagating along the characteristics (so that $x_1 - ct_1 > 0$), then the solution given by (7.16) is the same as the solution given by (7.9). This is because the characteristic lines passing through (x_1, t_1) intersect the positive x axis, and hence the domain of dependence is in the physical domain of the problem.

On the other hand, if (x_1, t_1) lies behind the waveform propagating along the characteristics (so that $x_1 - ct_1 \leq 0$), then the positive characteristic line passing through (x_1, t_1) will intersect the t axis (the physical boundary of our domain) before eventually intersecting the negative x axis. The effect here is that the wave *reflects* off the physical boundary and the reflected characteristic line intersects the positive x axis, resulting in the domain of dependence $ct_1 - x_1 < x < x_1 + ct_1$. See Figure 7.9 and then a specific example of these concepts in Figure 7.10.

The technique outlined analytically and supported geometrically is called *the method of reflections.* This explains why the only changes made in the first part of (7.16) involved "reflecting" $x - ct$ to obtain $ct - x$ in spots.

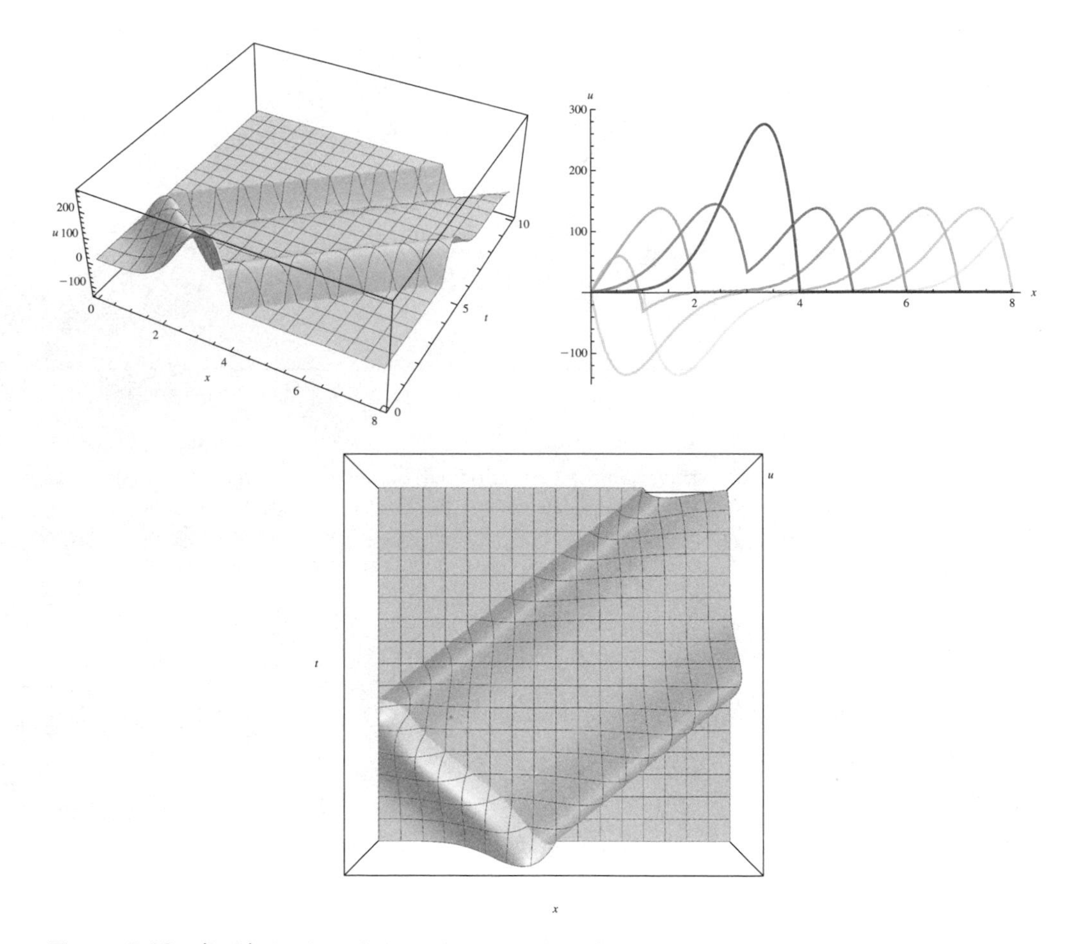

Figure 7.10: (Left) A plot of the solution surface for the semi-infinite d'Alembert solution of (7.12) with $c = 1$, $f(x) = x^5(4 - x)$, $0 < x < 4$ and zero otherwise. (Right) Time slices at $t = 0, 1, 2, 3, 4, 5$. Note how the left traveling wave reflects off the boundary at $x = 0$ to produce a negative wave, but one that still preserves the boundary condition. (Bottom) A view of the solution surface from above illustrates the reflection of the positive characteristic line.

Neumann Boundary Condition

If we impose a Neumann boundary condition at $x = 0$, the problem becomes

$$u_{tt} = c^2 u_{xx}, \qquad 0 < x < \infty,\ t > 0, \tag{7.17a}$$

$$u_x(0, t) = 0, \qquad t > 0, \tag{7.17b}$$

$$u(x, 0) = f(x), \qquad 0 < x < \infty, \tag{7.17c}$$

$$u_t(x, 0) = g(x), \qquad 0 < x < \infty. \tag{7.17d}$$

The strategy is similar to the Dirichlet case: start with the general solution (7.14) of the (7.17a), determine the appropriate modification $\tilde{F}$ of F when $x - ct \leq 0$, and retain the d'Alembert solution when $x - ct > 0$. The final solution is assembled from these pieces based on the sign of $x - ct$.

Since $u(x,t) = \tilde{F}(x - ct) + G(x + ct)$, the Neumann boundary condition (7.17b) yields $u_x(0,t) = \tilde{F}'(-ct) + G'(ct) = 0$. Letting $w := -ct$, we get $\tilde{F}'(w) = -G'(-w)$ and integration implies $\tilde{F}(w) = G(-w)$ (see Exercise 4). Since G is known from (7.8),

$$\begin{aligned} u(x,t) &= \tilde{F}(x - ct) + G(x + ct) \\ &= G(ct - x) + G(x + ct) \\ &= \frac{1}{2}[f(x+ct) + f(ct-x)] + \frac{1}{2c}\int_0^{ct-x} g(s)\,ds + \frac{1}{2c}\int_0^{x+ct} g(s)\,ds, \quad x - ct \le 0. \end{aligned} \tag{7.18}$$

When $x - ct > 0$, we use d'Alembert's solution (7.13). Combining these two pieces, we have the solution of (7.17):

$$u(x,t) = \begin{cases} \frac{1}{2}[f(x+ct) + f(ct-x)] + \dfrac{1}{2c}\displaystyle\int_0^{ct-x} g(s)\,ds + \dfrac{1}{2c}\displaystyle\int_0^{x+ct} g(s)\,ds, & x - ct \le 0, \\ \frac{1}{2}[f(x+ct) + f(x-ct)] + \dfrac{1}{2c}\displaystyle\int_{x-ct}^{x+ct} g(s)\,ds, & x - ct > 0. \end{cases} \tag{7.19}$$

We have shown that the Neumann boundary condition produces two intervals which together play the role of the domain of dependence for (x_1, t_1): $0 < x < ct_1 - x_1$ and $ct_1 - x_1 < x < x_1 + ct_1$. The only reason for separating them is to point out that in the first part of (7.19), $\int_0^{ct-x} g(s)\,ds$ is counted twice. See Figures 7.11 and 7.12.

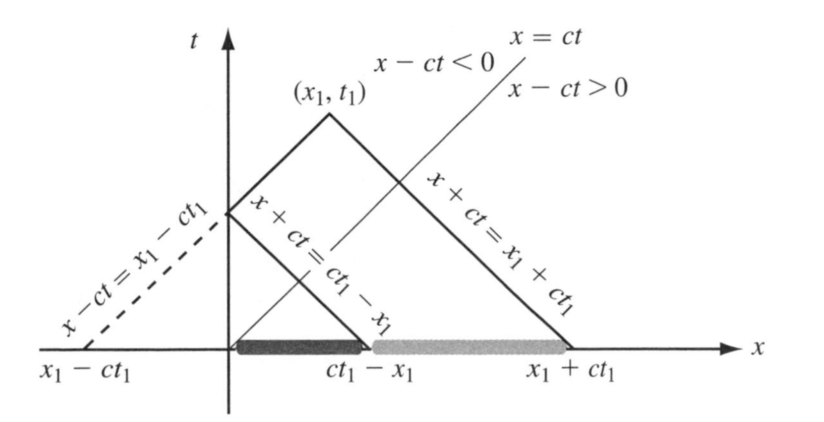

Figure 7.11: The geometry here is slightly different. When (x_1, t_1) is behind the waveform, the positive characteristic is again reflected off the physical boundary. The domain of dependence for (x_1, t_1) is $0 < x < x_1 + ct_1$ but $0 < x < ct_1 - x_1$ is counted twice in a sense.

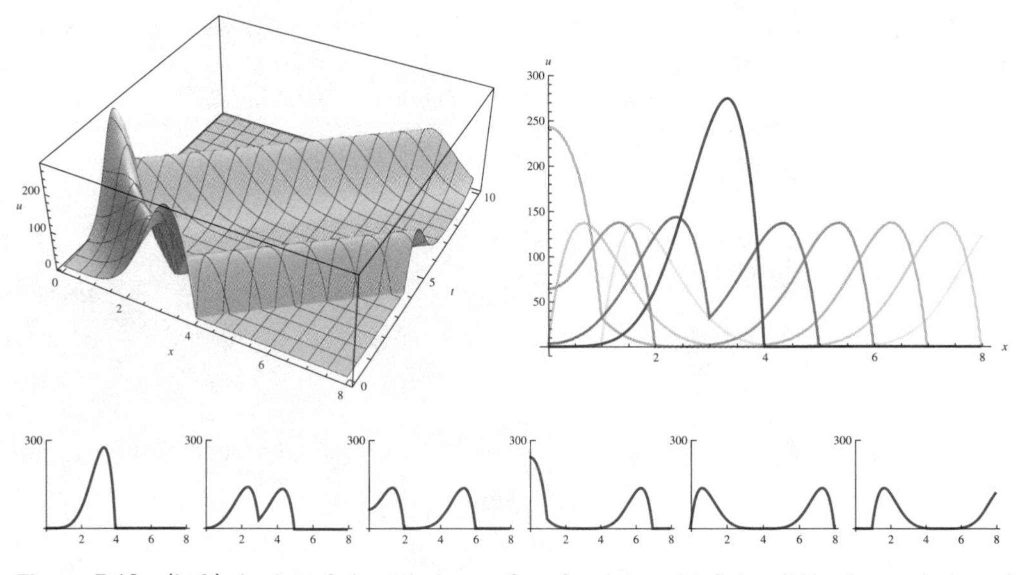

Figure 7.12: (Left) A plot of the solution surface for the semi-infinite d'Alembert solution of (7.17) with $c = 1$, $f(x) = x^5(4 - x)$, $0 < x < 4$ and zero otherwise. (Right) Time slices at $t = 0, 1, 2, 3, 4, 5$. The wave interactions with the boundary are different than in Figure 7.10.

Exercises

1. Give the physical interpretation of each part of (7.12). Justify the name *fixed end problem.*

2. Give the physical interpretation of each part of (7.17). Justify the name *free end problem.*

3. Write out the details of the three equalities leading to (7.15).

4. (a) In the Neumann example, explain why $\tilde{F}'(-ct) + G'(ct) = 0$ and $w := -ct$ imply $\tilde{F}(w) = G(-w)$.

   (b) Write out the details of the three equalities leading to (7.18).

   (c) Why did we not combine the two integrals in (7.18) as we did in (7.15)?

5. Consider (7.12) with $c = 1$, $f(x) = h(x - 1) - h(x - 3)$, where $h(x)$ is the unit step function, and $g(x) \equiv 0$.

   (a) Write out the appropriate semi-infinite d'Alembert solution for this problem and simplify.

   (b) Plot the solution surface and enough time snapshots to demonstrate the dynamics of the solution.

6. The sequence of time snapshots shown below is for a solution to a semi-infinite wave equation with zero initial velocity. Is there a Dirichlet or Neumann boundary condition? Justify your answer geometrically.

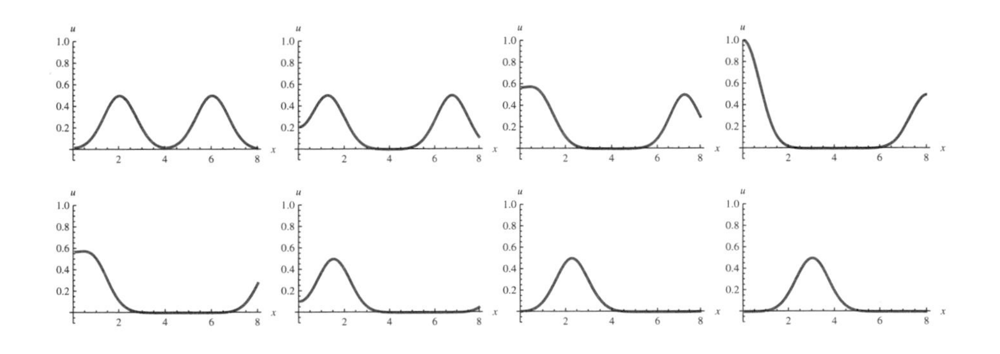

7. Consider (7.12) with $c = 1$, $f(x) = -\sin x$ for $\pi < x < 2\pi$ and zero elsewhere, and $g(x) = -0.25(h(x-\pi) - h(x-2\pi))$, where $h(x)$ is the unit step function.

   (a) Write out the appropriate semi-infinite d'Alembert solution for this problem and simplify.

   (b) Plot the solution surface and enough time snapshots to demonstrate the dynamics of the solution.

8. Consider (7.17) with $c = 1$, $f(x) = h(x-1) - h(x-3)$, where $h(x)$ is the unit step function, and $g(x) \equiv 0$.

   (a) Write out the appropriate semi-infinite d'Alembert solution for this problem and simplify.

   (b) Plot the solution surface and enough time snapshots to demonstrate the dynamics of the solution.

9. Consider (7.17) with $c = 1$, $f(x) = -\sin x$ for $\pi < x < 2\pi$ and zero elsewhere, and $g(x) = -0.25(h(x-\pi) - h(x-2\pi))$, where $h(x)$ is the unit step function.

   (a) Write out the appropriate semi-infinite d'Alembert solution for this problem and simplify.

   (b) Plot the solution surface and enough time snapshots to demonstrate the dynamics of the solution.

10. In a general problem of the form (7.12) or (7.17), how much time must pass until the waveform will no longer interact with the boundary? Justify your claim with a diagram like Figures 7.9 or 7.11.

11. Show that the solution (7.16) of (7.12) can also be obtained by starting with the infinite problem

$$\begin{aligned} u_{tt} &= c^2 u_{xx}, && -\infty < x < \infty,\ t > 0, \\ u(0,t) &= 0, && t > 0, \\ u(x,0) &= f_{\text{odd}}(x), && -\infty < x < \infty, \\ u_t(x,0) &= g_{\text{odd}}(x), && -\infty < x < \infty, \end{aligned}$$

applying d'Alembert's solution, and then restricting that solution to $x > 0$. This is called the *method of reflections.*

12. Show that the solution (7.19) of (7.17) can also be obtained by starting with the infinite problem

$$\begin{aligned} u_{tt} &= c^2 u_{xx}, && -\infty < x < \infty,\ t > 0, \\ u_x(0,t) &= 0, && t > 0, \\ u(x,0) &= f_{\text{even}}(x), && -\infty < x < \infty, \\ u_t(x,0) &= g_{\text{even}}(x), && -\infty < x < \infty, \end{aligned}$$

applying d'Alembert's solution, and then restricting that solution to $x > 0$. This is the method of reflections for a Neumann problem.

7.4 The Infinite Rod: The Method of Fourier Transforms

In this section, we introduce another powerful technique for solving PDEs on unbounded domains. It is reminiscent of the method of Laplace transforms from ODE theory.

Definition 7.1. The *Fourier transform of* $f(x)$, is given by

$$\mathcal{F}\{f(x)\} := F(\omega) = \int_{-\infty}^{\infty} f(x) e^{-i\omega x}\, dx, \quad -\infty < \omega < \infty. \tag{7.20}$$

The *inverse Fourier transform of* $F(\omega)$, is given by

$$\mathcal{F}^{-1}\{F(\omega)\} = f(x) = \frac{1}{2\pi} \int_{-\infty}^{\infty} F(\omega) e^{i\omega x}\, d\omega, \quad -\infty < x < \infty.$$

We call $f(x)$ and $F(\omega)$ *Fourier transform pairs.*

There are many slightly different ways to define the Fourier transform (and its inverse). The definitions above[2] are common, but in some sources you may see other variations, and it affects the precise statements of the theorems below.

[2]In *Mathematica* calculations, the definitions here require `FourierParameters->{1,-1}`.

We notice immediately from the definition that if $f(x)$ has a Fourier transform, then $f(x) \to 0$ as $x \to \pm\infty$; otherwise, the integral in (7.20) will diverge. The next example illustrates a rather simple function and how to compute its Fourier transform.

Example 7.4.1. Compute the Fourier transform of the pulse $f(x) = h(x+1)-h(x-1)$, where $h(x)$ is the unit step (or Heaviside) function.

From (7.20),

$$\begin{aligned}
\mathcal{F}\{f(x)\} &= \int_{-\infty}^{\infty} (h(x+1) - h(x-1))e^{-i\omega x}\, dx \\
&= \int_{-1}^{1} e^{-i\omega x}\, dx \\
&= \left.\frac{e^{-i\omega x}}{-i\omega}\right|_{x=-1}^{x=1}, \qquad \omega \neq 0 \\
&= \frac{2}{\omega}\left[\frac{e^{i\omega} - e^{-i\omega}}{2i}\right] \\
&= \frac{2}{\omega}\sin\omega,
\end{aligned}$$

where the last equality follows from the identity $\sin x = \frac{e^{ix}-e^{-ix}}{2i}$.

When $\omega = 0$,

$$\mathcal{F}\{f(x)\} = \int_{-\infty}^{\infty} (h(x+1) - h(x-1))\, dx = 2.$$

Since $\lim_{\omega\to 0} \frac{2\sin\omega}{\omega} = 2$, we conclude $F(\omega) = \frac{2\sin\omega}{\omega}$ is continuous for all ω and is the Fourier transform we seek. See Figure 7.13. ◊

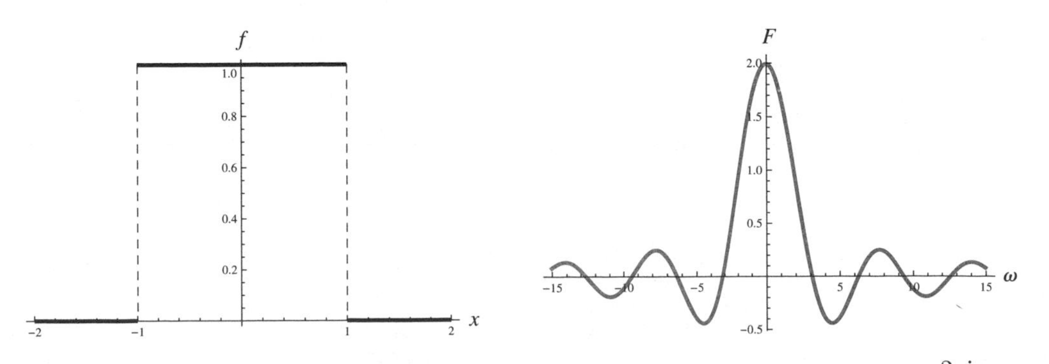

Figure 7.13: The function $f(x) = h(x+1) - h(x-1)$ and its Fourier transform $F(\omega) = \dfrac{2\sin\omega}{\omega}$.

Example 7.4.2. Find the Fourier transform of $f(x) = e^{-ax^2}$, $-\infty < x < \infty$, where $a > 0$ is constant.

By the definition of the Fourier transform,

$$F(\omega) = \mathcal{F}\{e^{-ax^2}\} = \int_{-\infty}^{\infty} e^{-ax^2} e^{i\omega x}\, dx.$$

Differentiating both sides with respect to ω,

$$\begin{aligned}
\frac{dF}{d\omega} &= \int_{-\infty}^{\infty} e^{-ax^2} \frac{d}{d\omega}\left(e^{-i\omega x}\right) dx \\
&= \int_{-\infty}^{\infty} e^{-ax^2} e^{-i\omega x}(-ix)\, dx \\
&= \frac{-i}{-2a} \int_{-\infty}^{\infty} -2axe^{-ax^2} e^{-i\omega x}\, dx \\
&= \frac{i}{2a} \int_{-\infty}^{\infty} \frac{d}{dx}\left(e^{-ax^2}\right) e^{-i\omega x}\, dx.
\end{aligned}$$

Integrating by parts and applying limits as $x \to \pm\infty$ (see Exercise 4), we obtain

$$\frac{dF}{d\omega} = \frac{i}{2a}\left[0 + i\omega \int_{-\infty}^{\infty} e^{-ax^2} e^{-i\omega x}\, dx\right] = \frac{\omega}{2a} F(\omega). \tag{7.21}$$

This is a separable ODE in the unknown $F(\omega)$; the solution is

$$F(\omega) = C \exp\left(\frac{-\omega^2}{4a}\right).$$

The constant is determined by evaluating $F(0)$ in the definition of the Fourier transform and then integrating in polar coordinates (see Exercise 14):

$$F(0) = \int_{-\infty}^{\infty} e^{-ax^2}\, dx = \sqrt{\frac{\pi}{a}}. \tag{7.22}$$

Combining these last two results, we obtain the transform

$$\mathcal{F}\{e^{-ax^2}\} = \sqrt{\frac{\pi}{a}} \exp\left(\frac{-\omega^2}{4a}\right). \tag{7.23}$$

See Figure 7.14. ◊

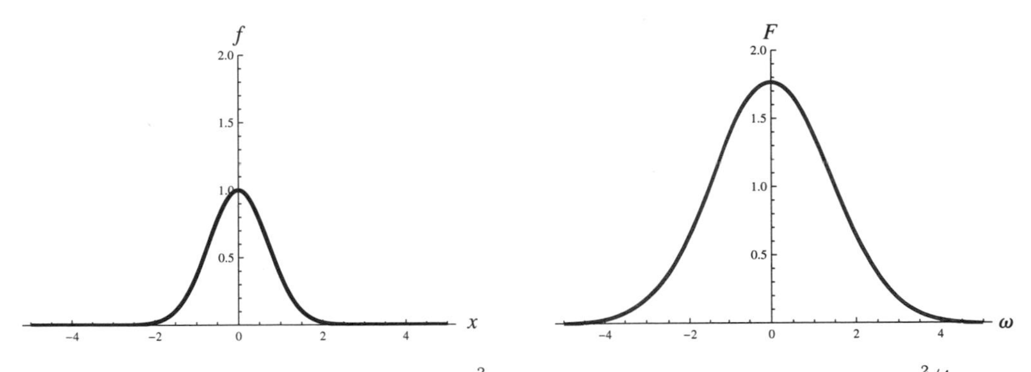

Figure 7.14: The function $f(x) = e^{-x^2}$ and its Fourier transform $F(\omega) = \sqrt{\pi}e^{-\omega^2/4}$.

Fourier transforms enjoy several properties similar to Laplace transforms, which we summarize below. We mainly focus on those properties which are useful in solving PDEs with Fourier transforms, but many others can be proved.

Theorem 7.1 (Operational Properties of Fourier Transforms).
Let $F(\omega)$ and $G(\omega)$ denote the Fourier transforms of $f(x)$ and $g(x)$, respectively.

(a) **Linearity:** *For $f, g \in L^2(-\infty, \infty)$, and any constants α, β,*

$$\mathcal{F}\{\alpha f(x) + \beta g(x)\} = \alpha\mathcal{F}\{f(x)\} + \beta\mathcal{F}\{g(x)\}.$$

(b) **Shift in the x Domain:** *If $f \in L^2(-\infty, \infty)$ and a is any constant, then*

$$\mathcal{F}\{f(x-a)\} = e^{-i\omega a}F(\omega).$$

(c) **Shift in the ω Domain:** *If $f \in L^2(-\infty, \infty)$ and a is any constant, then*

$$\mathcal{F}\{e^{iax}f(x)\} = F(\omega - a).$$

(d) **Transform of Derivatives:** *If $f^{(k)} \in L^2(-\infty, \infty)$ for $k = 0, 1, \ldots, n$, then*

$$\mathcal{F}\{f^{(n)}(x)\} = (i\omega)^n F(\omega), \ n = 0, 1, 2, \ldots.$$

(e) **Derivative of Transforms:** *If $f(x)$ and $x^n f(x)$ are both in $L^2(-\infty, \infty)$, then*

$$\frac{d^n F}{d\omega^n} = \mathcal{F}\{(-ix)^n f(x)\}.$$

(f) **Convolution Theorem:** *If $f, g \in L^2(-\infty, \infty)$, then*

$$\mathcal{F}^{-1}\{F(\omega)G(\omega)\} = f(x) * g(x) := \int_{-\infty}^{\infty} f(s)g(x-s)\,ds.$$

Theorem 7.1 enables us to compile tables of new Fourier transforms without having to appeal repeatedly to (7.20).

Example 7.4.3. Assuming we know the Fourier transform of $f(x)$, we can quickly compute $\mathcal{F}\{f(x)\cos(ax)\}$ for any constant a using Theorem 7.1(a),(c), and the identity $\cos x = \frac{1}{2}(e^{ix} + e^{-ix})$:

$$\begin{aligned}\mathcal{F}\{f(x)\cos(ax)\} &= \mathcal{F}\{f(x)\cdot\frac{1}{2}(e^{ix}+e^{-ix})\}\\ &= \frac{1}{2}\left[\mathcal{F}\{e^{iax}f(x)\} + \mathcal{F}\{e^{-iax}f(x)\}\right]\\ &= \frac{1}{2}\left[F(w-a)+F(w+a)\right].\end{aligned}$$

In particular, if $f(x) = e^{-2x^2}$, as in the last example, we quickly obtain

$$\mathcal{F}\{e^{-2x^2}\cos(3x)\} = \frac{1}{2}\cdot\sqrt{\frac{\pi}{2}}\left[\exp\left(\frac{-(w-3)^2}{8}\right)+\exp\left(\frac{-(w+3)^2}{8}\right)\right]. \tag{7.24}$$

See Figure 7.15. ◇

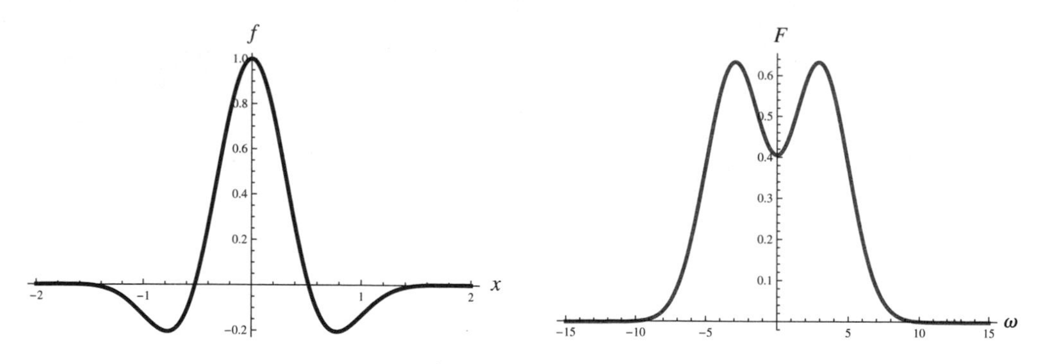

Figure 7.15: The function $f(x) = e^{-2x^2}\cos(3x)$ and its Fourier transform $F(\omega)$ given by (7.24).

Next, we demonstrate how the method of Fourier transforms is used to solve problems on infinite domains.

Example 7.4.4. Consider the initial-boundary value problem for the 1D heat equation given by

$$u_t = ku_{xx}, \qquad -\infty < x < \infty,\ t > 0, \tag{7.25a}$$

$$u(x,0) = f(x), \qquad -\infty < x < \infty. \tag{7.25b}$$

The Fourier transform of $u(x,t)$ with respect to the x variable leaves the t variable unaffected, so

$$\mathcal{F}\{u(x,t)\} = U(\omega,t).$$

Fourier transforming the PDE with respect to the x variable and using Theorem 7.1(d),

$$\begin{aligned}\mathcal{F}\{u_t(x,t)\} &= \mathcal{F}\{ku_{xx}(x,t)\}\\ U_t(\omega,t) &= -\omega^2 kU(\omega,t),\end{aligned}$$

while transforming the initial condition yields $\mathcal{F}\{u(x,0)\} = \mathcal{F}\{f(x)\} = F(\omega)$. Combining these two calculations, we see that $U(\omega,t)$ satisfies the initial value problem

$$\begin{aligned}U_t(\omega,t) &= -\omega^2 kU(\omega,t),\\ U(\omega,0) &= F(\omega).\end{aligned}$$

Regarding ω as a parameter and t as the independent variable, this is a separable ODE in the unknown $U(\omega,t)$. The solution is

$$U(\omega,t) = F(\omega)e^{-k\omega^2 t}.$$

Taking the inverse Fourier transform of both sides,

$$u(x,t) = \mathcal{F}^{-1}\{F(\omega)e^{-k\omega^2 t}\}.$$

Applying the Convolution Theorem to the right-hand side and using (7.23),

$$\begin{aligned}u(x,t) &= f(x) * \mathcal{F}^{-1}\{e^{-k\omega^2 t}\}\\ &= f(x) * \frac{1}{2\sqrt{k\pi t}} \exp\left(\frac{-x^2}{4kt}\right),\end{aligned} \tag{7.26}$$

or equivalently,

$$\boxed{u(x,t) = \frac{1}{2\sqrt{k\pi t}} \int_{-\infty}^{\infty} f(s) \exp\left(\frac{-(x-s)^2}{4kt}\right) ds.} \tag{7.27}$$

This integral representation of the solution to (7.25) is called the *fundamental solution of the heat equation.* The function

$$K(x,t) := \frac{1}{2\sqrt{k\pi t}} \exp\left(\frac{-x^2}{4kt}\right),$$

is called the *heat kernel.* The representation of the solution in (7.26) as a convolution of the initial function $f(x)$ with the heat kernel $K(x,t)$ is interesting in that it shows the role that the initial function $f(x)$ plays as well as the role the PDE plays (the heat kernel term) in the solution. The special type of "multiplication" given by the convolution ties these two parts together to form the solution of (7.25). See Figure 7.16. ◇

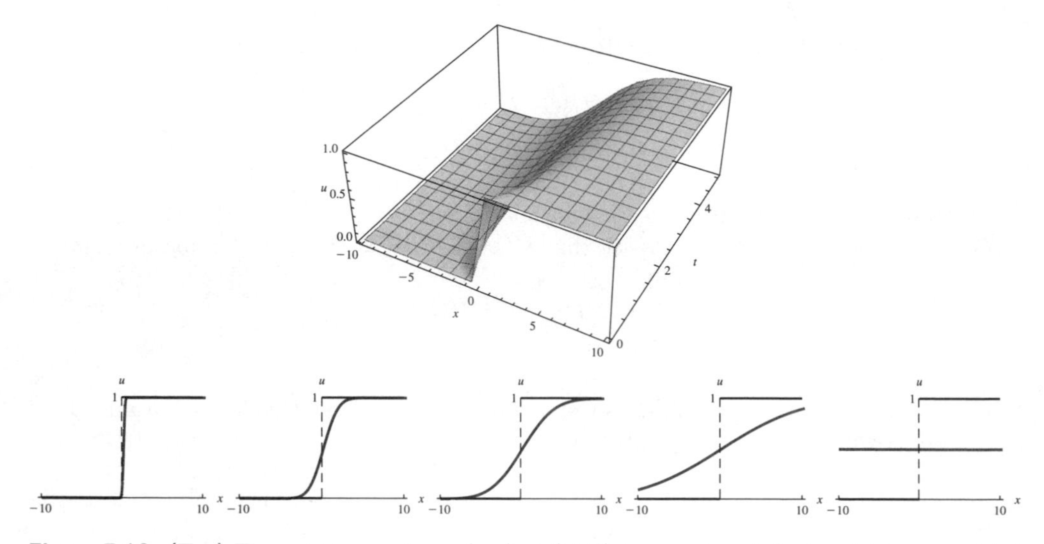

Figure 7.16: (Top) The solution surface $u(x,t)$ of (7.25) for $k=1$ and $f(x)=h(x)$, where $h(x)$ denotes the unit step function. (Bottom) Time slices of the solution for $t=0,1,5,30,10^{10}$.

Example 7.4.5. Consider the boundary value problem on the upper half plane,

$$u_{xx}+u_{yy}=0, \qquad -\infty<x<\infty,\ y>0, \tag{7.28a}$$

$$u(x,0)=f(x), \qquad -\infty<x<\infty. \tag{7.28b}$$

We will also impose the mathematical boundary condition

$$|u(x,y)| \text{ is bounded as } y\to\infty, \tag{7.29}$$

which is certainly physically realistic in any steady-state heat distribution or potential problem.

Taking the Fourier transform of the PDE with respect to the x variable (since it is x that ranges over $(-\infty,\infty)$),

$$\begin{aligned}-\omega^2U(\omega,y)+U_{yy}(\omega,y)&=0\\ U_{yy}-\omega^2U&=0.\end{aligned}$$

Viewing ω as a parameter, this is a second order constant coefficient ODE in the independent variable y. Its solution is

$$U(\omega,y)=A(\omega)e^{\omega y}+B(\omega)e^{-\omega y},$$

which must remain bounded as $y\to\infty$ in order for (7.29) to hold. But this requires $B(\omega)=0$ for $\omega>0$, and $A(\omega)=0$ for $\omega<0$. Therefore, we rewrite the general solution succinctly as

$$U(\omega,y)=C(\omega)\exp(-|\omega|y).$$

Figure 7.17: Siméon Poisson (1781–1840) of France worked mainly in applied problems whose solutions were rooted in differential equations. He published 300–400 papers on a wide variety of subjects, including potential theory, elasticity, electricity, and probability.

The initial condition (7.28b) transforms to $U(\omega, 0) = F(\omega)$, so $C(\omega) = F(\omega)$ and

$$U(\omega, y) = F(\omega)\exp(-|\omega|y).$$

Taking inverse Fourier transforms, applying the Convolution Theorem and Exercise 1(a), we get

$$u(x, y) = f(x) * \frac{y}{\pi(y^2 + x^2)},$$

or in terms of a convolution integral,

$$\boxed{u(x, y) = \frac{1}{\pi}\int_{-\infty}^{\infty} f(s)\frac{y}{y^2 + (x - s)^2}\, ds.} \tag{7.30}$$

This is known as *Poisson's Integral Formula for the half plane* (see Figure 7.17). It is a closed-form integral representation for the solution of the boundary value problem (7.28). In the same spirit as the fundamental solution of the heat equation, (7.30) can also be written as the convolution

$$u(x, y) = f(x) * P(x, y),$$

where

$$P(x, y) := \frac{y}{\pi(y^2 + x^2)},$$

is called the *Poisson kernel*. See Figure 7.18. ◊

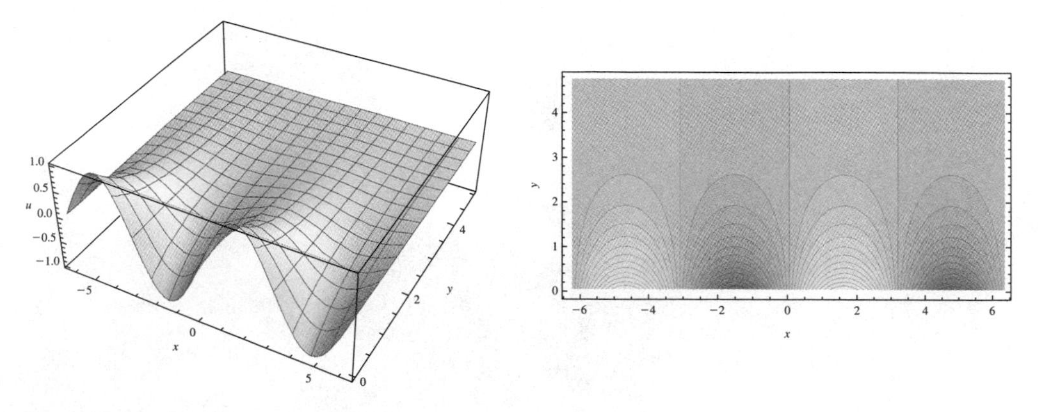

Figure 7.18: (Left) The solution surface $u(x, y)$ of (7.28) with $f(x) = \sin x$. (Right) A contour plot of the level curves of $u(x, y)$.

Exercises

1. Use the definition of $\mathcal{F}$ to compute the following:

   (a) $\mathcal{F}\{e^{-a|x|}\}$, $a > 0$

   (b) $\mathcal{F}\{f(x)\}$, where $f(x) = \begin{cases} |x|, & |x| < 1, \\ 0, & \text{else} \end{cases}$

   (c) $\mathcal{F}\{-h(x+1) + 2h(x) - h(x-1)\}$, where $h(x)$ is the unit step function

2. Plot the transform pairs $f(x)$ and $F(\omega)$ from Exercise 1. When $F(\omega)$ is complex-valued, plot $|F(\omega)|$.

3. Use Theorem 7.1 and/or known transforms to compute the following:

   (a) $\mathcal{F}\{xe^{-x^2}\}$

   (b) $\mathcal{F}\{f(x)\}$, where $f(x) = -6h(x-1) + 6h(x-3)$ and $h(x)$ is the unit step function

   (c) $\mathcal{F}^{-1}\left\{\dfrac{1}{1+\omega^2}\right\}$

   (d) $\mathcal{F}^{-1}\left\{\dfrac{2\sqrt{\pi}e^{-\omega^2/4}\sin\omega}{\omega}\right\}$

   (e) $\mathcal{F}^{-1}\left\{\dfrac{\sin(\omega-\pi)}{\omega-\pi}\right\}$

4. Write out the details of (7.21).

5. Although (7.25) has no physical boundary (and hence no physical boundary conditions), there are two *mathematical* boundary conditions implicitly imposed in this problem. What are they and where do they come into play in the solution of (7.25)?

6. The function

$$K(x,t) := \frac{1}{2\sqrt{k\pi t}} \exp\left(\frac{-x^2}{4kt}\right),$$

is called the *heat kernel* and plays an important role in the fundamental solution of the heat equation given by (7.27).

   (a) Let $k = 1$. Plot $K(x,t)$ for $-10 < x < 10$ and $0 < t < 5$.

   (b) Plot time snapshots of $K(x,t)$ for a sequence of t values with $t \to 0^+$.

   (c) Compute $\int_{-\infty}^{\infty} K(x,t)\,dx$.

   (d) What "function" from ODEs has the geometry described in (b) and (c)?

7. Consider (7.25) subject to the initial condition

$$f(x) = h(x+a) - h(x-a), \quad a > 0,$$

where $h(x)$ denotes the unit step function.

   (a) What is the physical interpretation of this problem?

   (b) Write out the solution as an appropriate convolution integral.

   (c) Plot the solution surface when $k = 1$ and $a = 1,\ 0.5,\ 0.1,\ 0.01$.

   (d) For each case in (c), plot the solution $u(x,t)$ for $t = 0.1,\ 0.01, 0.001$. This demonstrates that solutions of the heat equation have the property of *infinite speed of propagation*, which means that even if the initial temperature profile is zero except on the interval $-a < x < a$ (for very small a), the solution $u(x,t)$ is positive *for all* x at *any* (arbitrarily small) positive t values. In a sense, the heat which is concentrated on the tiny interval $-a < x < a$ at $t = 0$ becomes instantly propagated to all points on the entire domain for any $t > 0$.

8. Consider (7.25) subject to the initial condition

$$f(x) = \begin{cases} 0, & x < -1, \\ 1+x, & -1 < x < 0, \\ 1-x, & 0 < x < 1, \\ 0, & x > 1. \end{cases}$$

   (a) What is the physical interpretation of this problem?

   (b) Write out the solution as an appropriate convolution integral when $k = 1$.

(c) Plot the solution surface and enough time snapshots to display the dynamics of the solution.

(d) What is the steady-state temperature distribution of the rod?

9. Consider (7.25) subject to the initial condition $f(x) = x^3 e^{-x^2}$.

(a) What is the physical interpretation of this problem?

(b) Write out the solution as an appropriate convolution integral when $k = 1$.

(c) Plot the solution surface and enough time snapshots to display the dynamics of the solution.

(d) What is the steady-state temperature distribution of the rod?

10. (a) Use the method of Fourier transforms to solve the initial value problem

$$\begin{aligned} au_x + bu_t &= 0, & -\infty < x < \infty,\ t > 0, \\ u(x,0) &= f(x), & -\infty < x < \infty. \end{aligned}$$

(b) Let $f(x) = \frac{1}{1+x^2}$. Plot the solution surface and enough time snapshots to demonstrate the dynamics.

(c) Compare the solution found in (a) with the solution found in Section 1.1.

(d) What is the physical interpretation of this problem?

11. Consider the initial value problem

$$\begin{aligned} u_{tt} &= c^2 u_{xx}, & -\infty < x < \infty,\ t > 0, \\ u(x,0) &= f(x), & -\infty < x < \infty, \\ u_t(x,0) &= g(x), & -\infty < x < \infty. \end{aligned}$$

(a) Use the method of Fourier transforms to show

$$U(\omega,t) = F(\omega)\cos(c\omega t) + \frac{G(\omega)}{c\omega}\sin(c\omega t)$$

in the transform domain. Here, U, F, and G denote the Fourier transforms of u, f, and g, respectively.

(b) When $g(x) \equiv 0$, show that taking the inverse Fourier transform of (a) yields d'Alembert's solution, $u(x,t) = \frac{1}{2}[f(x-ct) + f(x+ct)]$. (Hint: Use the identity $\cos x = \frac{e^{ix}+e^{-ix}}{2}$.)

(c) Let $f(x) = h(x+a) + h(x-a)$ for $a = 1,\ 0.5,\ 0.1,\ 0.01$. Plot the solution surface in each case, taking $c = 1$. Compare the results here with those of Exercise 7(d).

(d) Prove (b) in another way, this time using the Convolution Theorem and the fact that $\mathcal{F}^{-1}\{\cos(c\,\omega\,t)\} = \frac{1}{2}\delta(ct-x) + \frac{1}{2}\delta(x-ct)$, where $\delta(x)$ is the Dirac delta distribution. (Hint: Use the *sifting property* of the Dirac delta: $\int_{-\infty}^{\infty} f(x)\delta(x-a)\,dx = f(a)$.)

12. (a) Use the method of Fourier transforms to solve the initial value problem with nonconstant coefficients

$$\begin{aligned} u_t &= t^2 u_{xx}, && -\infty < x < \infty,\ t > 0, \\ u(x,0) &= f(x), && -\infty < x < \infty. \end{aligned}$$

(b) Let $f(x) = h(x+1) - h(x-1)$. For $k_1 = k_2 = 1$, plot the solution surface and enough time snapshots to demonstrate the dynamics.

13. (a) Use the method of Fourier transforms to solve the initial value problem

$$\begin{aligned} u_t &= k_1 u_{xx} + k_2 u_x, && -\infty < x < \infty,\ t > 0, \\ u(x,0) &= f(x), && -\infty < x < \infty, \end{aligned}$$

where $k_1, k_2 > 0$ arc constants.

(b) Let $f(x) = h(x+1) - h(x-1)$. For $k_1 = k_2 = 1$, plot the solution surface and enough time snapshots to demonstrate the dynamics.

(c) What is the physical interpretation of this problem?

14. The integral in (7.22) appears in many branches of mathematics. Since there is no closed-form antiderivative for e^{-ax^2} in terms of elementary functions, we evaluate the integral as follows.

(a) Let $I = \int_{-\infty}^{\infty} e^{-ax^2}\,dx$. Then $I^2 = \int_{-\infty}^{\infty} e^{-ax^2}\,dx \cdot \int_{-\infty}^{\infty} e^{-ay^2}\,dy$. Convert this last integral to a double integral in polar coordinates using $r^2 = x^2 + y^2$, $dx\,dy = r dr\,d\theta$ to obtain

$$I^2 = \int_0^{2\pi} \int_0^{\infty} e^{-ar^2} r\,dr\,d\theta. \qquad (*)$$

(b) Evaluate $(*)$ to conclude $I = \sqrt{\pi/a}$.

15. **(Connection Between Fourier and Laplace Transforms)** The *Laplace transform* of $f(t)$ is given by

$$\mathscr{L}\{f(t)\}(s) = \int_0^{\infty} f(t) e^{-st}\,dt,$$

where s is a complex variable. We say $f(x)$ is *causal* if $f(x) = 0$ for all $x < 0$. Show that the Fourier transform of a causal function f is the Laplace transform of f evaluated along the imaginary axis in the ω domain; that is,

$$\mathcal{F}\{f(x)\}(\omega) = \mathscr{L}\{f(x)\}\Big|_{s=i\omega}.$$

Appendix

Power Series Solutions to ODEs

In this section, we briefly outline the methods and basic theory for obtaining power series solutions of ODEs of the form

$$y''(x) + p(x)y'(x) + q(x)y(x) = 0. \tag{A.1}$$

In contrast to the constant coefficient equations studied in Section 1.4, the theory here is more delicate. We begin with some definitions that play a fundamental role.

Definition A.1. A function $f(x)$ is *analytic at* $x = x_0$ if

$$f(x) = \sum_{k=0}^{\infty} a_k (x - x_0)^k,$$

for all x in some open interval about x_0.

For example, e^x, $\sin x$, and $\cos x$ are analytic for all $x_0 \in \mathbb{R}$. However, $\frac{1}{1-x}$ is analytic for $|x| < 1$, but not analytic at $x = 1$.

Power series solutions to (A.1) are classified according to the analyticity of the coefficient functions $p(x)$ and $q(x)$ in (A.1). This motivates the following definition.

Definition A.2. A point x_0 is called an *ordinary point* of (A.1) if p and q are analytic at x_0. If x_0 is not an ordinary point, it is called a *singular point* of (A.1).

Here are a few examples:

- The Cauchy-Euler equation $x^2y'' + xy' + y = 0$ has its only singular point at $x = 0$.

- Legendre's equation $(1 - x^2)y'' - 2xy' + n(n+1)y = 0$ has singular points at $x = \pm 1$.

- Chebyshev's equation $(1 - x^2)y'' - xy' + n^2y = 0$ has singular points at $x = \pm 1$.

- Bessel's equation $x^2y'' + xy' + (x^2 - n^2)y = 0$ has its singular point at $x = 0$.

In Section 4.3, we were able to compute valid power series solutions about $x = 0$ for Legendre's equation and Chebyshev's equation because $x = 0$ is an ordinary point for each. This is justified by the next theorem.

Theorem A.1 (Power Series Solutions about an Ordinary Point). *Suppose x_0 is an ordinary point of* (A.1). *Then* (A.1) *has two linearly independent solutions of the form*

$$y(x) = \sum_{k=0}^{\infty} a_k (x - x_0)^k.$$

Moreover, the radius of convergence of any such solution is at least as large as the distance (in the complex plane) from x_0 to the nearest (real or complex) singular point of (A.1).

However, the standard approach of looking for a power series outlined in Theorem A.1 fails on Bessel's equation (at least a power series solution about $x = 0$ fails) because $x = 0$ is a singular point of Bessel's equation. In Section 4.4, we illustrated how the Method of Frobenius does yield the power series solutions we seek. We clarify the theoretical picture with the following definition and theorem.

Definition A.3. A singular point x_0 of (A.1) is called a *regular singular point* provided

$$(x - x_0)p(x) \quad \text{and} \quad (x - x_0)^2 q(x)$$

are both analytic at x_0. Otherwise, x_0 is called an *irregular singular point.*

All four singular point examples above demonstrate regular singular points. In contrast, the equation $y'' + \frac{1}{x^3} y = 0$ has an irregular singular point at $x = 0$, since $x^2 \cdot \frac{1}{x^3} = \frac{1}{x}$ is not analytic at $x = 0$.

Theorem A.2 (Power Series Solutions: Regular Singular Point). *If x_0 is a regular singular point of* (A.1), *then there exists at least one series solution of the form*

$$y(x) = (x - x_0)^m \sum_{k=0}^{\infty} a_k (x - x_0)^k = \sum_{k=0}^{\infty} a_k (x - x_0)^{k+m}, \quad x > x_0.$$

Moreover, this series converges on $0 < x - x_0 < d$, where d is the distance (in the complex plane) from x_0 to the nearest other (real or complex) singular point of (A.1).

Selected Answers

1.1

1. (a) $u(x,t) = e^{-|x-t|}$
   (b)
   (c)

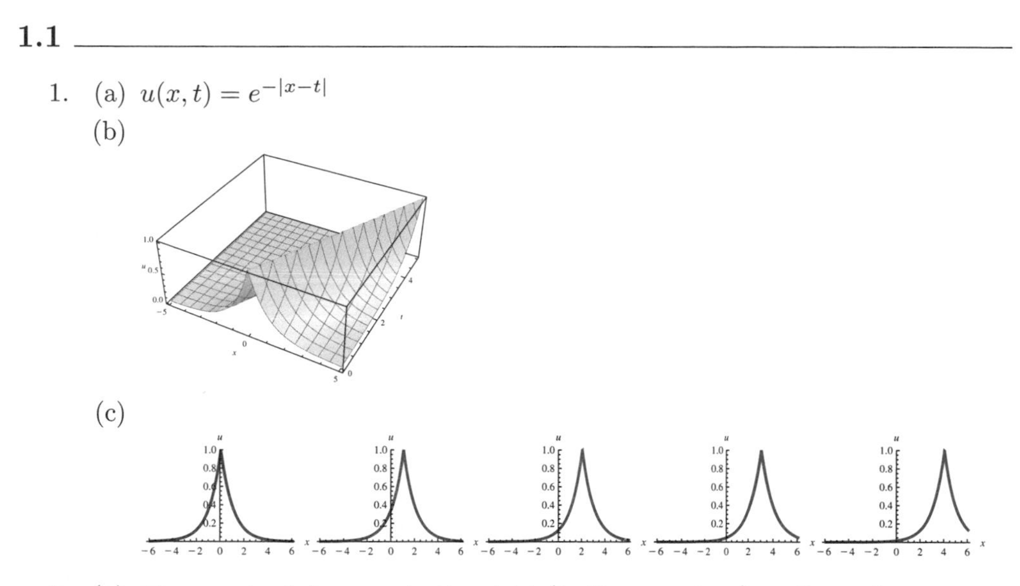

3. (b) The graph of f moves to the right (in the x-u plane) as time advances.
   (c)

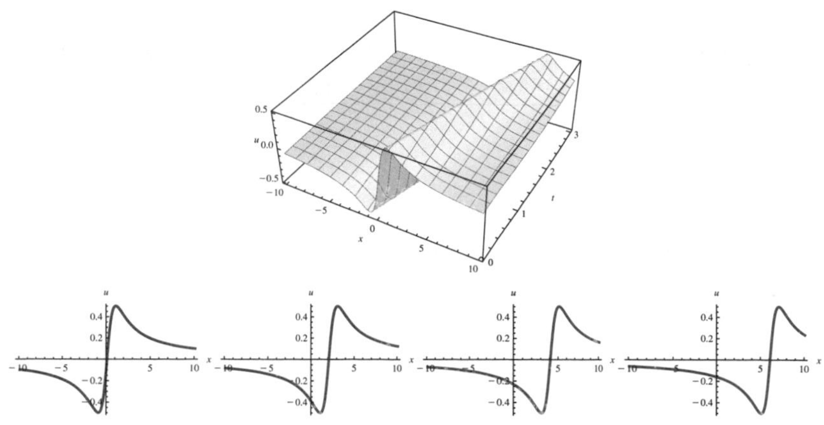

5. (b)

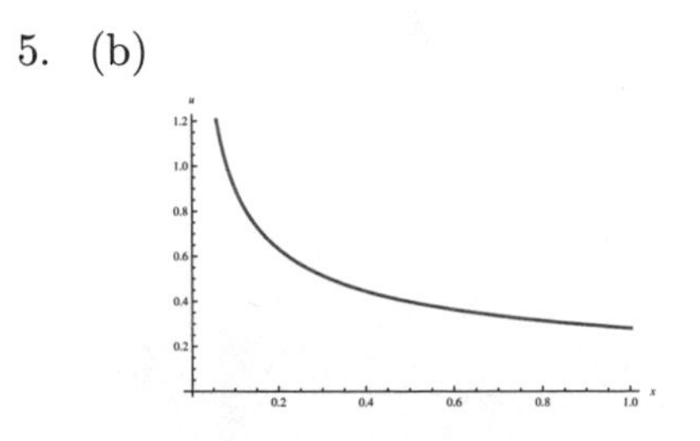

7. (a) For $u(r,\theta) = \ln r$, $u_r = \frac{1}{r}$, $u_{rr} = -r^{-2}$, $u_\theta = 0$, $u_{\theta\theta} = 0$ so $u_{rr} + \frac{1}{r}u_r + \frac{1}{r^2}u_{\theta\theta} = -r^{-2} + r^{-1} \cdot r^{-1} = -r^{-2} + r^{-2} = 0$.

   (b)

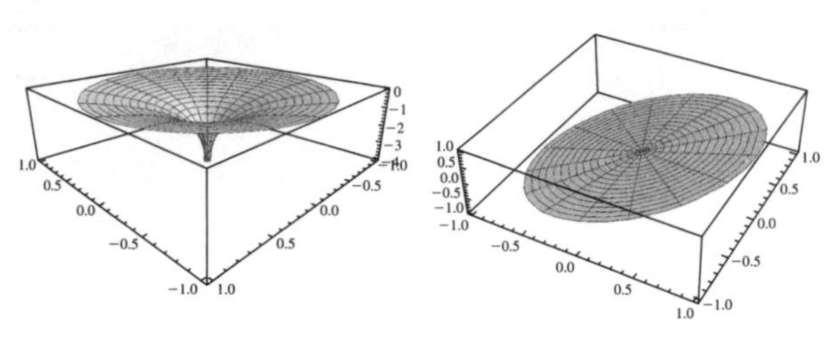

1.2

1. (a) $u(x,t)$ has units $\frac{[\text{quantity}]}{[\text{time}]^3}$; $\varphi(x,t)$ has units $\frac{[\text{quantity}]}{[\text{time}][\text{length}]^2}$; $f(x,t)$ has units $\frac{[\text{quantity}]}{[\text{time}][\text{length}]^3}$.

3. (a) $\varphi_x = uu_x + u_{xxx}$ so $\varphi = \frac{1}{2}u^2 + u_{xx}$; $f \equiv 0$

   (b) $\varphi_x = -u_x e^{-u}$ so $\varphi = e^{-u}$; $f \equiv 0$

   (c) $\varphi_x = \frac{u_x}{1+u^2}$ so $\varphi = \arctan u$; $f = |t|$

5. In the conservation law, an extra "rate out" term appears of the form $-\int_a^b cu(x,t)\,dx$ where $c > 0$ is a constant, resulting in the PDE $u_t = ku_{xx} - cu$.

7. (a) $u_{tt} = c^2 u_{xx} - g$

   (b) $u_{tt} = c^2 u_{xx} - du_t$

9. (a) parabolic

   (b) hyperbolic

   (c) elliptic

   (d) hyperbolic

1.3

1. $u_x(0,t) = 0$

3. (a) I (b) V (c) II (d) VI (e) VI (f) II (g) III (h) IV

5. (a) A linear function which passes through the point $(0, T_1)$ with slope T_2.
   (b) The constant function 100.

7. No, since the physical form at the left endpoint is is $u_x(0,t) = -(u(0,t)-(-100))$, resulting in $K < 0$.

1.4

1. (a) linear
   (b) nonlinear
   (c) linear
   (d) nonlinear

3. (a) $\mathscr{L}u = 0$ where $\mathscr{L}u := u'' + u - u^2$; nonlinear
   (b) $\mathscr{L}u = 0$ where $\mathscr{L}u := u' - ku$; linear, homogeneous
   (c) $\mathscr{L}u = \sin x$ where $\mathscr{L}u := a(x)u'' + b(x)u' + c(x)u$; linear, nonhomogeneous
   (d) $\mathscr{L}u = x$ where $\mathscr{L}u := uu'$; nonlinear

5. (a) $y = Ce^t$
   (b) $y = 2e^{-1}e^t$

7. (a) $y = x^2 \sin x + Cx^2$
   (b) $y = x^2 \sin x + \frac{1}{\pi^2}x^2$

9. (a) $y = c_1 e^{-3t} + c_2 te^{-3t}$
   (b) $y = e^{-3t} + 2te^{-3t}$
   (c) $y = e^{-3t} - te^{-3t}$

11. (a) $y = c_1 \cos 4t + c_2 \sin 4t$
    (b) $y = \cos 4t + \frac{1}{4} \sin 4t$
    (c) $y = -\cot(4) \cos 4t - \sin 4t$

13. (a) $y = c_1 x^{-3} + c_2 x^2$
    (b) $y = x^{-4}(c_1 \cos(\ln x) + c_2 \sin(\ln x))$

15. (a) $y = c_1 e^{5t} + c_2 e^{-5t}$
    (b) $y = \frac{e^5}{5(1+e^{10})} e^{5t} - \frac{e^5}{5(1+e^{10})} e^{-5t}$
    (c) $y = c_3 \cosh(5t) + c_4 \sinh(5t)$
    (d) $y = \frac{1}{5 \cosh 5} \sinh(5t)$
    (e) (d)!

2.1

1. (a) $\|\mathbf{u}\| = \sqrt{26}$, $\|\mathbf{v}\| = \sqrt{6}$
   (b) $\langle \mathbf{u}, \mathbf{v} \rangle \neq 0$ so not orthogonal
   (c) linearly independent
   (d) yes

3. (a) $1/\sqrt{2}$
   (b) no

5. (a) Any even function will work; for example, $u(t) = t^{2n}$ or $u(t) = \cos t$, since the integrand will then be odd and thus have zero integral.
   (b) $p(t) = at^2 + c$ works for any $a, c \in \mathbb{R}$
   (c) $\frac{p(t)}{\|p(t)\|}$ for any $p(t)$ from (b)

7. (a) $\mathscr{L}u = \lambda u$, $u'(0) = 0$, $u'(1) = 0$ where $\mathscr{L}u := -u''$
   (b) yes; $\lambda_0 = 0$, $u_0(x) = 1$ (up to a constant multiple)
   (c) none
   (d) $\lambda_n = (n\pi)^2$, $u_n(x) = \cos(\sqrt{\lambda_n}\, x)$, $n = 1, 2, \ldots$

9. (a) $\mathscr{L}u = \lambda u$, $u(0) = 0$, $u(1) = 0$ where $\mathscr{L}u := u'' + u'$.
   (b) $y'' + y' - \lambda y = 0$ has characteristic equation $r^2 + r - \lambda = 0$, with discriminant $1 + 4\lambda$.
   (c) $\lambda = -1/4$ results in $u(x) \equiv 0$, so that $-1/4$ is not an eigenvalue.
   (d) $\lambda > -1/4$ results in $u(x) \equiv 0$, so there are no such eigenvalues.
   (e) $\lambda < -1/4$ results in eigenvalues $\lambda_n = -\frac{4n^2\pi^2+1}{4}$, $n = 1, 2, \ldots$ with eigenfunctions $u_n(x) = e^{-x/2}\sin(\pi n x)$.

2.2

1. None at all.

3. (a) This is the 1D heat equation for a rod of length ℓ. The temperature at the left endpoint is fixed at zero. The right endpoint is insulated (i.e. zero flux). The initial temperature distribution in the rod is $f(x)$, $0 < x < \ell$.
   (b) $X'' + \lambda X = 0$, $X(0) = 0$, $X'(\ell) = 0$; $\quad T' + \lambda k T = 0$
   (c) $\lambda_n = \left(\frac{(2n-1)\pi}{2\ell}\right)^2$, $X_n(x) = \sin(\sqrt{\lambda_n}\, x)$, $n = 1, 2, \ldots$
   (d) $T_n(t) = C\exp(-\lambda_n k t)$, $n = 1, 2, \ldots$
   (e) $u(x, t) = \sum_{n=1}^{\infty} c_n \sin(\sqrt{\lambda_n}\, x)\exp(-\lambda_n k t)$

5. (a) $X'' + \lambda X = 0$, $X(-\ell) = X(\ell)$, $X'(-\ell) = X'(\ell)$; $\quad T' + \lambda k T = 0$

   (b) $\lambda_0 = 0$, $X_0(x) \equiv B$,
   $\lambda_n = \left(\frac{n\pi}{\ell}\right)^2$, $X_n(x) = a_n \cos(\sqrt{\lambda_n}\, x) + b_n \sin(\sqrt{\lambda_n}\, x)$, $n = 1, 2, \ldots$

   (c) $T_0(t) \equiv C$,
   $T_n(t) = C \exp(-\lambda_n k t)$, $n = 1, 2, \ldots$

   (d) $u(x,t) = c_0 + \sum_{n=1}^{\infty} \left[a_n \cos(\sqrt{\lambda_n}\, x) + b_n \sin(\sqrt{\lambda_n}\, x)\right] \exp(-\lambda_n k t)$

2.3

1. (a)
$$\begin{aligned}
&\int_0^{\ell} \sin(n\pi x/\ell)\sin(m\pi x/\ell)\, dx \\
&= \int_0^{\ell} \left[\frac{1}{2}\cos\left(\frac{n\pi x}{\ell} - \frac{m\pi x}{\ell}\right) - \frac{1}{2}\cos\left(\frac{n\pi x}{\ell} + \frac{m\pi x}{\ell}\right)\right] dx \\
&= \int_0^{\ell} \left[\frac{1}{2}\cos\left(\frac{(n-m)\pi x}{\ell}\right) - \frac{1}{2}\cos\left(\frac{(n+m)\pi x}{\ell}\right)\right] dx \\
a &= \frac{1}{2}\int_0^{\ell} \cos\left(\frac{(n-m)\pi x}{\ell}\right) dx - \frac{1}{2}\int_0^{\ell} \cos\left(\frac{(n+m)\pi x}{\ell}\right) dx \\
&= \frac{\ell}{2(n-m)\pi}\sin\left(\frac{(n-m)\pi x}{\ell}\right)\Bigg|_0^{\ell} - \frac{\ell}{2(n+m)\pi}\sin\left(\frac{(n+m)\pi x}{\ell}\right)\Bigg|_0^{\ell} \\
&= \frac{\ell}{2(n-m)\pi}[\sin((n-m)\pi) - \sin(0)] - \frac{\ell}{2(n+m)\pi}[\sin((n+m)\pi) - \sin(0)] \\
&= 0 \quad \text{since } n, m \text{ are integers}
\end{aligned}$$

   (b) Similar to (a).

3. Argue as in Exercise 1, using the given identity, and remembering that $\sin x$ is odd and $\cos x$ is even.

5. (a)
$$\begin{aligned}
a_0 &= \int_0^1 x\, dx = 1/2, \\
a_n &= \int_0^1 x\cos(n\pi x)\, dx = \frac{-1 + (-1)^n}{n^2\pi^2}, \; n = 1, 2, \ldots \\
b_n &= \int_0^1 x\sin(n\pi x)\, dx = \frac{(-1)^{n+1}}{n\pi}, \; n = 1, 2, \ldots
\end{aligned}$$

(b)

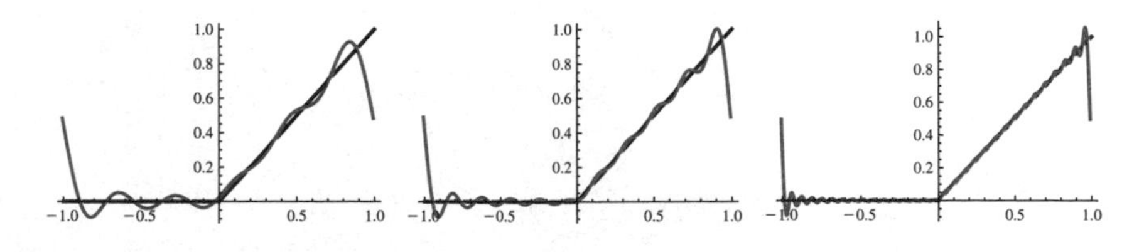

7. (a) $a_n = \dfrac{2}{\ell} \displaystyle\int_0^\ell f(x) \sin(\sqrt{\lambda_n}\, x)\, dx,\ n = 1, 2, \ldots,$

$$b_n = \frac{2}{\sqrt{\lambda_n}\, c\, \ell} \int_0^\ell g(x) \sin(\sqrt{\lambda_n}\, x)\, dx,\ n = 1, 2, \ldots$$

(b)

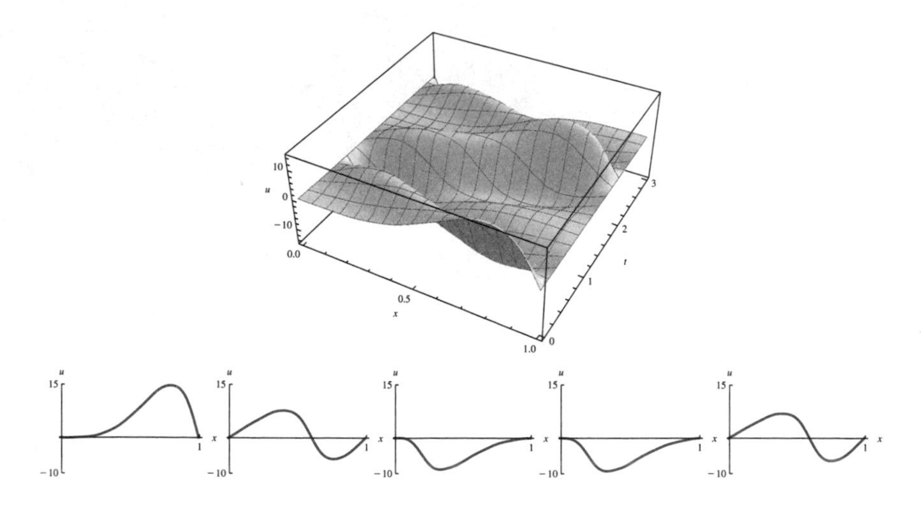

(c) Oscillates indefinitely.

9. (a) $a_n = \dfrac{2}{\ell} \displaystyle\int_0^\ell f(x) \cos(\sqrt{\lambda_n}\, x)\, dx,\ n = 1, 2, \ldots;$

$$b_n = \frac{2}{\sqrt{\lambda_n}\, c\, \ell} \int_0^\ell g(x) \cos(\sqrt{\lambda_n}\, x)\, dx,\ n = 1, 2, \ldots$$

(b)

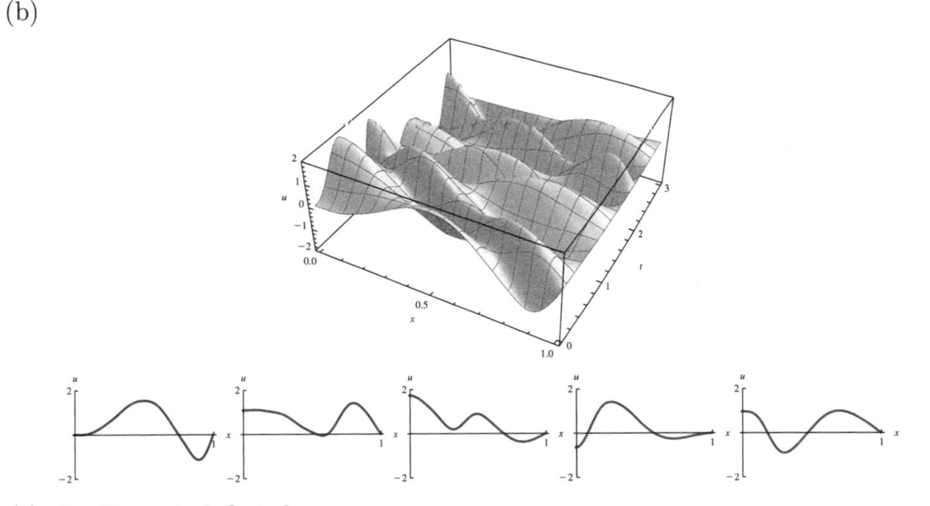

(c) Oscillates indefinitely.

11. The average value (in the calculus sense) of $f(x)$ on the interval $0 < x < \ell$.

2.4

1. Suppose $X'(a) + c_1X(a) = 0$ and $X'(b) + c_2X(b) = 0$ since every homogeneous Robin boundary condition can be written in this form. Then $X'(a) = -c_1X(a)$ and $X'(b) = -c_2X(b)$. Substitute these values for $X'(a)$ and $X'(b)$ in the right-hand side of (2.26) and simplify to get zero.

3. The right-hand side of (2.26) simplifies to $2(X_1'(a)X_2(a) - X_1(a)X_2'(a))$ in this case, which is not necessarily zero.

5. (a) Yes
   (b) Yes

7. (a) No integration is necessary. Since φ_n and φ_m are nonzero on disjoint intervals for $n \neq m$, their product is zero so that the integral of the product is also zero, so $\varphi_n \perp \varphi_m$.
   (b) $c_n = \dfrac{\langle f, \varphi_n \rangle}{\langle \varphi_n, \varphi_n \rangle} = \dfrac{\int_0^1 xe^{-5x}\varphi_n(x)\,dx}{\int_0^1 \varphi_n^2(x)\,dx}$, $n = 1, 2, \ldots, 10$

9. Rewrite the norms in terms of inner products and use the fact that this is an orthogonal family.

11. (a) Directly compute the inner products and show that they are zero after expanding using Euler's formula.
    (b) Similar to previous orthogonality arguments in the text.
    (c) Direct computation.

2.5

1. (a) $X(1) = 0$: right endpoint is held constant at 0.
   $X'(0) + X(0) = 0$: Robin boundary condition which can be restated as $X'(0) = -(X(0) - 0)$, so it is a physically unrealistic condition since $K < 0$.
   (b) $X_0(x) = x - 1$ (up to a constant multiple)
   (c) This results from the fact that $\tanh p = p$ has no nontrivial solutions.
   (d) $\lambda_n = p_n^2$ where p_n, $n = 1, 2, \ldots$ is the nth positive solution of $\tan p = p$.
   (f)

n	λ_n	Eigenfunction $y_n(x)$
1	20.191	$-4.493\cos(4.493x) + \sin(4.493x)$
2	59.680	$-7.725\cos(7.725x) + \sin(7.725x)$
3	118.900	$-10.904\cos(10.904x) + \sin(10.904x)$
4	197.858	$-14.066\cos(14.066x) + \sin(14.066x)$
5	296.554	$-17.221\cos(17.221x) + \sin(17.221x)$

3. (a) Left: The string is in a frictionless, vertical track attached to a spring which provides a restoring force in accordance with Hooke's Law when stretched. Right: The string is fixed at a height of zero. The last two conditions represent the position of the string, along with its velocity at each point, at time zero.
   (b) $X'' + \lambda X = 0, \quad X'(0) = X(0), \quad X(1) = 0; \qquad T'' + \lambda T = 0$
   (c) The eigenvalues are $\lambda_n = p_n^2$ where p_n is the nth positive solution of $p = -\tan p$, $n = 1, 2, \ldots$. The corresponding eigenfunctions are (up to a constant multiple) $X_n(x) = p_n \cos(p_n x) + \sin(p_n x)$, $n = 1, 2, \ldots$.
   (d) $T_n(t) = a_n \cos(\sqrt{\lambda_n}\, t) + b_n \sin(\sqrt{\lambda_n}\, t)$, $n = 1, 2 \ldots$
   (e) $u(x,t) = \sum_{n=1}^{\infty} \left[\sqrt{\lambda_n} \cos(\sqrt{\lambda_n}\, x) + \sin(\sqrt{\lambda_n}\, x)\right] \left[a_n \cos(\sqrt{\lambda_n}\, t) + b_n \sin(\sqrt{\lambda_n}\, t)\right]$
   (f)

$$a_n = \frac{\int_0^1 f(x) \left(\sqrt{\lambda_n} \cos(\sqrt{\lambda_n}\, x) + \sin(\sqrt{\lambda_n}\, x)\right)\, dx}{\int_0^1 \left(\sqrt{\lambda_n} \cos(\sqrt{\lambda_n}\, x) + \sin(\sqrt{\lambda_n}\, x)\right)^2\, dx}, \qquad n = 1, 2, \ldots,$$

$$b_n = \frac{1}{\sqrt{\lambda_n}} \frac{\int_0^1 g(x) \left(\sqrt{\lambda_n} \cos(\sqrt{\lambda_n}\, x) + \sin(\sqrt{\lambda_n}\, x)\right)\, dx}{\int_0^1 \left(\sqrt{\lambda_n} \cos(\sqrt{\lambda_n}\, x) + \sin(\sqrt{\lambda_n}\, x)\right)^2\, dx}, \qquad n = 1, 2, \ldots$$

5. (a) Second line: Physically unrealistic convection between the left endpoint and a surrounding medium fixed at 0°. Third line: Physically realistic convection between the right endpoint and a surrounding medium fixed at 0°. Last line: The initial temperature distribution of the wire.

(b) $X'' + \lambda X = 0,\ X'(0) + X(0) = 0,\ X'(1) + X(1) = 0$
$T' + \lambda T = 0$

(d) $\lambda_{-1} = -1$ with eigenfunction $X_{-1}(x) = \sinh x - \cosh x$.

(e) $\lambda_n = (n\pi)^2$ for $n = 1, 2, \ldots$, with corresponding eigenfunctions $X_n(x) = \sin(n\pi x) - n\pi \cos(n\pi x)$.

(f) $T_0(t) \equiv C$, $T_n(t) = Ce^{-\lambda t}$, $n = 1, 2, \ldots$

(g) $u(x,t) = a_{-1}e^t(\sinh x - \cosh x) + \sum_{n=1}^{\infty} a_n e^{-\lambda_n t}\left(\sin(\sqrt{\lambda_n}\, x) - \sqrt{\lambda_n}\cos(\sqrt{\lambda_n}\, x)\right)$

(h)

$$a_{-1} = \frac{\int_0^1 f(x)\,(\sinh x - \cosh x)\,dx}{\int_0^1 (\sinh x - \cosh x)^2\,dx},$$

$$a_n = \frac{\int_0^1 f(x)\,(\sin(n\pi x) - n\pi\cos(n\pi x))\,dx}{\int_0^1 (\sin(n\pi x) - n\pi\cos(n\pi x))^2\,dx}, \qquad n = 1, 2, \ldots$$

2.6

1. (a)

$$\begin{aligned} u_t &= ku_{xx}, && 0 < x < \ell, t > 0, \\ u(0,t) &= 100, && t > 0, \\ u_x(\ell,t) &= 0, && t > 0, \\ u(x,0) &= 1000x(1-x) + 100, && 0 < x < \ell \end{aligned}$$

(b)

$$\begin{aligned} v'' &= 0, && 0 < x < \ell, \\ v(0) &= 100, \\ v'(\ell) &= 0, \end{aligned}$$

which has the solution $v(x) = 100$.

(c)

$$\begin{aligned} w_t &= kw_{xx}, && 0 < x < \ell,\ t > 0, \\ w(0,t) &= 0, && t > 0, \\ w_x(\ell,t) &= 0, && t > 0, \\ w(x,0) &= f(x), && 0 < x < \ell, \end{aligned}$$

where $w(x,t) = \sum_{n=1}^{\infty} c_n \sin\left(\sqrt{\lambda_n}\, x\right) e^{-\lambda_n kt}$, $\lambda_n = \left(\frac{(2n-1)\pi}{2\ell}\right)^2$,

$$c_n = \frac{\int_0^1 f(x)\sin\left(\sqrt{\lambda_n}x\right)\,dx}{\int_0^1 \sin^2\left(\sqrt{\lambda_n}x\right)\,dx}, \qquad n = 1, 2, \ldots.$$

(d) $u(x,t) = v(x) + w(x,t) = 100 + \sum_{n=1}^{\infty} c_n \sin\left(\sqrt{\lambda_n}\, x\right) e^{-\lambda_n kt}$

(f) The "steady-state" solution is in fact a physically steady-state, since for large values of t, the exponential term takes over and all the terms in the sum go to zero, leaving $u(x,t) \to 100$ as $t \to \infty$.

3. (a)

$$\begin{aligned} u_t &= k u_{xx}, && 0 < x < \ell, t > 0, \\ u(0,t) &= 100, && t > 0, \\ -u_x(\ell,t) &= u(\ell,t) - 500, && t > 0, \\ u(x,0) &= f(x), && 0 < x < \ell \end{aligned}$$

(b)

$$\begin{aligned} v'' &= 0, && 0 < x < \ell, \\ v(0) &= 100, \\ -v'(\ell) &= v(\ell) - 500, \end{aligned}$$

which has the solution $v(x) = \frac{400}{\ell+1}x + 100$.

(c)

$$\begin{aligned} w_t &= k w_{xx}, && 0 < x < \ell,\ t > 0, \\ w(0,t) &= 0, && t > 0, \\ -w_x(\ell,t) &= w(\ell,t), && t > 0, \\ w(x,0) &= f(x) - v(x), \end{aligned}$$

which has solution

$$w(x,t) = \sum_{n=1}^{\infty} a_n e^{-\lambda_n kt} \sin(\sqrt{\lambda_n}\, x),$$

where $\lambda_n = p_n^2$, p_n is the nth positive solution of $p = -\tan p\ell$, $n = 1, 2, \ldots,$ and

$$a_n = \frac{\int_0^\ell (f(x) - v(x))\sin(\sqrt{\lambda_n}\, x)\,dx}{\int_0^\ell \sin^2(\sqrt{\lambda_n}\, x)\,dx}, \qquad n = 1, 2, \ldots$$

(d) $u(x,t) = w(x,t) + v(x)$

(f) As $t \to \infty$, the exponential term takes over in each term of the sum, so that $u(x,t) \to 200x + 100$. So in the limit, $u(x,t)$ is physically steady-state, with the temperature varying linearly from 100 at the left end to 300 at the right end.

5. (a) Heat diffusion in a 1D rod of length ℓ; the temperature at the left endpoint is maintained at $f(t)$ degrees; the right endpoint convects in a physically realistic way with a surrounding medium kept at temperature $f_2(t)$; the initial temperature distribution in the rod is $g(x)$.

(b) $v_1(t) = f_1(t)$, $v_2(t) = \dfrac{\ell f_2(t) + f_1(t)}{\ell + 1}$

(c) $v(x,0) = f_1(0) + \dfrac{f_2(0) - f_1(0)}{\ell + 1} x$

3.1

1. (a) $(fg)(-x) = f(-x)g(-x) = f(x) \cdot (-g(x) = -f(x)g(x) = -(fg)(x)$ so fg is odd.

(b) $(fg)(-x) = f(-x)g(-x) = f(x)g(x) = (fg)(x)$ so fg is even.

(c) $(fg)(-x) = f(-x)g(-x) = (-f(x)) \cdot (-g(x)) = f(x)g(x) = (fg)(x)$ so fg is even.

3. (b) $b_{99} = 0$

(c) $a_{99} = 0$

(d) 1; 0; it depends on how f is extended to $-1 < x < 0$.

5. (a)

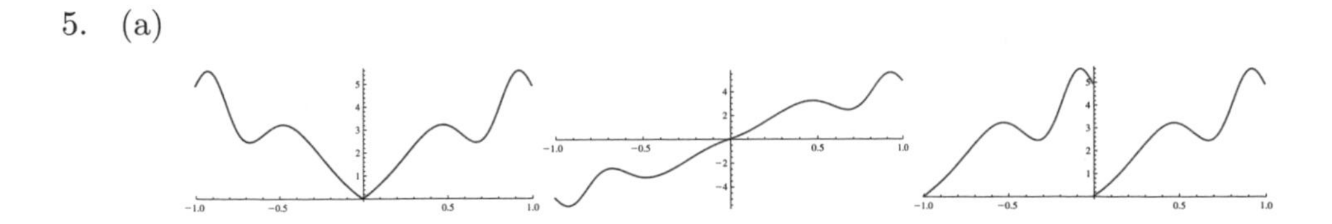

(b)

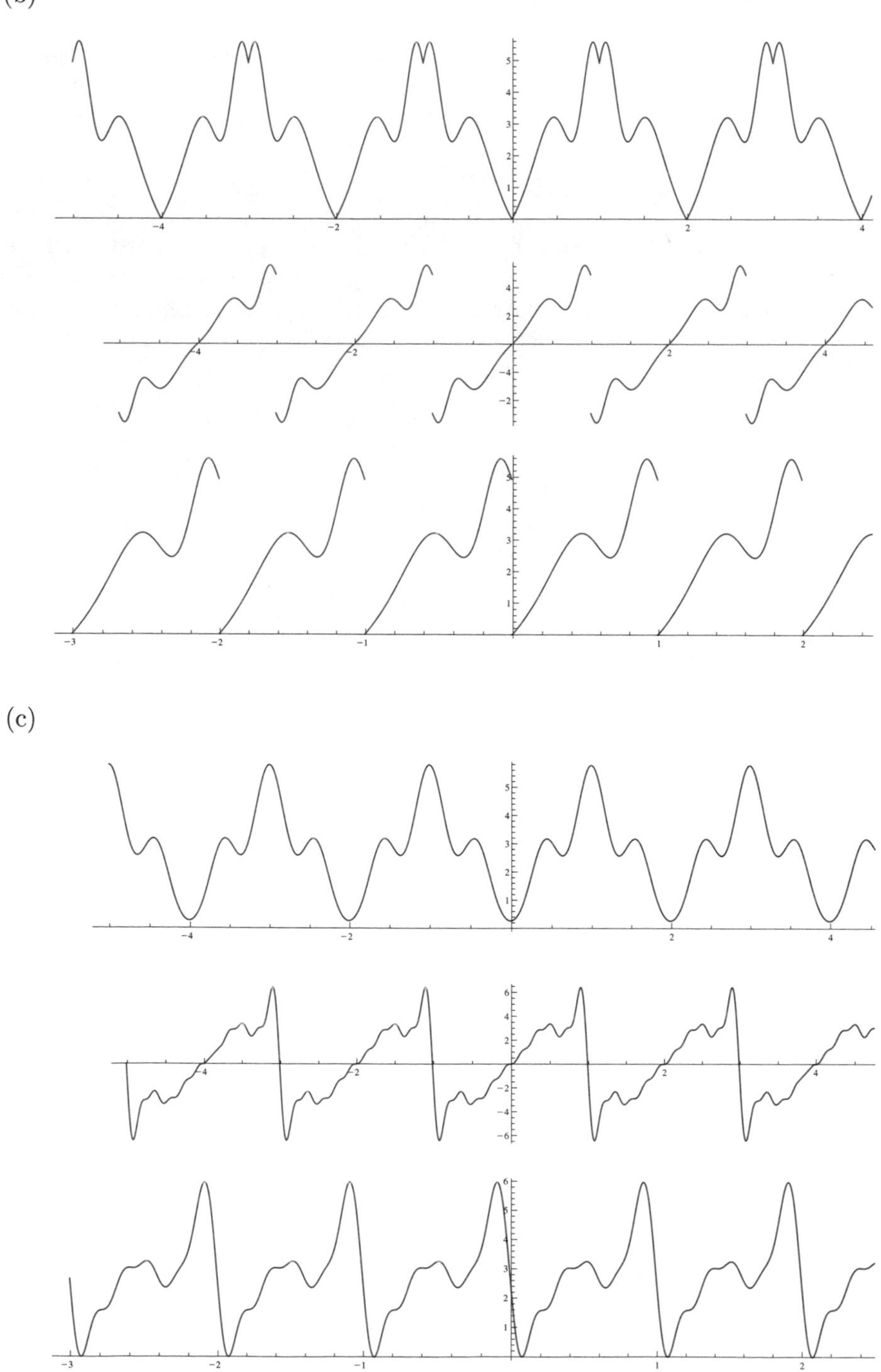

(c)

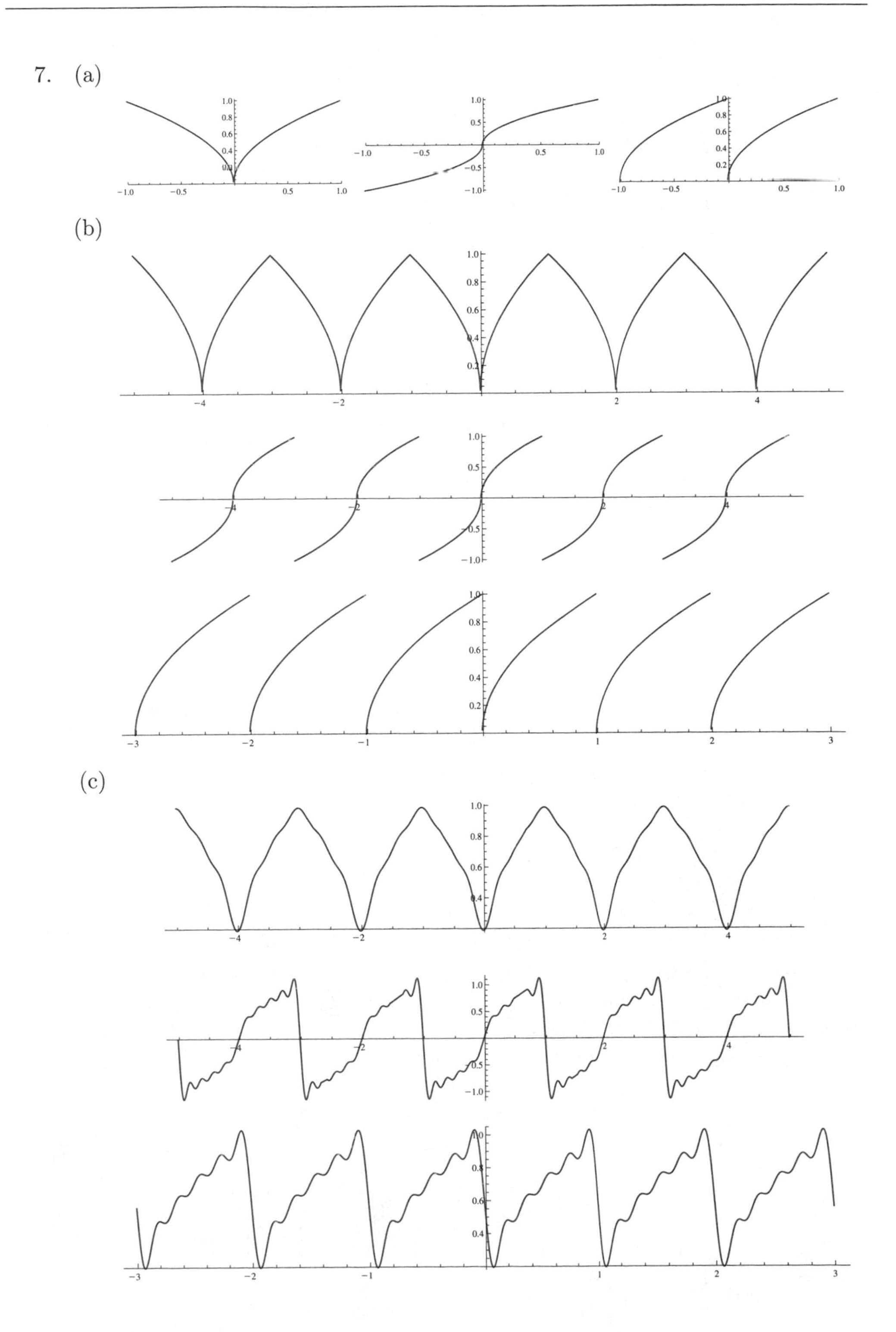
7. (a)
(b)
(c)

9. (a) $b_n = \dfrac{2}{3}\displaystyle\int_0^3 (1+2x)\sin\left(\frac{n\pi}{3}x\right)\,dx$

11. Fourier cosine series; Fourier sine series; full Fourier series

13. (a) f must be continuous on $[0,\ell]$.

    (b) f must be continuous on $[0,\ell]$, and $f(0)=f(\ell)=0$.

3.2

1. (a)

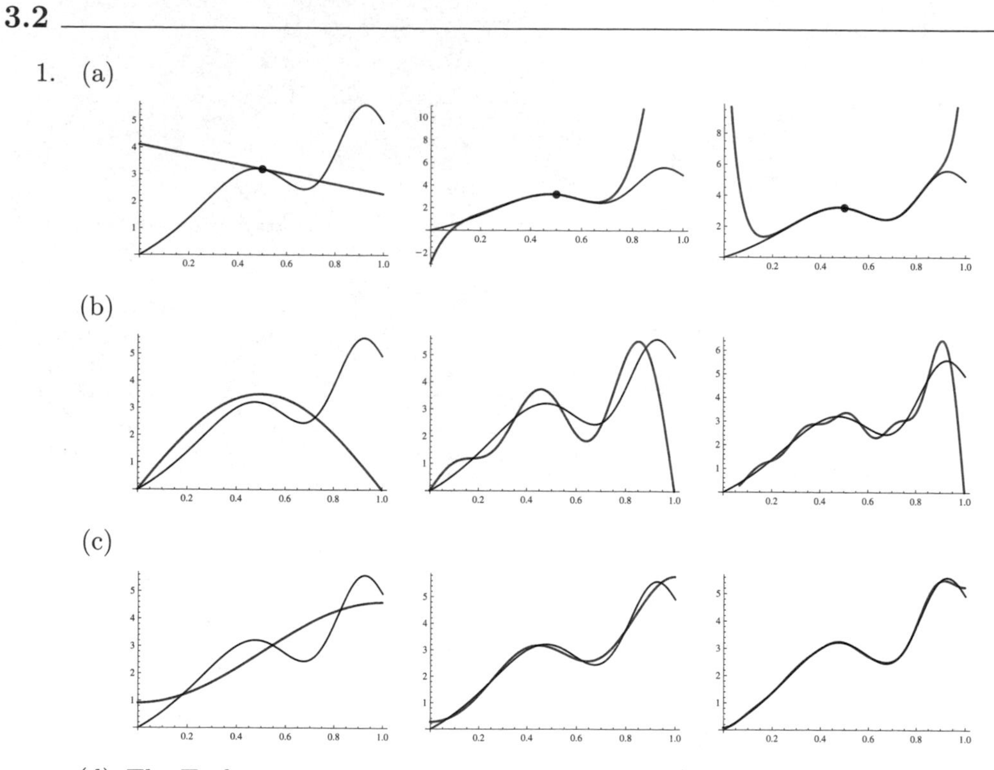

   (d) The Taylor approximations are best near $x=1/2$, and differ from the given function more and more away from $x=1/2$. The Fourier approximations, on the other hand, provide a better approximation over the entire interval rather than at one particular point.

3. (a)

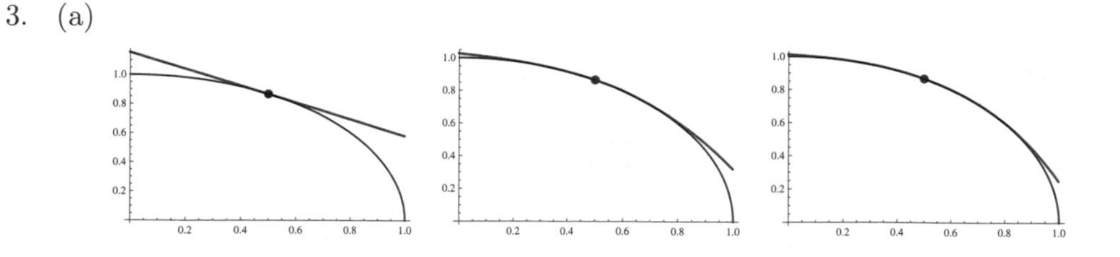

(b)

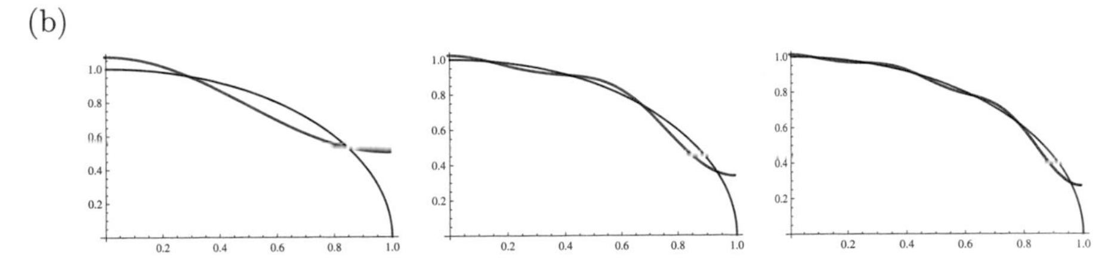

(c) Both are good approximations, but note how the Taylor series is striving to approximate the function at $x = 1/2$, whereas the Fourier series is striving to approximate the behavior of the given function over the entire interval.

7. (a) $x_0 - R < x < x_0 + R$, where R is the radius of convergence of the Taylor series for $f(x)$ at x_0.

   (b) $x_0 - R < x < x_0 + R$, where R is the radius of convergence of the Taylor series for $f(x)$ at x_0.

3.3

1. (a) $p(0.25) \approx 0.2932$, $p(0.5) \approx 0.7154$, $p(0.75) \approx 1.3674$

   (b) ≈ 4.9249 at $x = 1$

   (c) ≈ 1.5893.

3. (c)

N	$\max\limits_{0\le x\le 3} \lvert f(x) - T_N(x)\rvert$	$\max\limits_{0\le x\le 3} \lvert f(x) - F_N(x)\rvert$
1	1.71748	1
5	4.54949	1
10	12.2732	1

   (d)

N	$\sqrt{\int_0^3 \lvert\text{erf}(x) - T_N(x)\rvert^2\,dx}$	$\sqrt{\int_0^3 \lvert\text{erf}(x) - F_N(x)\rvert^2\,dx}$
1	1.39602	0.564403
5	2.18542	0.332003
10	4.11986	0.240524

5. (a) Show that $|1 - x^N - 1| \to 0$ as $N \to \infty$. Use the fact that $0 < x < 1$.

   (b) Show that $\max\limits_{0\le x\le 1} |1 - x^N - 1| \not\to 0$ as $N \to \infty$.

   (c) Show that $\int_0^1 |1 - x^N - 0|^2\,dx \to 0$ as $N \to \infty$.

3.4

1. (b) The Pointwise Convergence Theorem says that the Fourier series converges to the average of the left- and right-hand limits. In this case, both limits are 3.

   (c) No (explain why).

   (d) Yes (explain why).

3. (a) ℓ^2

   (b) 0

   (c) It depends on the extension chosen. It is the average of the left and right limits of the extension of f at $\pm\ell$.

5. (a) Converges for all three; can differentiate to get f'.

   (b) Converges pointwise and L^2, but not uniformly. Cannot conclude that we can differentiate to get f'.

   (c) Cannot conclude that it either does or does not converge pointwise or uniformly; does converge in L^2. Cannot conclude that we can differentiate to get f'.

   (d) Cannot conclude that it either does or does not converge pointwise. Does not converge uniformly but does in L^2. Cannot conclude that we can differentiate to get f'.

   (e) Converges for all three; can differentiate to get f' except at $x = 0$ (where f' is undefined).

   (f) Cannot conclude that it converges pointwise; does not converge uniformly or in L^2. Cannot conclude that we can differentiate to get f'.

   (g) Converges in all three; can differentiate to get f' except at $x = -\pi$, 0, or π, where f is not differentiable.

7. (a) For example, extend $f(x)$ to $[-1, 1]$ via $f_{\text{even}}(x)$.

   (b) e

9. (a) Any values of m and b will work.

   (b) $m = e^{-1} - 1$, $b = 1$

   (c) Any values of m and b will work.

11. (a) Apply the Weierstrass M-Test with $M_n = \frac{4}{\pi} \cdot \frac{1}{n^2}$. Then $\sum_{n=1}^{\infty} M_n$ converges since $\sum_{n=1}^{\infty} \frac{1}{n^2}$ converges.

    (c) Use (b) with the given a_n, b_n to conclude yes.

3.5

1. (a) (IP1)–(IP4) all follow directly, with some calculation, from the corresponding properties of the dot product.

   (b) (IP1)–(IP3) are all trivial properties of integrals. (IP4) follows from the fact that $u(x)^2 \geq 0$ so that its integral is nonnegative, and is strictly positive unless $u(x) \equiv 0$.

   (c) No. Take $u(x) \equiv 1$, $v(x) \equiv -1$, $w(x) \equiv 1$. Then (IP2) fails.

3. (a) and (d) are in $L^2[0,1]$, while (b) and (c) are not.

5. (a) $c_n = \dfrac{\langle f, P_n \rangle}{\langle P_n, P_n \rangle} = \dfrac{\int_{-1}^{1} |x| P_n(x)\, dx}{\int_{-1}^{1} P_n^2\, dx}$, $n = 0, 1, 2$, which yields the coefficients $c_0 = \frac{1}{2}$, $c_1 = 0$, $c_2 = \frac{5}{8}$.

7. (a)

N	1	2	3	5	10	19	20
E_N	0.33423	0.27436	0.23601	0.19042	0.13862	0.10188	0.09937

   The first method seemed slightly faster to compute. 20 terms are required in order for the RMS error to be less than 0.1.

   (b)

N	1	2	3	5	10
E_N	0.12248	0.05875	0.03702	0.01904	0.00727

   The first method seemed slightly faster to compute. Two terms are required in order for the RMS error to be less than 0.1.

9. (a) Taking $\sin(n\pi x/\ell)$ to be the complete orthogonal family, apply Parseval's equality and then evaluate $\|\sin(n\pi x/\ell)\|^2 = \ell/2$. The result follows.

   (b) Taking $\cos(n\pi x/\ell)$ to be the complete orthogonal family, apply Parseval's equality and then evaluate $\|\cos(n\pi x/\ell)\|^2 = \ell/2$ (except for $n = 0$, where the norm is ℓ). The result follows.

   (c) Similar to the first two parts, using $\{\sin(n\pi x/\ell)\}_{n=1}^{\infty} \cup \{\cos(n\pi x/\ell)\}_{n=0}^{\ell}$ as the complete orthogonal family.

13. No. Parseval's equality in this case gives the harmonic series on the right-hand side, which diverges, while the left-hand side must be finite since $f \in L^2[-\ell, \ell]$.

15. (a) (IP2) and (IP3) are proved exactly as in Exercise 1(a). (IP1′) and (IP4) follow from basic properties of complex conjugation.

   (b) (IP2) and (IP3) are proved exactly as in Exercise 1(b). (IP1′) follows from basic properties of complex conjugation. (IP4) follows from the fact that the integrand of the integral defining $\langle u, u \rangle$ is $u(x)\overline{u(x)} = \|u(x)\|^2$.

3.6

1. (a) False. It refers specifically to the persistent overshoot of the approximations in the neighborhood of a jump discontinuity.

   (b) True, assuming the full Fourier series converges pointwise. Fourier *series* means summing *infinitely* many terms, but the Gibbs phenomenon only occurs when summing a *finite* number of terms in the series.

   (c) False. The Gibbs phenomenon only occurs in the neighborhood of a jump discontinuity, and the Uniform Convergence Theorem says that functions with a jump discontinuity *cannot* converge uniformly.

   (d) False. Lanczos sigma factors reduce but do not eliminate the overshoots.

   (e) False. Lanczos sums do not even converge pointwise to $f(x)$ at the jump discontinuity. Also, see the reasoning in (c).

5. (a)

N	$F_N(x)$	$\lvert F_N(x) - f(x)\rvert$
10	1.28361	0.28361
20	1.22952	0.22952
40	1.20382	0.20382
60	1.19544	0.19544
80	1.19129	0.19129

   It appears that the overshoot is approaching the expected value of 0.17898.

   (c) Note that $S_N(x)$ does not have a local maximum near zero; its first positive critical point is at $x \approx 2.69469$. So we will start the table at $N = 20$.

	Standard Sums, $F_N(x)$		**Lanczos Sums, $S_N(x)$**		**Comparison**	
N	overshoot	L^2 error	overshoot	L^2 error	overshoot %	L^2 error %
20	0.22952	0.356692	0.11032	0.439097	51.9367	23.1027
40	0.20382	0.252290	0.06561	0.310506	67.8086	23.0753
60	0.19544	0.206004	0.05141	0.253530	73.6942	23.0703
80	0.19129	0.178408	0.04441	0.219564	76.7842	23.0685

   (d) Using Lanczos sums continues to increase the L^2 error by about 23%.

7. (c) $N = 7$

   (d) With $N = 7$, the error is ≈ 1.85194, and $\frac{1.85194}{\pi} \approx 0.58949$.

   (e) Because this Taylor series has an infinite radius of convergence.

9. (a) Evaluate $D_N(-\theta)$ using the formula from Exercise 8(c).

4.1

3. (a) $p(x) = e^x$, $q(x) = 0$, $w(x) = e^x$

   $\frac{1}{e^x}\left((e^x y')'\right) + \lambda y = 0,\ 0 < x < 1;\ y(0) = 0,\ y(1) = 0$

   (b) Yes

   (c) Yes. From (b), the operator is a regular Sturm-Liouville operator, and these are symmetric by Theorem 4.2.

   (d) $\lambda_n = \frac{4n^2\pi^2+1}{4}$, $y_n = e^{-x/2}\sin(n\pi x)$, $n = 1, 2, \ldots$

   The eigenvalues are real and form an unbounded, increasing sequence. The eigenfunctions form a complete orthogonal family for $L^2_w[0,1]$. Each eigenfunction is unique up to a constant multiple. We can use the coefficient formulas in Theorem 4.3 to compute eigenfunction expansions with this set of basis functions.

   (e) $x = \sum_{n=1}^{\infty} c_n e^{-x/2}\sin(n\pi x)$ where $c_n = 2\int_0^1 x e^{x/2}\sin(n\pi x)\,dx$

   (f) $N = 55$

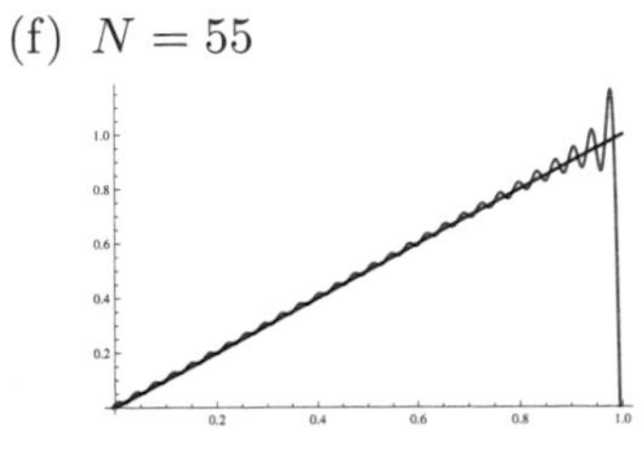

5. (a) $p(x) = x$, $q(x) = 0$, $w(x) = 1/x$

   $\frac{1}{1/x}\left[(xy')'\right] + \lambda y = 0,\ 1 < x < 2;\ y(1) = 0,\ y'(2) = 0$

   (b) Yes

   (c) Yes. From (b), the operator is a regular Sturm-Liouville operator, and these are symmetric by Theorem 4.2.

   (d) $\lambda_n = \left(\frac{(2n+1)\pi}{\ln 4}\right)^2$, $y_n = \sin\left(\sqrt{\lambda_n}\,\ln x\right)$, $n = 0, 1, 2\ldots$

   (e) $f(x) = \sum_{n=0}^{\infty} c_n \sin\left(\sqrt{\lambda_n}\,\ln x\right)$ where $c_n = \frac{\int_1^2 f(x)\sin\left(\sqrt{\lambda_n}\,\ln x\right)\frac{1}{x}\,dx}{\int_1^2 (\sin\left(\sqrt{\lambda_n}\,\ln x\right))^2\frac{1}{x}\,dx}$

   (f) $N = 14$

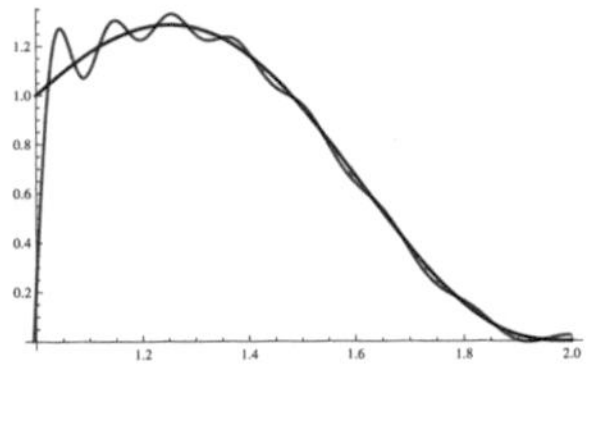

7. (a) $p(x) = 1$, $q(x) = -2$, $w(x) = 1$
   $\left[(y')' - 2y\right] + \lambda y = 0$, $0 < x < \pi$; $y(0) = y(\pi)$, $y'(0) = y'(\pi)$
   (b) No, the boundary conditions are not in the required form.
   (c) Yes
   (d) Yes, periodic Sturm-Liouville operators are symmetric by Theorem 4.4.
   (e) $\lambda_0 = 2$, $y_0(x) = 1$;
   $\lambda_n = 4n^2 + 2$, $y_n(x) = A\cos(2nx) + B\sin(2nx)$, $n = 1, 2, \ldots$
   (f) The orthogonal expansion is given by Theorem 4.3(d), with coefficients

$$a_0 = \frac{\pi}{8}, \qquad a_n = \frac{2}{\pi}\int_{\pi/2}^{\pi} f(x)\cos(2nx)\,dx, \qquad b_n = \frac{2}{\pi}\int_{\pi/2}^{\pi} f(x)\sin(2nx)\,dx$$

   (g) $N = 39$

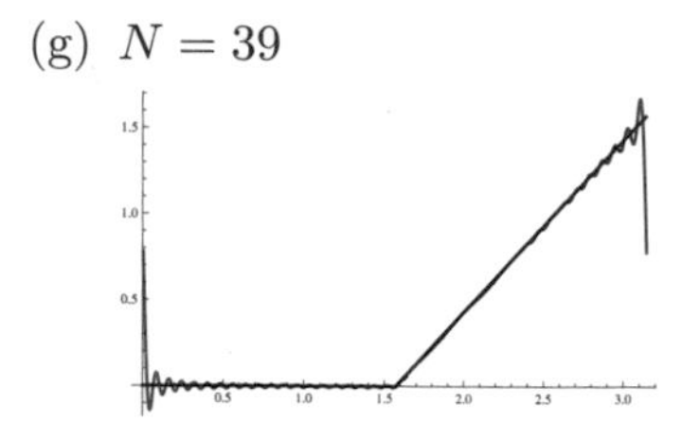

9. Expand the right-hand side of (4.8) or (4.9); periodicity causes the expansion to reduce to zero.

11. There are no eigenvalues. This does not contradict Theorem 4.5 because the boundary conditions are not regular or periodic.

13. Since $\mathcal{S}$ is a real operator (i.e., has real coefficients), it follows that $\mathcal{S}\bar{y} = \bar{\mathcal{S}}\bar{y} = \overline{\mathcal{S}y}$. But $\overline{\mathcal{S}y} = \overline{-\lambda y} = -\bar{\lambda}\bar{y}$ by basic properties of complex conjugation.

15. The theory is analogous to the theory of eigenvalues and eigenvectors of real symmetric matrices: any such matrix has n real eigenvalues and the eigenvectors form an orthogonal system and thus a basis for $\mathbb{R}^n$; this matches parts (a)–(c) of Theorem 4.3. Part (d) corresponds exactly to the linear algebra theorem regarding representation of a vector in the basis given by eigenvectors.

4.2

1. (a)
   - $w(x) = x$, $p(x) = x^2$, $q(x) = 0$;

$$\frac{1}{x}\left[\left(x^2 y'\right)'\right] + \lambda y = 0, \quad 0 < x < 1$$

   - The problem is singular at $x = 0$ because $p(0) = 0$, but also because $w(0) = 0$.

- The modified boundary condition at $x = 0$ is y, y' bounded as $x \to 0^+$. The boundary condition at $x = 1$ remains unchanged.

(b)
- $w(x) = \frac{1}{(x-1)^2}$, $p(x) = \dfrac{1}{1-x}$, $q(x) = \dfrac{-1}{(x-1)^2}$;

$$(x-1)^2 \left[\left(\frac{1}{1-x} y' \right)' - \frac{1}{(x-1)^2} y \right] + \lambda y = 0, \quad 0 < x < 1$$

- The problem is singular at $x = 1$ because p, q, and w become infinite as $x \to 1^-$.
- The modified boundary condition at $x = 1$ is y, y' bounded as $x \to 1^-$. The boundary condition at $x = 0$ remains unchanged.

(c)
- $w(x) = \frac{1}{(1-x)^{3/2}\sqrt{1+x}}$, $p(x) = \dfrac{\sqrt{1+x}}{\sqrt{1-x}}$, $q(x) = \frac{1}{(1-x)^{3/2}\sqrt{1+x}}$;

$$(1-x)^{3/2}\sqrt{1+x} \left[\left(\frac{\sqrt{1+x}}{\sqrt{1-x}} y' \right)' + \frac{1}{(1-x)^{3/2}\sqrt{1+x}} y \right] + \lambda y = 0$$

- The problem is singular at $x = \pm 1$ since p, q, w become infinite there.
- The modified boundary conditions are y, y' bounded as $x \to 1^-$ and as $x \to -1^+$.

(d)
- $w(x) = \sin x$, $p(x) = \sin x$, $q(x) = \csc x$;

$$\csc x \left[(y' \sin x)' + \csc x \, y \right] + \lambda y = 0, 0 < x < \pi; \quad y(0) = 0, y(\pi) = 0$$

- The problem is singular at $x = 0$ and $x = \pi$ since $p(x)$ and $w(x)$ are zero at these points (and q becomes infinite at these points).
- The modified boundary condition at $x = 0$ is y, y' bounded as $x \to 0^+$. The modified boundary condition at $x = \pi$ is y, y' bounded as $x \to \pi^-$.

3. Every $\lambda > 0$ is an eigenvalue, with corresponding eigenfunction $\sin\left(\sqrt{\lambda}\, x\right)$.

5. (a) The Sturm-Liouville form is $x\left[(xy')'\right] + \lambda y = 0,\ 0 < x < 1;\ y(0) = 0,\ y(1) = 0$. The only singular point is $x = 0$ since $p(0) = 0$ and $w(x) \to \infty$ as $x \to 0^+$.

   (b) The modified boundary condition at $x = 0$ is y, y' remain bounded as $x \to 0^+$. The boundary condition at $x = 1$ remains unchanged.

   (c) Every $\lambda > 0$ is an eigenvalue, with corresponding eigenfunctions $\sin\left(\sqrt{\lambda}\, \ln x\right)$.

7. $\alpha\beta = -1$

4.3

3. (a) $x^3 + |x| = \sum_{n=0}^{\infty} c_n P_n(x)$ where $c_n = \dfrac{2n+1}{2} \displaystyle\int_{-1}^{1} \left(x^3 + |x|\right) P_n(x)\, dx$.

(b)

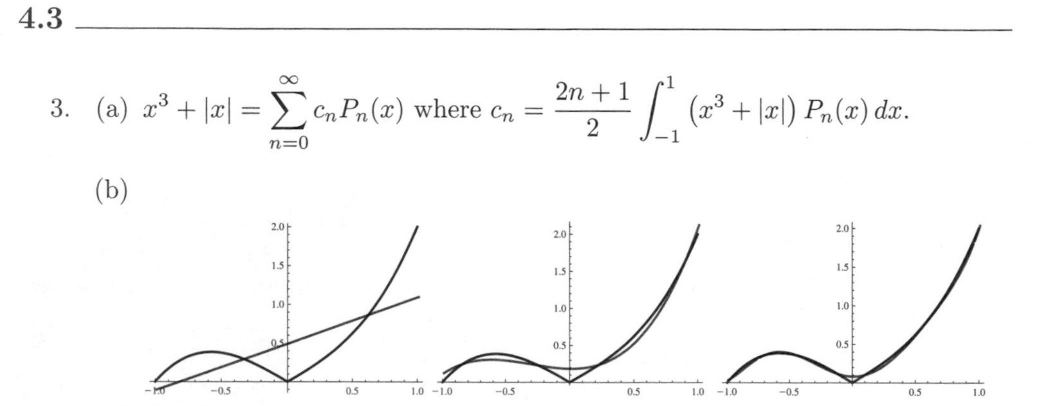

(c) Note that all three are plotted on the same scale:

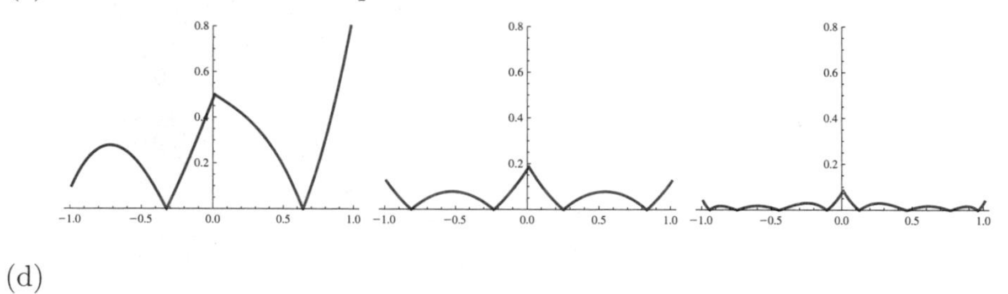

(d)

N	Uniform error	L^2_w error
1	0.9	0.4608
3	0.1875	0.1021
6	0.0854	0.0319

(e) $N = 4$

(f) Converges in all three senses.

5. (a) $e^{-x}\sin(2\pi x) = \sum_{n=0}^{\infty} c_n P_n(x)$ where $c_n = \dfrac{2n+1}{2} \displaystyle\int_{-1}^{1} e^{-x}\sin(2\pi x) P_n(x)\, dx$.

(b)

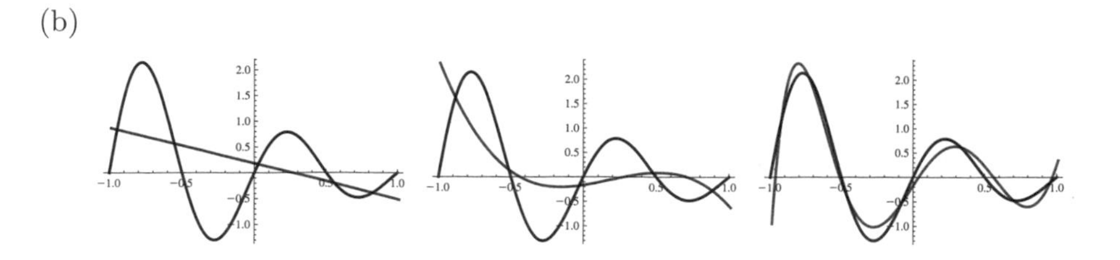

(c) Note that all three are plotted on the same scale:

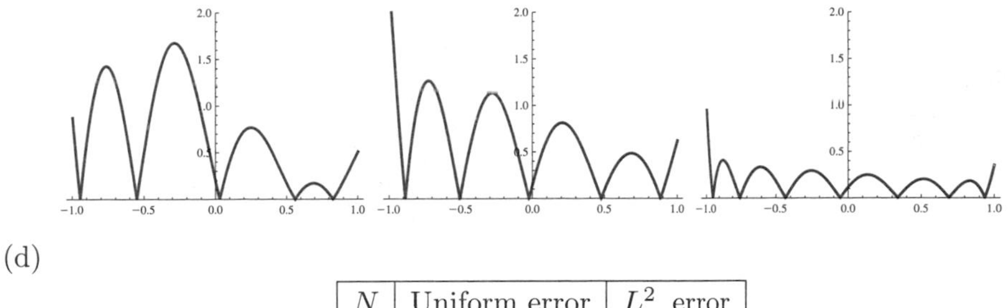

(d)

N	Uniform error	L^2_w error
1	1.6775	1.1761
3	2.3302	1.0133
6	0.9437	0.2963

(e) $N = 8$

(f) Converges pointwise and in the L^2_w sense, but not uniformly.

7. (a) $a_{k+2} = \dfrac{(k-n)(k+n)}{(k+1)(k+2)} a_k,\ k = 0, 1, 2, \ldots.$

(b) The recurrence associates all even-numbered terms and all odd-numbered terms. The terms become zero once $k = n$. Thus for a given n, one series terminates and one does not. The nonterminating series is rejected as in the text. Thus we expect the even-numbered polynomials to consist of only even powers and the odd-numbered ones to consist of only odd powers.

(c) The limit of the ratio of successive terms in either sequence is x^2, so the radius of convergence is 1. At the endpoints, note that the signs of all terms after $k = n + 1$ are the same, so the sum is unbounded, so it cannot satisfy (4.32).

(d) Apply the recurrence starting with $a_n = 2^{n-1}$ and compute.

(e) See part (b).

9. (a) $\sqrt{x+1} = \sum_{n=0}^{\infty} c_n T_n(x)$ where $c_n = \dfrac{\int_{-1}^{1} \sqrt{x+1}\, T_n(x)(1-x^2)^{-1/2}\, dx}{\int_{-1}^{1} T_n^2(x)(1-x^2)^{-1/2}\, dx}$.

(b)

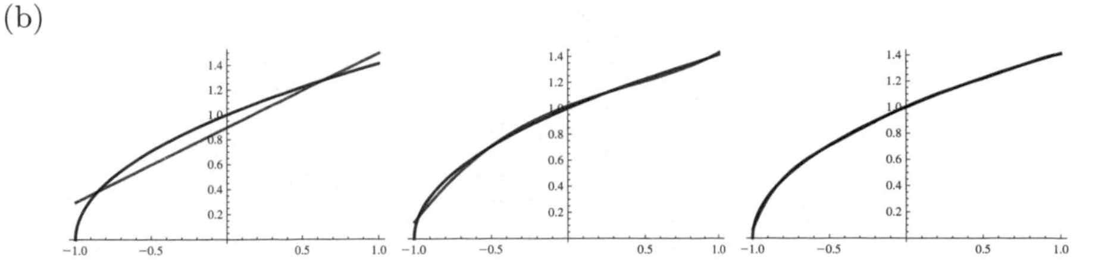

(c) Note that all three are plotted on the same scale:

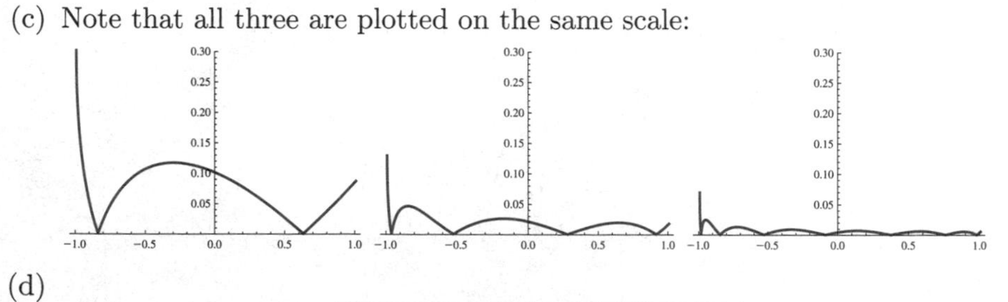

(d)

N	Uniform error	L^2_w error
1	0.3001	0.1710
3	0.1286	0.0494
6	0.0243	0.0196

(e) $N = 2$

(f) Converges in all three senses.

11. (a) $\sqrt{1-x^2} = \sum_{n=0}^{\infty} c_n T_n(x)$ where $c_n = \dfrac{\int_{-1}^{1} T_n(x)\,dx}{\int_{-1}^{1} T_n(x)^2\,dx}$. Note that the $\sqrt{1-x^2}$ term cancels with the weight function. Since $T_n(x)$ is odd for n odd, $c_n = 0$ for n odd.

(b)

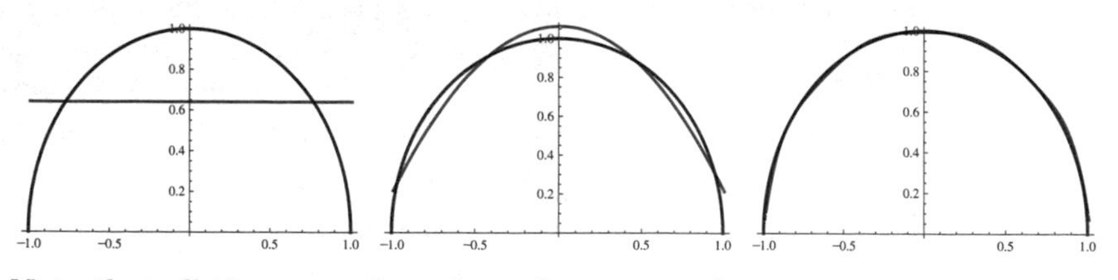

(c) Note that all three are plotted on the same scale:

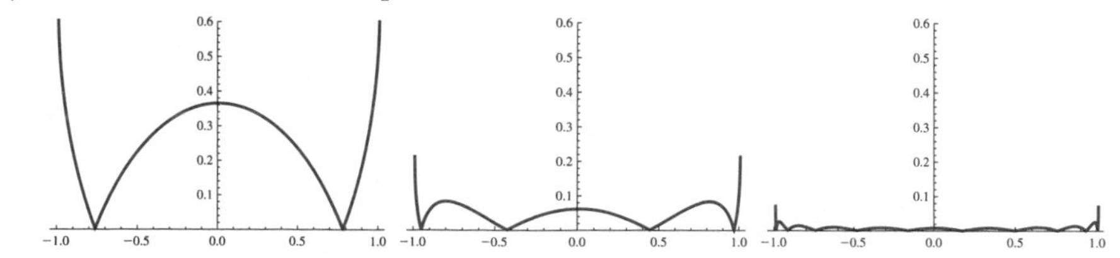

(d)

N	Uniform error	L^2_w error
1	0.6366	0.5455
3	0.2122	0.1209
6	0.0324	0.0349

(e) $N = 4$

(f) Converges in all three senses.

13. (a) $xe^{-5x} = \sum_{m=1}^{\infty} c_m J_1\left(\sqrt{\lambda_{1m}}\,x\right)$ where $c_m = \dfrac{\int_0^1 xe^{-5x} J_1\left(\sqrt{\lambda_{1m}}\,x\right) x\,dx}{\int_0^1 J_1\left(\sqrt{\lambda_{1m}}\,x\right)^2 x\,dx}$.

(b)

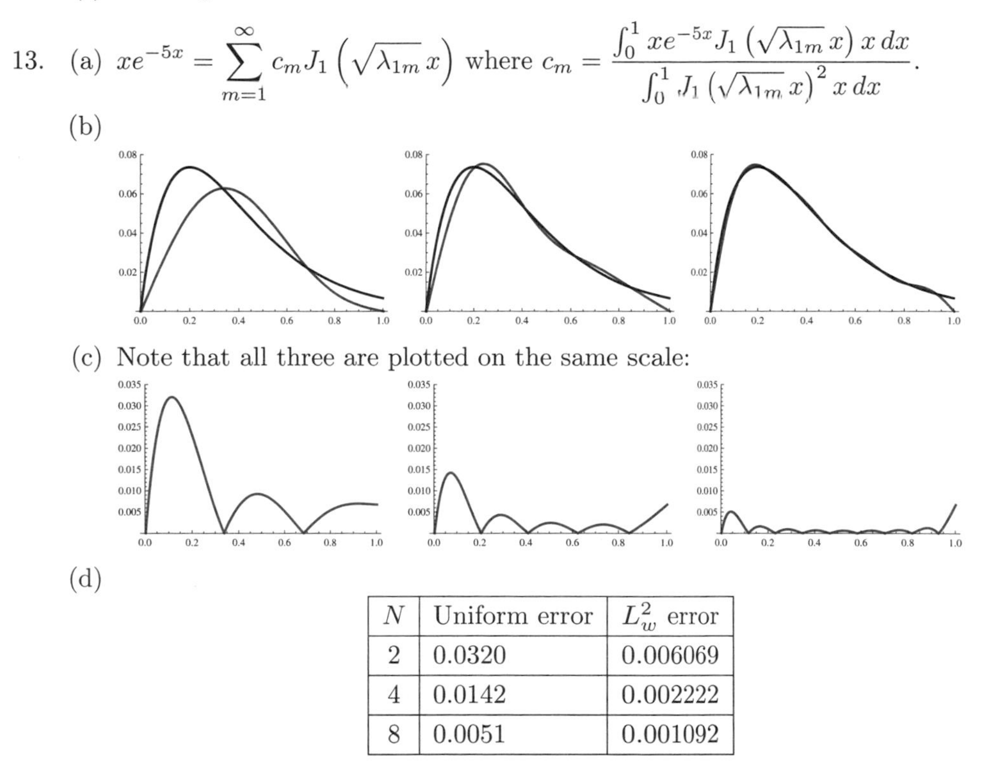

(c) Note that all three are plotted on the same scale:

(d)

N	Uniform error	L^2_w error
2	0.0320	0.006069
4	0.0142	0.002222
8	0.0051	0.001092

(e) $N = 1$

(f) Converges in all three senses.

15. Substitute $x = \frac{\xi}{\ell}$; the range is transformed from $0 < x < 1$ to $0 < \xi < \ell$. Compute $y'(x)$ and $y''(x)$ in terms of ξ, substitute, and simplify. The formula for the eigenvalues follows. To see that the eigenfunctions are as required, write $y_{nm}(\xi) = J_n(z_{nm}\xi/\ell)$; then

$$y'_{nm}(\xi) = J'_n(z_{nm}\xi/\ell)x'(\xi) = \frac{1}{\ell}J'_n(z_{nm}\xi),$$
$$y''_{nm}(\xi) = \frac{1}{\ell}J''_n(z_{nm}\xi)x'(\xi) = \frac{1}{\ell^2}J''_n(z_{nm}\xi)$$

Substituting these into the differential equation involving ξ reduces it to the original equation, so it is zero and these are indeed the eigenfunctions.

4.4

1. (a) $\dfrac{1}{x}\left[(xy'(x))' - \dfrac{p^2}{x}y(x)\right] + y = 0$

(b) $\int_0^1 J_p(ax)J_p(bx)x\,dx = 0$, where a and b are distinct roots of J_p.

3. (a) The general solution is $y(x) = c_1 J_{1/2}(x) + c_2 J_{-1/2}(x)$. The first three nonzero terms for $J_{1/2}(x)$ are

$$\frac{1}{2^{1/2}\Gamma(3/2)}x^{1/2} - \frac{1}{2^{5/2}\Gamma(5/2)}x^{5/2} + \frac{1}{2^{11/2}\Gamma(7/2)}x^{9/2}$$

and for $J_{-1/2}(x)$ they are

$$\frac{1}{2^{-1/2}\Gamma(1/2)}x^{-1/2} - \frac{1}{2^{3/2}\Gamma(3/2)}x^{3/2} + \frac{1}{2^{9/2}\Gamma(5/2)}x^{7/2}$$

(b) The general solution is $y(x) = c_1 J_{\sqrt{\pi}}(x) + c_2 J_{-\sqrt{\pi}}(x)$. The first three nonzero terms for $J_{\sqrt{\pi}}(x)$ are

$$\frac{1}{2^{\sqrt{\pi}}\Gamma(1+\sqrt{\pi})}x^{\sqrt{\pi}} - \frac{1}{2^{2+\sqrt{\pi}}\Gamma(2+\sqrt{\pi})}x^{2+\sqrt{\pi}} + \frac{1}{2^{5+\sqrt{\pi}}\Gamma(3+\sqrt{\pi})}x^{4+\sqrt{\pi}}$$

and for $J_{-\sqrt{\pi}}$ they are

$$\frac{1}{2^{-\sqrt{\pi}}\Gamma(1-\sqrt{\pi})}x^{-\sqrt{\pi}} - \frac{1}{2^{2-\sqrt{\pi}}\Gamma(2-\sqrt{\pi})}x^{2-\sqrt{\pi}} + \frac{1}{2^{5-\sqrt{\pi}}\Gamma(3-\sqrt{\pi})}x^{4-\sqrt{\pi}}$$

5. (a)

p	Roots				
0	2.40483	5.52008	8.65373	11.79153	14.93092
1	3.83171	7.01559	10.17347	13.32369	16.47063
2	5.13562	8.41724	11.61984	14.79595	17.95982

(b)

p	Roots				
0	0.89358	3.95768	7.08605	10.22235	13.36110
1	2.19714	5.42968	8.59601	11.74915	14.89744
2	3.38424	6.79381	10.02348	13.20999	16.37897

9. We compute the limit of the ratio between successive terms of the series

$$\begin{aligned}\lim_{k\to\infty}\left|\frac{a_{k+1}}{a_k}\right| &= \lim_{k\to\infty}\left|\frac{(-1)^{k+1}x^{2(k+1)+p}}{2^{2(k+1)+p}(k+1)!(k+p+1)!}\cdot\frac{2^{2k+p}k!(k+p)!}{(-1)^k x^{2k+p}}\right| \\ &= \lim_{k\to\infty}\left|\frac{1}{4(k+1)(k+p+1)}x^2\right| = 0\end{aligned}$$

so that the radius of convergence is infinite.

4.5

1. (a) The determinant of the matrix formed by the four vectors is nonzero, so they are four linearly independent vectors, and therefore form a basis for $\mathbb{R}^4$.

   (b)

$$\mathbf{u}_1 = \begin{bmatrix} 2 \\ 3 \\ 5 \\ 7 \end{bmatrix}, \quad \mathbf{u}_2 = \begin{bmatrix} 133/29 \\ 98/29 \\ 28/29 \\ -100/29 \end{bmatrix}, \quad \mathbf{u}_3 = \begin{bmatrix} -1160/3939 \\ 2940/1313 \\ -10610/3939 \\ 4130/3939 \end{bmatrix}, \quad \mathbf{u}_4 = \begin{bmatrix} 704/265 \\ -528/265 \\ 396/265 \\ 308/265 \end{bmatrix}$$

   (c) $\langle \mathbf{v}_1, \mathbf{v}_4 \rangle = 817$; $\text{proj}_{\mathbf{v}_3}(\mathbf{v}_2) = \begin{bmatrix} 2139/185 \\ 2697/185 \\ 2883/185 \\ 93/5 \end{bmatrix}$; $\|\mathbf{v}_1\| = \sqrt{87}$

3. (a) For example, $\langle \sin x, \sin(2x) \rangle = \int_0^{\pi/2} \sin(x)\sin(2x)\,dx = \frac{2}{3}$, so $\sin x$ and $\sin 2x$ are not orthogonal.

   (b)

$$\begin{aligned}
u_1 &= \sin x, \\
u_2 &= \sin(2x) - \frac{8\sin x}{3\pi}, \\
u_3 &= \sin(3x) + \frac{48\sin(x)(4 - 3\pi\cos(x))}{5\left(9\pi^2 - 64\right)}, \\
\frac{u_1}{\|u_1\|} &= \frac{2\sin(x)}{\sqrt{\pi}}, \\
\frac{u_2}{\|u_2\|} &= \frac{4\sin(x)(3\pi\cos(x) - 4)}{\sqrt{\pi\left(9\pi^2 - 64\right)}}, \\
\frac{u_3}{\|u_3\|} &= \frac{2\left(192\sin(x) - 72\pi\sin(2x) + 5\left(9\pi^2 - 64\right)\sin(3x)\right)}{\sqrt{\pi\left(9\pi^2 - 64\right)\left(225\pi^2 - 2176\right)}}
\end{aligned}$$

   (c)

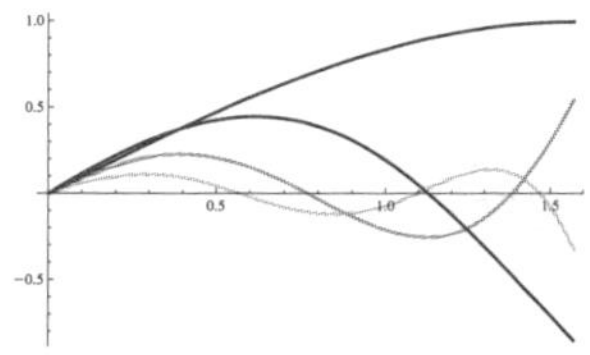

5. Let $v_n = \cos nx$ for $n = 0, 1, 2, \ldots$.

   (a) For example, $\langle v_0, v_1 \rangle_w = \int_0^{\pi} x\cos x\,dx = -2$, so v_0, v_1 are not orthogonal.

(b)

$$\begin{aligned}
u_0 &= v_0 = 1,\\
u_1 &= \cos(x) + \frac{4}{\pi^2},\\
u_2 &= \frac{40\left(\pi^2\cos(x)+4\right)}{9\left(\pi^4-32\right)} + \cos(2x),\\
\frac{u_0}{\|u_0\|} &= \frac{\sqrt{2}}{\pi},\\
\frac{u_1}{\|u_1\|} &= \frac{2(4+\pi^2\cos x)}{\pi\sqrt{\pi^4-32}},\\
\frac{u_2}{\|u_2\|} &= \frac{80\pi^2\cos(x)+18\left(\pi^4-32\right)\cos(2x)+320}{\pi\sqrt{(\pi^4-32)\left(81\pi^4-4192\right)}}
\end{aligned}$$

7. (a) Use $\langle x^m, x^n\rangle = \begin{cases} 0, & m+n \text{ odd},\\ \frac{2}{m+n+1}, & m+n \text{ even},\end{cases}$ and then Gram-Schmidt.

   (b) Compute the norms of the first four Legendre polynomials. Note that we want $\frac{P_n}{\|P_n\|} = \frac{\mathbf{u}_n}{\|\mathbf{u}_n\|}$ to compute the scaling factors.

   (c) No, since $L^2[-1,1]$ is infinite-dimensional, Gram-Schmidt tells us only that they are orthogonal and linearly independent, not that they form a basis.

9. (a) Compute using the weighted inner product of x^m and x^n with $w(x) = e^{-x^2}$.

   (b) Compute the norms of the first four Hermite polynomials. Note that we want $\frac{H_n}{\|H_n\|} = \frac{\mathbf{u}_n}{\|\mathbf{u}_n\|}$ to find the scaling factors.

   (c) No, since $L^2_w(-\infty,\infty)$ is infinite-dimensional, Gram-Schmidt tells us only that they are orthogonal and linearly independent, not that they form a basis.

11. Since the underlying L^2_w spaces are infinite dimensional, the Gram-Schmidt procedure makes no claim about *completeness* of the resulting orthogonal family. The Sturm-Liouville theory does.

5.1

1. Writing $\nabla = \langle \frac{\partial}{\partial x}, \frac{\partial}{\partial y}, \frac{\partial}{\partial z}\rangle$, ∇f looks like "scalar multiplication" by f while $\nabla\cdot\mathbf{F}$ looks like the "dot product" of ∇ and $\mathbf{F}$.

3. (a) $\nabla f = \langle 2xy^3 + y\cos(xy), 3x^2y^2 + x\cos(xy)\rangle$
   $\Delta f = 2y^3 + 6x^2y - (x^2+y^2)\sin(xy)$

(b) $\operatorname{grad} f = \left\langle \frac{x}{\sqrt{x^2+y^2+z^2}}, \frac{y}{\sqrt{x^2+y^2+z^2}}, \frac{z}{\sqrt{x^2+y^2+z^2}} \right\rangle$

$\nabla^2 f = \Delta f = \frac{2}{\sqrt{x^2+y^2+z^2}}$

5. (b) $\nabla \cdot \mathbf{F} = 0$

   (c) The divergence is zero since this is a purely circulating field, and the amount of fluid coming into a point equals the amount going out, as can be seen from the sizes of the arrows on any circle around the origin.

   (d) No, none exists. (We will see why in Exercise 12(a).)

   (e)

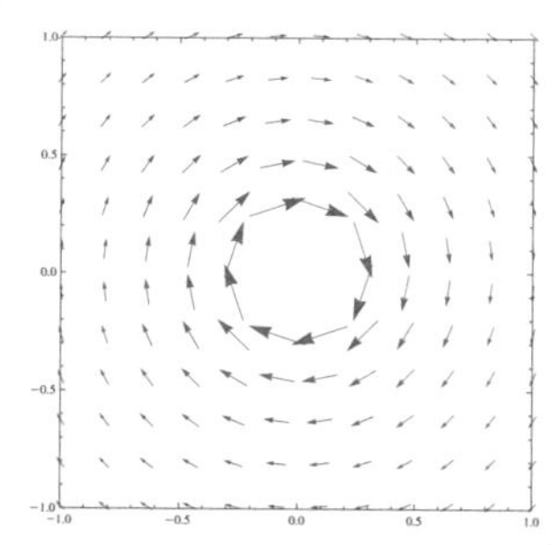

7. (a) Compute each side independently and compare.

   (b) No. For example, let $\mathbf{a} = \langle 1, 0, 0 \rangle$ and $\mathbf{b} = \mathbf{c} = \langle 0, 0, 1 \rangle$.

9. (a), (b), (d), (f) are meaningful; of these, (b) and (d) are always zero if the components of $\mathbf{F}$ have continuous second partials. (c) and (e) are not meaningful.

11. Use Stokes' Theorem.

13. (a) $\frac{\partial F_1}{\partial y} = \frac{\partial F_2}{\partial x} = 2x + 3y^2$

    (b) $f(x, y) = x^2 y + xy^3 + h(y)$

    (c) $f(x, y) = x^2 y + xy^3 + y^2 + C$

15. $f(x, y, z) = xy + \frac{1}{4} z^4 + C$

17. Use the Divergence Theorem.

19. (a) Use the Divergence Theorem with $\mathbf{F} = \mathbf{\Phi}$.

    (b) Use Stokes' Theorem with $\mathbf{F} = \mathbf{n} \times \mathbf{k}$.

5.2

1. $\nabla \cdot \mathbf{n}$ is the directional derivative in the outward normal direction, but flux, $\mathbf{\Phi}$, is *minus* $\nabla \cdot n$ so that the heat flow will obey Fourier's Law. Thus, the flux and the directional derivative in the outward normal direction have *opposite* signs.

3. Show that $\mathbf{a} \cdot (\mathbf{b} \times \mathbf{c}) = \begin{vmatrix} a_1 & a_2 & a_3 \\ b_1 & b_2 & b_3 \\ c_1 & c_2 & c_3 \end{vmatrix}$. Then the other scalar triple products are the same matrix with an even number of rows exchanged.

5. Most of the steps are obvious. The second step results from exchanging time and spatial partial derivatives and assuming mixed partials are equal; the sixth step results from the fact that $\nabla \cdot \mathbf{E} = 0$ by (5.10b).

7. $\dfrac{\partial \rho}{\partial t} = \dfrac{\partial}{\partial t}(\nabla \cdot \mathbf{D}) = \nabla \cdot \dfrac{\partial \mathbf{D}}{\partial t} = \nabla \cdot (\nabla \times \mathbf{H} - \mathbf{J}) = -\nabla \cdot \mathbf{J}$ since $\nabla \cdot (\nabla \times \mathbf{H}) = 0$ for any smooth vector field.

5.3

1.

Boundary	**Partial Notation**
$x = 0,\ 0 < y < b$	$u_x(0, y, t) = K(u(0, y, t) - T/K),\ K > 0$
$x = a,\ 0 < y < b$	$-u_x(a, y, t) = R$
$y = 0,\ 0 < x < a$	$u_y(x, 0, t) = R$
$y = b,\ 0 < x < a$	$u(x, b, t) = T$

3. (a) The top and right edges are insulated while the bottom and left edges are maintained at a temperature of zero.
   (b)
   - Radiating along the left edge.
   - At the right edge, physically realistic convection with an outside medium which is fixed at R_1 degrees.
   - Along the bottom edge, physically unrealistic convection with an outside medium which is fixed at R_2 degrees.
   - Top edge is maintained at a constant temperature of T.

5.4

1. (a) 1 initial condition (IC), 2 boundary conditions (BCs)
   (b) 4 BCs
   (c) 2 ICs, 4 BCs
   (d) 2 ICs, 4 BCs

(e) 4 BCs for each variable; 12 total BCs

(f) 6 BCs

(g) 6 BCs

(h) 1 IC, 6 BCs

(i) 2 ICs, 2 BCs

3. Green's First Identity can be rewritten as

$$\iint_D v\Delta u\, dA = -\iint_D \nabla v \cdot \nabla u\, dA + \int_{\partial D} v\frac{\partial u}{\partial \mathbf{n}}\, ds.$$

Here, $\Delta u = u_{xx} + u_{yy}$ is analogous to a second derivative in integration by parts, while $\nabla v \cdot \nabla u = \langle v_x, v_y\rangle \cdot \langle u_x, u_y\rangle$ is analogous to the product of first derivatives. The last term is analogous to v times the first derivative; note that that integral is along the boundary of D, which corresponds to evaluating at the boundary points a and b in the integration by parts formula.

5. If u is a solution, Green's First Identity with $v = 1$ gives $\int_{\partial\Omega} \frac{\partial u}{\partial \mathbf{n}}\, ds = \iint_\Omega \Delta u\, dA$, which is the same as $\int_{\partial\Omega} g\, ds = \iint_\Omega f\, dA$. This is simply a restatement of the Divergence Theorem for fluid flow: let $\mathbf{F} = \nabla u$ and rewrite the result using the given conditions.

5.5

1. (a) $u(x, y) = \sum_{n=1}^{\infty} [a_n \cosh(n\pi x) + b_n \sinh(n\pi x)] \sin(n\pi y)$,

$$a_n = 2\int_0^1 y \sin(n\pi y)\, dy,$$

$$b_n = \frac{2\int_0^1 (-y)\sin(n\pi y)\, dy - a_n \cosh(n\pi)}{\sinh(n\pi)}, \quad n = 1, 2, \ldots,$$

(b) The temperature is held fixed at 0 at the top and bottom, and held at the given function at the left and right.

(c)

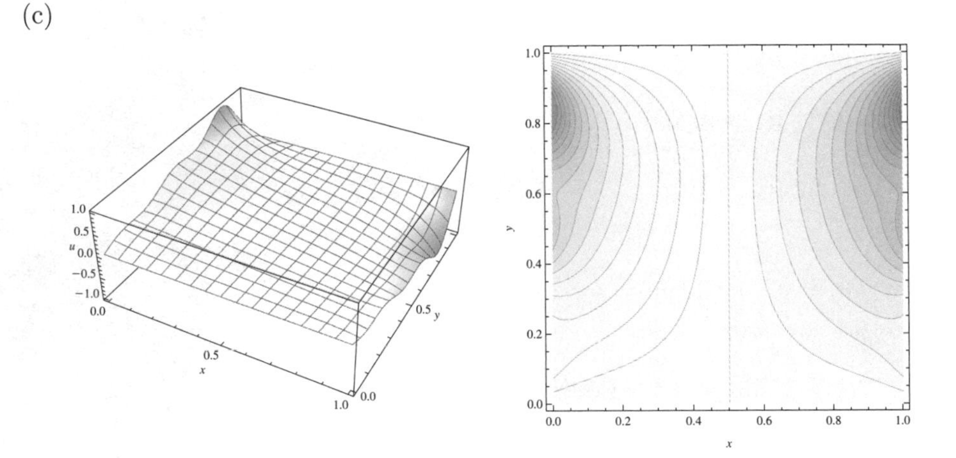

3. (a) $u(x, y) = b_0 x + \sum_{n=1}^{\infty} b_n \sinh(nx) \cos(ny),$

$$b_0 = \frac{2}{\pi^2} \int_0^{\pi} \cos(y^2)\, dy,$$

$$b_n = \frac{2}{\pi \sinh(n\pi)} \int_0^{\pi} \cos(y^2) \cos(ny)\, dy, \quad n = 1, 2, 3, \ldots$$

(b) The steady-state temperature along the left edge is 0 and along the right edge is $\cos(y^2)$. The top and bottom edges of the plate are insulated.

(c)

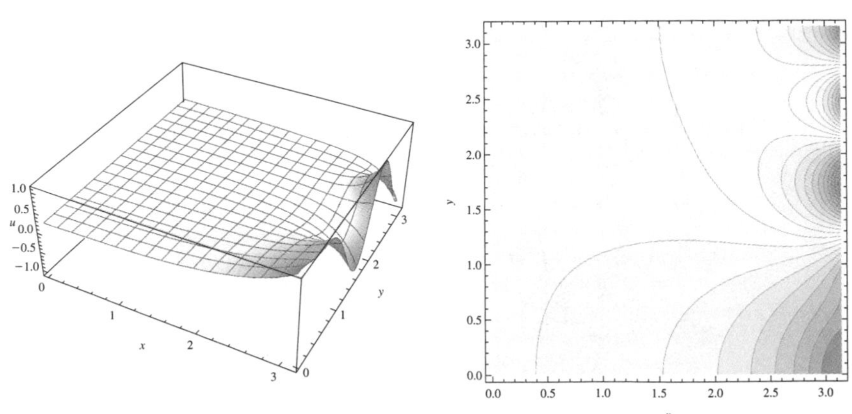

5. Homogenize the boundary conditions by letting $u(x,y) := v(x,y) + w(x,y)$ where v homogenizes the boundary conditions for x and w those for y.

   The solution of the v problem is

$$\begin{aligned} v(x,y) &= \sum_{n=1}^{\infty} \left[a_n \cosh\left(\sqrt{\lambda_n}\, y\right) + b_n \sinh\left(\sqrt{\lambda_n}\, y\right)\right] \sin\left(\sqrt{\lambda_n}\, x\right), \\ \lambda_n &= \left(\frac{n\pi}{a}\right)^2, \quad n = 1, 2, \ldots, \\ a_n &= \frac{2}{a} \int_0^a p(x) \sin(\sqrt{\lambda_n}\, x)\, dx, \\ b_n &= \frac{2}{a} \cdot \frac{\int_0^a q(x) \sin(\sqrt{\lambda_n}\, x)\, dx - \cosh(b\sqrt{\lambda_n}) \int_0^a p(x) \sin(\sqrt{\lambda_n}\, x)\, dx}{\sinh(b\sqrt{\lambda_n})}. \end{aligned}$$

   The solution of the w problem is

$$\begin{aligned} w(x,y) &= \sum_{n=1}^{\infty} \left[c_n \cosh\left(\sqrt{\mu_n}\, x\right) + d_n \sinh\left(\sqrt{\mu_n}\, x\right)\right] \sin\left(\sqrt{\mu_n} y\right), \\ \mu_n &= \left(\frac{n\pi}{b}\right)^2, \quad n = 1, 2, \ldots, \\ c_n &= \frac{2}{b} \int_0^b f(y) \sin(\sqrt{\mu_n} y)\, dy, \\ d_n &= \frac{2}{b} \cdot \frac{\int_0^b g(y) \sin(\sqrt{\mu_n} y)\, dy - \cosh(a\sqrt{\mu_n}) \int_0^b f(y) \sin(\sqrt{\mu_n} y)\, dy}{\sinh(a\sqrt{\mu_n})}. \end{aligned}$$

5.6

1. (a)

$$\begin{aligned} u(x,y,t) &= \sum_{n=1}^{\infty}\sum_{m=1}^{\infty} \left(c_{nm} \cos(c\sqrt{\lambda_{nm}}\, t) + d_{nm} \sin(c\sqrt{\lambda_{nm}}\, t)\right) \sin(\sqrt{\mu_n}\, x) \sin(\sqrt{\nu_m}\, y), \\ \mu_n &= \left(\frac{n\pi}{a}\right)^2, \quad n = 1, 2, \ldots, \\ \nu_m &= \left(\frac{m\pi}{b}\right)^2, \quad m = 1, 2, \ldots, \\ \lambda_{nm} &= \mu_n + \nu_m, \\ c_{nm} &= \frac{\int_0^b \int_0^a f(x,y) \sin(\sqrt{\mu_n}\, x) \sin(\sqrt{\nu_m}\, y)\, dx\, dy}{\int_0^b \int_0^a \sin^2(\sqrt{\mu_n}\, x) \sin^2(\sqrt{\nu_m}\, y)\, dx\, dy}, \\ d_{nm} &= \frac{\int_0^b \int_0^a g(x,y) \sin(\sqrt{\mu_n}\, x) \sin(\sqrt{\nu_m}\, y)\, dx\, dy}{c\sqrt{\lambda_{nm}} \int_0^b \int_0^a \sin^2(\sqrt{\mu_n}\, x) \sin^2(\sqrt{\nu_m}\, y)\, dx\, dy} \end{aligned}$$

(b) The boundary conditions say that the edges of the sheet are held fixed at a height of 0 over time. The initial conditions give the initial position and velocity of each point in the sheet's interior.

(c)

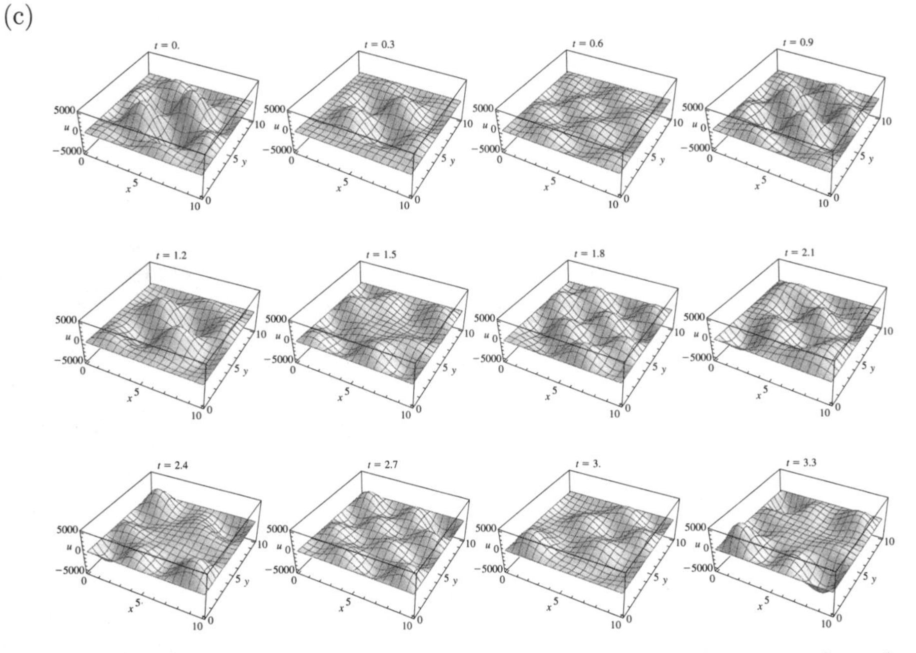

(d) Both from the periodic nature of the equation and the plots, it is clear that the position of the sheet over time is periodic as well; the period seems to be somewhere around $t = 3$. The initial and boundary conditions appear to be satisfied.

(e) $\sqrt{\int_0^{10} \int_0^{10} (f(x,y) - F_5(x,y,0))^2 \; dx \, dy} \approx 512.805$

6.1

1. (a)

$$u(r,\theta) = \sum_{n=1}^{\infty} c_n r^{\sqrt{\lambda_n}} \sin\left(\sqrt{\lambda_n}\,\theta\right),$$

$$\lambda_n = \left(\frac{n\pi}{\theta_0}\right)^2, \quad n = 1, 2, \ldots,$$

$$c_n = \frac{2}{\theta_0 \rho^{n\pi/\theta_0}} \int_0^{\theta_0} f(\theta) \sin\left(n\pi\theta/\theta_0\right) \, d\theta$$

(b) u is zero along the radial boundaries of the wedge and is given by $f(\theta)$ along the circular boundary. There is also a mathematical boundary condition, that R, R' remain bounded as $r \to 0^+$.

(c) $\langle \sin\left(\sqrt{\lambda_n}\,\theta\right), \sin\left(\sqrt{\lambda_m}\,\theta\right)\rangle = 0,\ n \neq m$

(d) $N = 8$

(e)

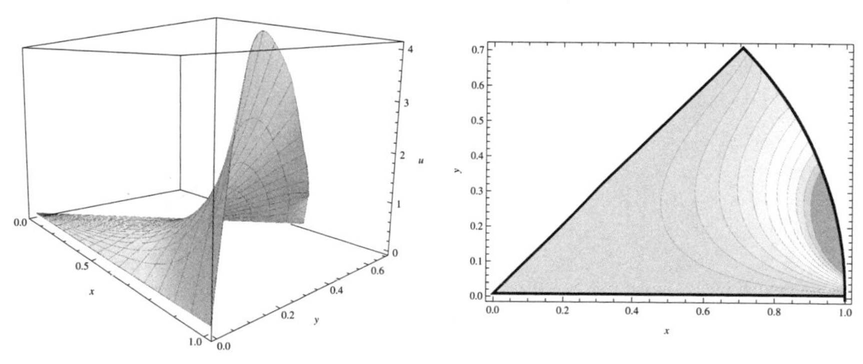

3. (a)

$$u(r,\theta) = c_0 + d_0 \ln r + \sum_{n=1}^{\infty} \left[c_n r^n + d_n r^{-n}\right]\left[a_n \cos(n\theta) + b_n \sin(n\theta)\right],$$

$$c_0 = \frac{1}{2\pi}\int_{-\pi}^{\pi} f(\theta)\,d\theta,$$

$$d_0 = 0,$$

and the products $a_n c_n$, $a_n d_n$, $b_n c_n$, and $b_n d_n$ are determined from the four equations

$$a_n c_n \rho_1^n + a_n d_n \rho_1^{-n} = \frac{1}{\pi}\int_{-\pi}^{\pi} f(\theta)\cos(n\theta)\,d\theta$$

$$b_n c_n \rho_1^n + b_n d_n \rho_1^{-n} = \frac{1}{\pi}\int_{-\pi}^{\pi} f(\theta)\sin(n\theta)\,d\theta$$

$$n a_n c_n \rho_2^{n-1} - n a_n d_n \rho_2^{-n-1} = 0$$

$$n b_n c_n \rho_2^{n-1} - n b_n d_n \rho_2^{-n-1} = 0$$

(b) $u(\rho_1,\theta) = f(\theta)$ and $u_r(\rho_2,\theta) = 0$ are physical boundary conditions since they are conditions on the physical boundary of the domain (i.e, the inner and outer radius of the annulus). The mathematical boundary conditions implicitly present are the periodic boundary conditions $u(r,-\pi) = u(r,\pi)$, $u_\theta(r,-\pi) = u_\theta(r,\pi)$.

(c) They are eigenfunctions of a regular Sturm-Liouville problem.

(d) $N = 3$

(e)

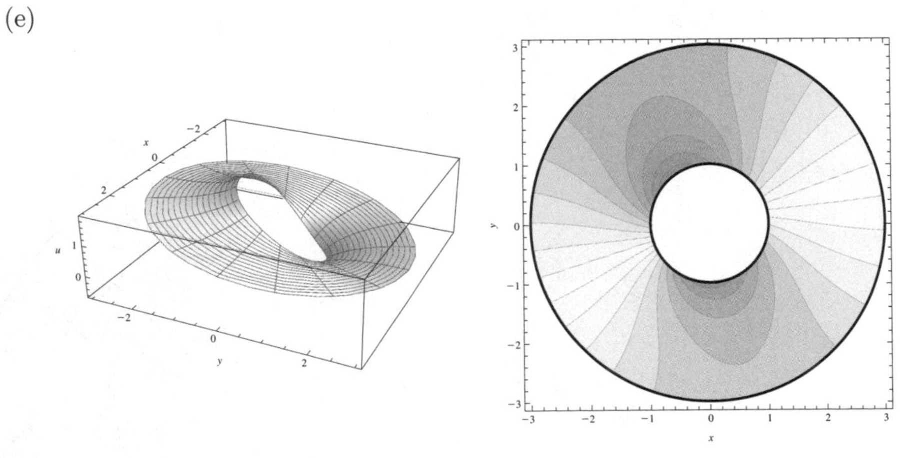

(f) The steady-state temperature distribution in an annulus with inner radius ρ_1 and outer radius ρ_2. The temperature of the inner boundary is kept at $f(\theta)$ degrees while the outer boundary is kept at $0°$.

5. In all four exercises, the maximum and minimum values of the solution occur on the boundary of the domain, and nowhere inside.

7. Each boundary condition must be independent of one coordinate or the other.

9.

$$u(r,\theta) = \sum_{n=1}^{\infty} \left[a_n r^{4n} + b_n r^{-4n}\right] \sin(4n\theta),$$

$$a_n = b_n = \frac{8T \int_0^{\pi/4} \sin(4n\theta)\, d\theta}{\pi(16^{-n} + 16^n)} = \frac{2T(1-(-1)^n)}{n\pi(16^{-n}+16^n)}, \quad n = 1, 2, \ldots$$

6.2

3. $25°$

5. (a) 4 (b) -2 (c) 1

6.3

1. (a)

$$u(r,\theta,t) = \sum_{m=1}^{\infty}\sum_{n=1}^{\infty} A_{nm} J_{2n}(\sqrt{\lambda_{nm}}\, r) \cos(\sqrt{\lambda_{nm}}\, ct) \sin(2n\theta),$$

$$A_{nm} = \frac{\int_0^{\rho}\int_0^{\pi/2} f(r,\theta) J_{2n}\left(\sqrt{\lambda_{nm}}\, r\right) \sin(2n\theta) r\, d\theta\, dr}{\int_0^{\rho}\int_0^{\pi/2} J_{2n}^2\left(\sqrt{\lambda_{nm}}\, r\right) \sin^2(2n\theta) r\, d\theta\, dr}, \quad n, m = 1, 2, \ldots,$$

where $\lambda_{nm} = (z_{nm}/\rho)^2$ and z_{nm} is the mth positive zero of $J_{2n}(x)$.

(b)

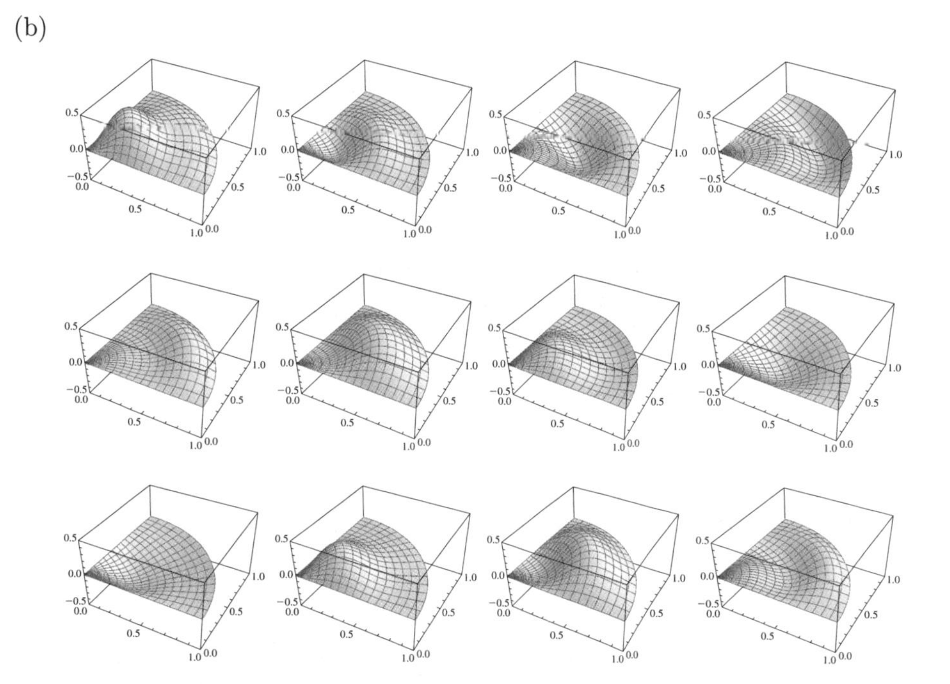

(c) This is a vibrating wedge problem in which all three edges of the wedge are held at 0 over time, the initial position of the wedge is given by $f(r, \theta)$, and the initial velocity is zero.

3.

$$u(r, \theta, t) = \sum_{m=1}^{\infty} \sum_{n=1}^{\infty} a_{nm} J_n(\sqrt{\lambda_{nm}}\, r) \exp(-k\lambda_{nm} t) \sin(n\theta),$$

$$\lambda_{nm} = (z_{nm}/\rho)^2, \quad m, n = 1, 2, \ldots,$$

$$a_{nm} = \frac{\int_0^{\pi} \int_0^{\rho} f(r, \theta) J_n(\sqrt{\lambda_{nm}}\, r) \sin(n\theta) r\, dr\, d\theta}{\int_0^{\pi} \int_0^{\rho} J_n^2(\sqrt{\lambda_{nm}}\, r) \sin^2(n\theta) r\, dr\, d\theta},$$

where z_{nm} is the mth positive zero of $J_n(x)$.

6.4

3. (a) This is a steady-state temperature problem on a half-cylinder of radius ρ and height ℓ. The first five boundary conditions say that the temperature on all sides (including the flat side) as well as the bottom of the cylinder are fixed at 0. The final boundary condition gives the heat distribution at the top of the cylinder.

(c)

$$u(r,\theta,z) = \sum_{m=1}^{\infty}\sum_{n=1}^{\infty} a_{nm} J_n\left(\sqrt{\lambda_{nm}}\, r\right) \sin(n\theta) \sinh\left(\sqrt{\lambda_{nm}}\, \ell\right),$$

$$\lambda_{nm} = (z_{nm}/\rho)^2, \quad m, n = 1, 2, \ldots,$$

where z_{nm} is the mth positive zero of $J_n(x)$.

(d) $a_{nm} = \dfrac{1}{\sinh(\sqrt{\lambda_{nm}}\ell)} \cdot \dfrac{\int_0^\rho \int_0^\pi J_n(\sqrt{\lambda_{nm}}\, r) \sin(n\theta) f(r,\theta) r \, d\theta \, dr}{\int_0^\rho \int_0^\pi J_n^2(\sqrt{\lambda_{nm}}\, r) \sin^2(n\theta) r \, d\theta \, dr}$, $m, n = 1, 2, \ldots$

(e)

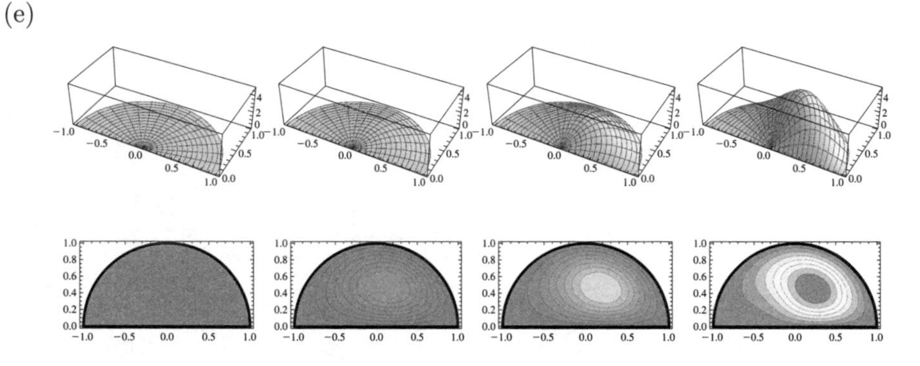

(f) $N = 2$ provides an excellent approximation to the system along the top half-disk, and the model does seem to conform to the other boundary conditions of the problem.

5. (a) This problem would model the steady-state temperature distribution in a cylinder of radius ρ and height ℓ. The first boundary condition specifies that the temperature along the sides of the cylinder is held at 0; the remaining boundary conditions specify the heat distribution along the bottom and top of the cylinder in terms of $f(r,\theta)$ and $g(r,\theta)$.

(b)

$$u(r,\theta,z) = \sum_{n=0}^{\infty}\sum_{m=1}^{\infty} J_n(\sqrt{\lambda_{nm}}\, r) \left[a_{nm}\cos(n\theta) + b_{nm}\sin(n\theta)\right]$$
$$\times \left[c_{nm}\cosh(\sqrt{\lambda_{nm}} z) + d_{nm}\sinh(\sqrt{\lambda_{nm}} z)\right]$$

where $\lambda_{nm} = (z_{nm}/\rho)^2$ and z_{nm} is the mth positive zero of $J_n(x)$.

(c) The products of coefficients (which are all that is needed) can be found by solving the following set of equations:

$$a_{0m}c_{0m} = \frac{\int_0^\rho \int_{-\pi}^{\pi} J_0(\sqrt{\lambda_{0m}}r) f(r,\theta) r \, d\theta \, dr}{\int_0^\rho \int_{-\pi}^{\pi} J_0^2(\sqrt{\lambda_{0m}}r) r \, d\theta \, dr}$$

$$a_{0m}c_{0m}\cosh(\sqrt{\lambda_{0m}}\ell) + a_{0m}d_{0m}\sinh(\sqrt{\lambda_{0m}}\ell) = \frac{\int_0^\rho \int_{-\pi}^{\pi} J_0(\sqrt{\lambda_{0m}}r) g(r,\theta) r \, d\theta \, dr}{\int_0^\rho \int_{-\pi}^{\pi} J_0^2(\sqrt{\lambda_{0m}}r) r \, d\theta \, dr}$$

$$a_{nm}c_{nm} = \frac{\int_0^\rho \int_{-\pi}^{\pi} J_n(\sqrt{\lambda_{nm}}\, r)\cos(n\theta) f(r,\theta) r \, d\theta \, dr}{\int_0^\rho \int_{-\pi}^{\pi} J_n^2(\sqrt{\lambda_{nm}}\, r)\cos^2(n\theta) r \, d\theta \, dr}$$

$$a_{nm}c_{nm}\cosh(\sqrt{\lambda_{nm}}\ell) + a_{nm}d_{nm}\sinh(\sqrt{\lambda_{nm}}\ell) = \frac{\int_0^\rho \int_{-\pi}^{\pi} J_n(\sqrt{\lambda_{nm}}\, r)\cos(n\theta) g(r,\theta) r \, d\theta \, dr}{\int_0^\rho \int_{-\pi}^{\pi} J_n^2(\sqrt{\lambda_{nm}}\, r)\cos^2(n\theta) r \, d\theta \, dr}$$

$$a_{nm}d_{nm} = \frac{\int_0^\rho \int_{-\pi}^{\pi} J_n(\sqrt{\lambda_{nm}}\, r)\sin(n\theta) f(r,\theta) r \, d\theta \, dr}{\int_0^\rho \int_{-\pi}^{\pi} J_n^2(\sqrt{\lambda_{nm}}\, r)\sin^2(n\theta) r \, d\theta \, dr}$$

$$b_{nm}c_{nm}\cosh(\sqrt{\lambda_{nm}}\ell) + b_{nm}d_{nm}\sinh(\sqrt{\lambda_{nm}}\ell) = \frac{\int_0^\rho \int_{-\pi}^{\pi} J_n(\sqrt{\lambda_{nm}}\, r)\sin(n\theta) g(r,\theta) r \, d\theta \, dr}{\int_0^\rho \int_{-\pi}^{\pi} J_n^2(\sqrt{\lambda_{nm}}\, r)\sin^2(n\theta) r \, d\theta \, dr},$$

for $n, m = 1, 2, \ldots$ and where $b_{0m} = 0$ for all m.

(e) $N = 2$ is not sufficient to get a good match between the solution and the model. In particular, the model is not close to $g(r,\theta)$ at $z = 1$. This is likely because there is a discontinuity on the circle $r = 1$, $z = 1$ between the boundary condition at $z = 1$ and the boundary condition along the sides. The model does show that the sides of the cylinder are held fixed at zero, and the bottom is a reasonable approximation to $f(r,\theta)$.

6.5

5. (a) $\left[((1-x^2)y')' - \frac{m^2}{1-x^2}y\right] + \lambda y = 0$

   (b) y, y' bounded as $x \to \pm 1$

7. (a) $r^2R'' + 2rR' - \lambda R = 0$, $\rho < r < \infty$

   $\Phi'' + \cot\varphi\,\Phi' + \lambda\Phi = 0$, $0 < \varphi < \pi$

   (b) The mathematical boundary condition at $\rho = \infty$ is that R, R' are bounded as $r \to \infty$. This eliminates the r^n terms in the R solution.

(c)

$$u(r,\varphi) = \sum_{n=0}^{\infty} c_n r^{-(n+1)} P_n(\cos\varphi),$$

$$c_n = \rho^{n+1} \frac{\int_0^\pi f(\varphi) P_n(\cos\varphi)\sin\varphi\,d\varphi}{\int_0^\pi P_n^2(\cos\varphi)\sin\varphi\,d\varphi}, \qquad n = 0, 1, 2, \ldots.$$

9. (a) The key is to realize that the Φ equation is still the associated Legendre's equation, except here $0 < \varphi < \pi/2$. Then the boundary conditions are $\Phi(\pi/2) = 0$ and Φ, Φ' bounded as $\varphi \to \frac{\pi}{2}^-$. Using the hint, we get

$$u(r,\varphi) = \sum_{n=0}^{\infty} c_{2n+1} r^{2n+1} P_{2n+1}(\cos\varphi),$$

$$c_{2n+1} = \frac{1}{\rho^{2n+1}} \frac{\int_0^{\pi/2} f(\varphi) P_{2n+1}(\cos\varphi)\sin\varphi\,d\varphi}{\int_0^{\pi/2} P_{2n+1}^2(\cos\varphi)\sin\varphi\,d\varphi}, \qquad n = 0, 1, 2, \ldots.$$

(d) 0.241513

11. (a)

$$u(r,\theta,\varphi) = \sum_{n=0}^{\infty} \sum_{m=-n}^{n} c_{2n+1,m} r^{2n+1} Y_{2n+1}^m(\theta,\varphi),$$

$$c_{2n+1,m} = \frac{2}{\rho^{2n+1}} \int_0^{2\pi} \int_0^{\pi/2} f(\theta,\varphi) \overline{Y_{2n+1}^m(\theta,\varphi)} \sin\varphi\,d\varphi\,d\theta, \quad n = 0, 1, 2, \ldots.$$

(b) See Figures 6.17 and 6.18.

(d) 13.3233

13. (c) 2.01943

7.1

3. (a) From Figure 7.2, the peaks of the waves move left (or right) one x unit every t unit.

   (b) The peaks of the waves will move left (or right) by c units every time unit, since for example when $t = 1$, $u(x,t) = f(x-c) + g(x+c)$ so that the peaks will occur at $x = \pm c$. Thus for $c < 1$ the waves will spread more slowly than for $c = 1$, while for $c = 3 > 1$ they will spread more rapidly.

5. (a) $u(x,t) = \frac{1}{2}[f(x-t) + f(x+t)]$, which consists of two copies of one period of the sine wave which overlap to varying degrees depending on t; for $t \geq \pi$, they are nonoverlapping, while at $t = 0$ they are superimposed.

(b)

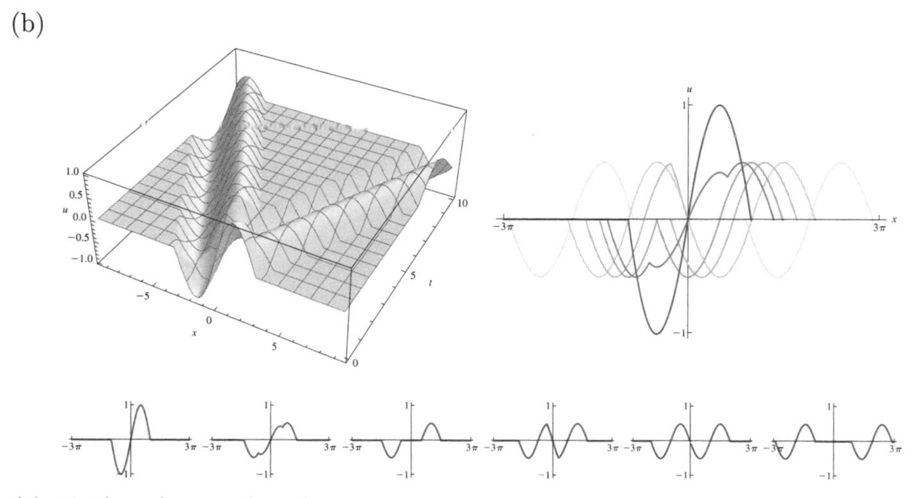

(c) If $f(x+t)$ and $f(x-t)$ were both sinusoidal everywhere, their sum would be as well. However, since they are both zero outside of a single period, there is a "corner" when each of them becomes zero. This corner is reflected in the "kinks", which results from the fact that $u(x,t)$ is then composed solely of one of the two functions. By $t=\pi$, the waves have completely separated, and the kink is gone.

7. (a) $u(x,t)=\dfrac{1}{2}\left(\begin{cases}0 & x<-t\\ x+t & x>-t\end{cases}-\begin{cases}0 & x<t\\ x-t & x>t\end{cases}\right)$

(b)

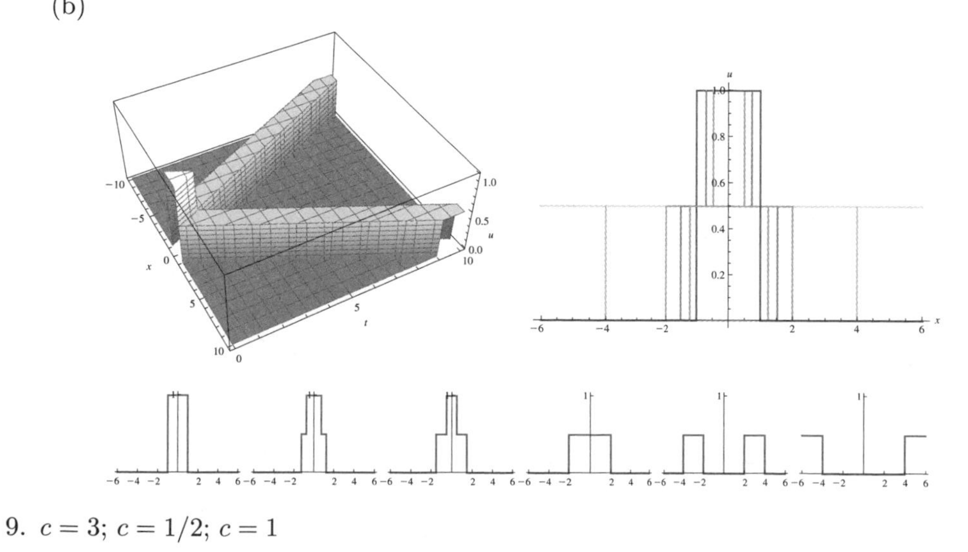

9. $c=3$; $c=1/2$; $c=1$

11. (a) Compute u_t+u_x in terms of derivatives of U using the chain rule.

(b) Integrate U_η with respect to η and re-interpret in x-t coordinates.

7.2

1. (a) $t \geq \ell/c$

   (b) $t \geq (b-a)/2c$

3. $u(0,2) = 2$; $u(0,4) = 0$; $u(5,5) = 1$; $u(10,6) = 0$; $u(-5,3) = 1$

7.3

1. This is the wave equation in a semi-infinite string. The left endpoint is maintained at a height of zero, hence the name *fixed end problem.* The initial displacement of the string is $f(x)$ and the initial velocity of the string is $g(x)$.

5. (a) $u(x,t) = \begin{cases} \frac{1}{2}\left[h(x+t-1) - h(x+t-3) - h(t-x-1) + h(t-x-3)\right], & x-t \leq 0 \\ \frac{1}{2}\left[h(x+t-1) - h(x+t-3) + h(x-t-1) + h(x-t-3)\right], & x-t > 0 \end{cases}$

7. (a) $u(x,t) = \begin{cases} \frac{1}{2}\left[f(x+t) - f(t-x)\right] + \frac{1}{2}\int_{t-x}^{x+t} g(s)\,ds, & x-t \leq 0 \\ \frac{1}{2}\left[f(x+t) + f(x-t)\right] + \frac{1}{2}\int_{x-t}^{x+t} g(s)\,ds, & x-t > 0 \end{cases}$

   (b)

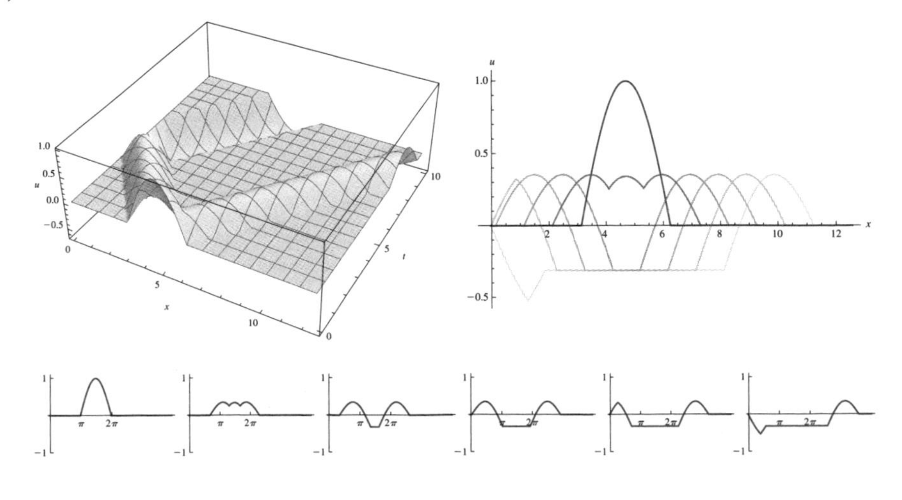

9. (a) $u(x,t) = \begin{cases} \frac{1}{2}\left[f(x+t) + f(t-x)\right] + \frac{1}{2}\int_{0}^{t-x} g(s)\,ds + \frac{1}{2}\int_{0}^{x+t} g(s)\,ds, & x-t \leq 0 \\ \frac{1}{2}\left[f(x+t) + f(x-t)\right] + \frac{1}{2}\int_{x-t}^{x+t} g(s)\,ds, & x-t > 0 \end{cases}$

(b)

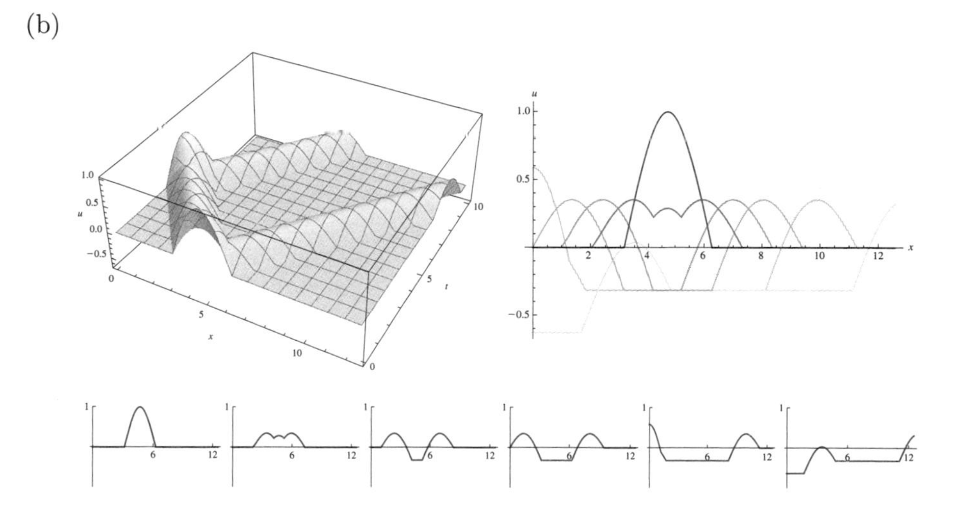

11. Use the definition of the odd extension.

7.4

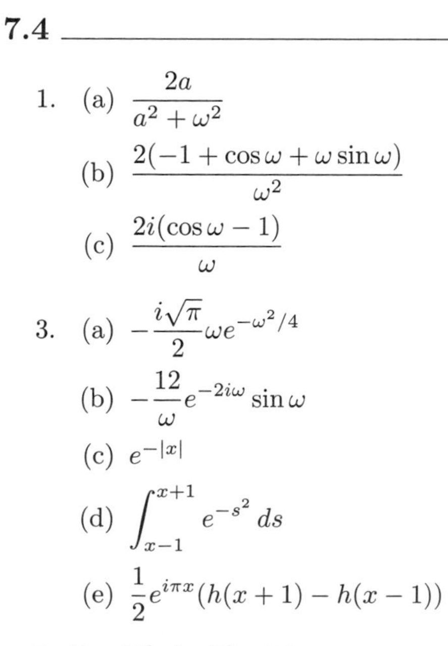

1. (a) $\dfrac{2a}{a^2+\omega^2}$

   (b) $\dfrac{2(-1+\cos\omega+\omega\sin\omega)}{\omega^2}$

   (c) $\dfrac{2i(\cos\omega-1)}{\omega}$

3. (a) $-\dfrac{i\sqrt{\pi}}{2}\omega e^{-\omega^2/4}$

   (b) $-\dfrac{12}{\omega}e^{-2i\omega}\sin\omega$

   (c) $e^{-|x|}$

   (d) $\displaystyle\int_{x-1}^{x+1} e^{-s^2}\,ds$

   (e) $\dfrac{1}{2}e^{i\pi x}(h(x+1)-h(x-1))$

5. For $\mathcal{F}\{u(x,t)\}$ with respect to x to exist, we must have $u(x,t)\to 0$ as $x\to\pm\infty$. Similarly, for $\mathcal{F}\{u_t(x,t)\}$ to exist, we must have $u(x,t)\to 0$ as $t\to\infty$.

7. (a) This is a heat diffusion problem on an "infinite" rod with the initial heat distribution centered on an interval of width $2a$.

   (b) $u(x,t)=\dfrac{1}{2\sqrt{k\pi t}}\displaystyle\int_{-a}^{a}\exp\left(\dfrac{-(x-s)^2}{4kt}\right)ds$

(c)

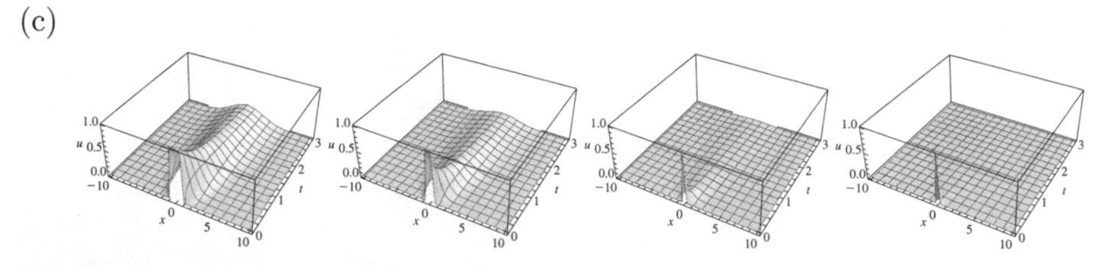

9. (a) This is a heat diffusion problem over an infinite rod where the initial temperature distribution of the system is $x^3 e^{-x^2}$.

   (b) $u(x,t) = \frac{1}{2\sqrt{k\pi t}} \int_{-\infty}^{\infty} x^3 \exp\left(-x^2 + \frac{-(x-s)^2}{4kt}\right) ds$

   (c)

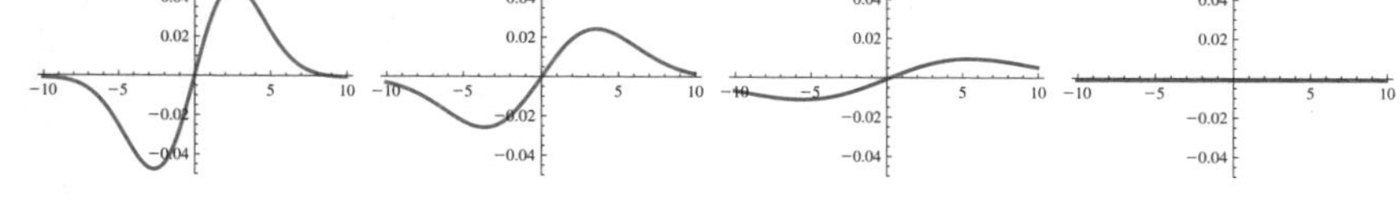

   (d) Zero

13. (a) $u(x,t) = \displaystyle\int_{-\infty}^{\infty} f(x + k_2 t - s) K(s,t)\, ds$, where $k = k_1$ in the heat kernel.

   (c) This is the transport-diffusion equation, see equation (1.6). It describes the concentration of a diffusing substance in a flowing fluid.

Photo and Illustration Credits

Fig 1.7: INTERFOTO / Alamy

Fig 1.10: Bettmann/CORBIS

Fig 2.1: akg-images

Fig 3.12: Bettmann/CORBIS

Fig 4.1: SPL / Photo Researchers, Inc.

Fig 4.2: INTERFOTO / Alamy

Fig 4.4: RIA Novosti / Alamy

Fig 4.11: Bettmann/CORBIS

Fig 5.2: Adapted from Jon Rogawski, *Calculus: Early Transcendentals*, Second Edition, ©2012 by W.H. Freeman and Company.

Fig 5.3: Adapted from Jon Rogawski, *Calculus: Early Transcendentals*, Second Edition, ©2012 by W.H. Freeman and Company.

Fig 5.4: Adapted from Jon Rogawski, *Calculus: Early Transcendentals*, Second Edition, ©2012 by W.H. Freeman and Company.

Fig 5.5: Bettmann/CORBIS

Fig 5.7: Bettmann/CORBIS

Fig 5.9: Bettmann/CORBIS

Fig 6.1: Jon Rogawski, *Calculus: Early Transcendentals*, Second Edition, ©2012 by W.H. Freeman and Company.

Fig 6.7: Jon Rogawski, *Calculus: Early Transcendentals*, Second Edition, ©2012 by W.H. Freeman and Company.

Fig 6.11: Jon Rogawski, *Calculus: Early Transcendentals*, Second Edition, ©2012 by W.H. Freeman and Company.

Fig 7.1: Bettmann/CORBIS

Fig 7.17: Bettmann/CORBIS

Index